Two Manuals Combined into One Book

AF885451

NBC Decontamination

And

Military Role in Civil Disturbance Operations

Department of Defense

Pentagon
Publishing

***FM 3-5**
***MCWP 3-37.3**

Field Manual
No. 3-5

Headquarters,
Department of the Army

Marine Corps Warfighting Publication (MCWP)
3-37.3

Commandant,
US Marine Corps

Washington, DC, 28 July 2000

NBC DECONTAMINATION

Contents

DESTRUCTION RESTRICTIONS: Approved for public release; distribution is unlimited.

* This publication supersedes Field Manual (FM) 305/Fleet Marine Force Manual (FMFM) 11-10, 17 November 1993.

Preface

FM 3-5 integrates the nuclear, biological, and chemical (NBC) fundamentals published in FM 3-100 and implements Standardized North Atlantic Treaty Organization Agreement (STANAG) 2426.

The extent and timing of decon depends on the tactical situation, the mission, the area of contamination, and the decon resources available. Survivability and reducing the effect of any chemical threat are the ultimate goals of decon. This manual provides detailed guidance on conducting decon operations performed by chemical and nonchemical personnel. Individual soldiers and leaders must be familiar with the basic procedures and concepts in this manual.

Appendix A contains an English to metric measurement conversion chart.

The proponent of this publication is Headquarters TRADOC. To submit changes for improving this publication use Department of the Army (DA) Form 2028 (Recommended Changes to Publications and Blank Forms) and forward to Commandant, US Army Chemical School, ATTN: ATSN-CMZ, Fort Leonard Wood, Missouri 65473-8926.

Unless this publication states otherwise, nouns and pronouns do not refer exclusively to men.

Chapter 1

Introduction

The use of NBC weapons creates unique residual hazards that may require decon. In addition to the deliberate use of these weapons, collateral damage, natural disasters, and industrial emitters may require decon. The presence of contamination generally reduces the effectiveness of our combat power. Contamination forces us into protective equipment that degrades our ability to perform individual and collective tasks.

CONTAMINATION FORMS

1-1. The following are the different forms of contamination:

- Solids – radioactive particles, biological spores, or dusty agents that could appear as a fine dust.
- Liquids – liquid droplets that fall like rain. Droplets can range from thick and sticky to the consistency of water.
- Vapors or gases – created by bursting munitions or generators. These clouds are affected by the weather and can cover large areas.
- Aerosols – fine liquids or solid particles suspended in the air. They behave much like vapors.

CONTAMINATION-HAZARDS TRANSMISSION

1-2. Contamination hazards can be transmitted in the following manner:

- **Transfer**. Anything that touches a surface covered with liquid or solid contamination will tend to pick up that contamination and move it from one surface to another.
- **Spread**. Touching a surface covered with liquid chemical agent can spread contamination on the same surface, thereby, increasing the size of the contaminated area.
- **Vapor**. Vapors can be carried through the air in the form of a dust, atomized liquids (aerosols), or true gases. Vapors in an open/outdoor area disperse rapidly, so there is no need to decon.
- **Desorption**. Liquid-chemical contamination absorbs into porous surfaces. Once absorbed, it begins to desorb or give off gas; that is, low levels of vapor pass out of the contaminated surface into the air and can be transferred to any surface that contacts it, including bare skin.
- **Radiation**. Radiation is given off by radioactive dust or dirt, most of which appears as fallout. For decon purposes, radiation can be thought of as a solid.

AGENT CLASSIFICATION

1-3. Depending on the length of time agents will be a hazard, they are classified as—

- Nonpersistent – an immediate threat that lasts a few minutes. They rarely require decon.
- Persistent – takes a longer time to act and may last for days. In a protected environment, these agents can last for long periods of time. All agents are affected to some extent by the weather. Even radiological particles can "rain out" of the air and form hot spots on the ground. Biological organisms are greatly affected by sunlight. Chemical agents can be decontaminated by the weather. The duration of a hazard is a complex estimation that is based on numerous factors which include the—
  - Type of contamination.
  - Contamination density and droplet size.
  - Temperature.
  - Wind speed.
  - Sunlight.
  - Humidity and rain.
  - Composition of the contaminated surface.
  - Type of soil and terrain.

DECONTAMINANTS

1-4. Decontaminants are described as follows:

- Natural – weather (rain, wind, and humidity).
- Standard – supertropical bleach (STB) and decontaminating solution number 2 (DS2).
- Nonstandard – soaps and detergents.

1-5. For more information on the types of decontaminants, see Appendix B.

DECON CONCEPTS

1-6. Whenever soldiers are unable to avoid contamination and have to use protective measures, decon is necessary to allow them to remove their protective gear and resume normal operations. Weathering is the most desirable means of decon; however, time and operational needs may not permit that option.

1-7. This chapter describes when, where, and how much to decon. Protective clothing (mission-oriented protective posture [MOPP] gear), protective equipment, and collective protective shelters (CPSs) offer only a temporary solution. Decon is the removal, destruction, or naturalization of contamination. If you become contaminated, some decon must occur as soon as possible.

DECON DECISIONS

1-8. The decision to decon is a risk assessment and is made within the context of the mission, enemy, terrain, troops, time available, and civilian consideration (METT-TC) and the resources available. Decon must be considered if the contamination levels exceed the negligible risk levels as follows:

- Chemical and biological contamination causes mild incapacitation in 5 percent or less of unprotected soldiers operating for 12 continuous hours within 1 meter of contamination. For the chemical-agent monitor (CAM), this equates to a one-bar reading at a distance of 1 inch from the surface.
- Radiological contamination can cause mild incapacitation in 2.5 percent or less of unprotected, nonpreviously exposed soldiers. This equates to a reading of no more than 0.33 centigray per hour (cGy/hr).

1-9. MOPP-gear exchange provides excellent protection against field concentrations of agents; however, wearing the gear causes performance degradation. Decon is performed to restore the normal operating tempo, but the logistical support that is required to keep soldiers in MOPP gear impacts operations. Therefore, decon should be conducted as soon as practical. Table 1-1 provides comparison data for decon levels/techniques.

Table 1-1. Comparison Data for Decon Levels/Techniques

Levels	Techniques[1]	Best Start Time	Performed by	Advantages
Immediate	Skin decon Personal wipe down Operator's spray down	Before 1 minute Within 15 minutes	Individual Individual or crew	Stops agent from penetrating.
Operational	MOPP-gear exchange[2] Vehicle washdown[3]	Within 6 hours	Unit Battalion crew or decon platoon (-)	Provides possible temporary relief from MOPP4. Limits liquid agent spread.
Thorough	DED and DAD DTD	When mission allows reconstitution	Decon platoon Unit	Provides probable long-term MOPP reduction with minimum risk.

[1]The techniques become less effective the longer they are delayed.
[2]Performance degradation and risk assessment must be considered when exceeding 6 hours. See FM 3-4 for battle-dress overgarment (BDO) risk assessment.
[3]Vehicle washdown is most effective if started within 1 hour.

DECON PRINCIPLES

1-10. Decon immediately for a chemical agent on the skin. Perform higher levels of decon as a result of the risk assessment. Personnel should—

- Decon as soon as possible. The sooner the contamination is removed, the sooner MOPP levels can be reduced and combat power can be restored.
- Decon only what is necessary. Weathering is the least costly method of decon. Expend resources where they count.
- Decon as far forward as possible. While METT-TC dependent, performing decon as close to the point as possible reduces the spread of contamination and minimizes any transfer hazard.
- Decon by priority. The commander identifies which items are most critical to the mission. Restoring combat power and reestablishing operating tempo are key.

DECON LEVELS

1-11. The three levels of decon operations are immediate, operational, and thorough (see Table 1-1, page 1-3).

IMMEDIATE

1-12. Skin decon is a soldier's basic survival skill and should be performed within 1 minute of being contaminated using the skin decontaminating kit (SDK).

1-13. Personal wipe down should be performed within 15 minutes. This is done to remove contamination from individual equipment by using an individual equipment decon kit (IEDK). Use detector paper or a CAM to locate the agent. For radiological contamination, use a radiac set to locate it and then brush, wipe, or shake it off.

1-14. Operators' spray down should be done within 15 minutes. Operators use the on-board decon apparatus to decon surfaces that they must touch or contact to operate the equipment. Radiological contamination in the form of dust particles may be wiped, scraped, or brushed off.

OPERATIONAL

1-15. A MOPP-gear exchange should be performed within 6 hours of being contaminated when thorough decon cannot be done. Soldiers will continue to wear MOPP gear and the operating tempo will be reduced.

1-16. Vehicle washdown should be performed within 6 hour of being contaminated when the mission does not permit a thorough decon. This process removes gross contamination and limits the spread of it.

THOROUGH

1-17. Detailed equipment decon (DED) and detailed aircraft decon (DAD) restore items so that they can be used without MOPP gear. Normally, the DED and the DAD are conducted as part of a reconstitution or during breaks in combat operations. These operations require support from a chemical decon unit.

1-18. Detailed troop decon (DTD) normally takes place in conjunction with DED/DAD. The contaminated unit conducts this process and supports the DED/DAD operations.

CHEMICAL-AGENT-RESISTANT COATING (CARC)

1-19. Army equipment is painted with CARC, which precludes the absorption of chemical agents and, thus, facilitates decon. By preventing penetration of an agent, contamination is easier to remove. Thickened agents are still a problem, but detergent and water should remove them.

1-20. The proper CARC paint colors should be used to paint bumper numbers, spot painting, and so forth. Do not use other paints in lieu of CARC.

WEATHER EFFECTS

1-21. If METT-TC allows, leaving the contamination alone is the easiest method of decon. However, marking, reporting, and periodically rechecking the contamination are required.

TEMPERATURE

1-22. The higher the temperature, the faster the rate of evaporation. Contamination persistency decreases as the temperature rises. It has no effect on radiological contamination.

WIND

1-23. Aeration aids in decon. Agents are dispersed by the wind, thus reducing the concentration.

HUMIDITY AND RAIN

1-24. Moisture tends to break down chemical agents but does so slowly. Heavy rain physically removes contamination; however, contaminated runoff may build up in drainage areas. Biological agents dehydrate (dry out) in low humidity, thus reducing their persistency. Rain may prevent the desorption of aerosols and leach contamination into the soil.

SUNLIGHT

1-25. Sunlight hastens the evaporation and decomposition of agents. Ultraviolet and infrared radiation in sunlight rapidly kills most biological agents.

TIME

1-26. Only time will neutralize and destroy radiological contamination. Other techniques merely shield or remove the problem.

DECON IN COMBAT

1-27. Combat effectiveness is degraded by MOPP. Decon can remove the contamination and restore combat power. The detection and warning network is vital to ensure that soldiers and equipment are protected when contamination is encountered. Considerations of decon in combat are—

- Immediate decon. It saves lives and permits the use of individual equipment and key systems.
- Operational decon. It reduces the spread and the level of contamination. In some cases, when combined with weathering, MOPP levels may be reduced without further decon.
- Thorough decon. It removes the unit from the fight but allows it to return with restored effectiveness.

1-28. Figure 1-1 depicts the drop in effectiveness as the affected unit and soldiers react.

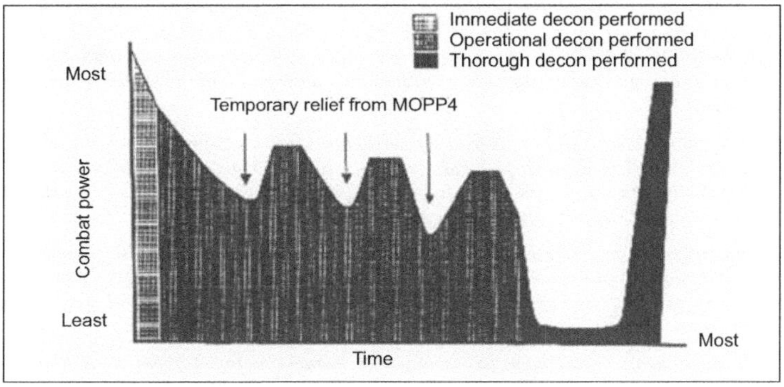

Figure 1-1. How Decon Affects Combat Effectiveness

Decon-Operation Scenario

You are the commander of the 155-millimeter Self-Propelled Howitzer Battery. Your unit is hit by a persistent nerve agent. Soldiers react by taking cover and going to MOPP4. The warning system provided sufficient alert and only a few have to conduct skin decon; most contamination is on the outside of your howitzers and vehicles.

The ongoing fire mission continues with minimal disruption, and the enemy is being detected. Your well-trained soldiers complete a personal wipe down and an operator's spray down as soon as possible. During this period, you are capable of providing fire support and moving, as necessary.

The temperature is 55°F, and humidity is low. While the work rate is strenuous, your troops drink water and suffer no heat casualties. However, after several hours, your response times to the fire mission are getting longer and soldiers have not eaten in more than 12 hours.

Upon making a decision to decon, you request support from your battalion. Coordination is made for an operational decon en route to an alternate position. You have 2 ½ hours to complete the movement and be in a firing position. The alternate position is 20 minutes away, with the decon site about halfway.

You send the advance party to meet the supporting power-driven decon-equipment (PDDE) crew and organize the site. The crew sets up and runs the vehicle washdown while the advance party, with battery supply personnel, prepares the MOPP-gear exchange.

The battalion chemical noncommissioned officer (NCO) directs the vehicle washdown as you rotate one platoon at a time through the site. The firing platoons go quickly as it is critical to have the artillery firing tables available. You complete the operation and are in position to fire at the prescribed time. Your soldiers are in MOPP4 because a vapor hazard still exists from the residual contamination. The advance party found the new area clear of contamination, and ammo resupply commences.

As sections begin improving their positions, gun crews one and two leave one soldier on each gun and move the rest of the soldiers 65 feet upwind. After a check on contamination proves negative, the battery NBC NCO directs unmasking procedures using two crew members. When no symptoms show up, the crews eat and get relief from the masks. The wind direction is monitored to ensure that the soldiers stay upwind of any vapors desorbing from their equipment.

Soldiers on the guns rotate into the clean area for rest and relief. Those returning to the guns assume MOPP4.

Chapter 2

Immediate Decon

Once a soldier is aware of chemical or biological contamination on his bare skin, he initiates immediate decon techniques, without command, by using his personal SDK. He decontaminates his hood, mask, gloves, and weapon using the IEDK or an additional SDK. To remove radiological contamination from equipment and personnel, brush it off and/or use soap and water.

NOTE: Throughout this manual, SDK refers to the M291 kit and IEDK refers to the M295 kit.

SKIN DECON

2-1. Start the skin-decon techniques within 1 minute of becoming contaminated. Some toxic chemical agents, especially nerve agents, kill in minutes.

CHEMICAL

2-2. Use the SDKs within 1 minute of contamination (see Figure 2-1). Instructions for use are listed on the outside of the kit itself, on the individual packet within the kit, and in STP 21-1-SMCT.

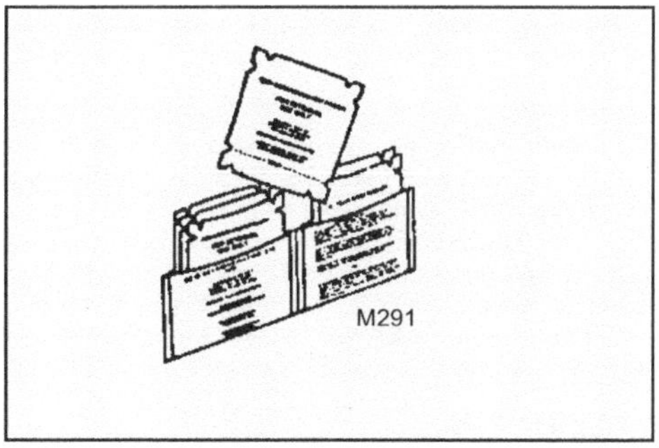

Figure 2-1. M291 SDK

2-3. If an SDK is not available, chemical contamination may be pinch-blotted from the skin with a cloth and flushed with water from a canteen. Soap, if available, can also be used to wash the agent from the skin. Washing with soap and water (or hot water) is the next best method for toxic-agent decon if SDKs are not available, but this method is not as effective as using the decon kits.

BIOLOGICAL

2-4. Washing with soap and water removes nearly all biological agents from the skin. A 0.5 percent chlorine (calcium hypochlorite [HTH] or household bleach) solution is an effective biological decontaminant. See Appendix B for details on how to make this solution.

RADIOLOGICAL

2-5. To remove radiological dust particles, brush, wash, or wipe them off.

PERSONAL WIPE DOWN

2-6. The personal wipe-down techniques are most effective when done within 15 minutes of being contaminated. The CAM and/or M8/M9 detector paper is used to detect and monitor equipment.

NOTE: Very high concentrations of DS2 can cause a CAM response in the G mode.

2-7. Wipe down your mask, hood, gloves, and other essential gear. Do not attempt to remove chemical contamination from your protective overgarment unless there are obvious clumps of the agent. In this case, scrape off the material. Brush off radiological or frozen chemical-agent contamination. See FM 3-4 for detailed information on your overgarment's protective qualities.

CHEMICAL

2-8. Decon individual equipment using the IEDKs. Wearing your Kevlar® helmet protective cover will prevent or reduce the adsorption of any liquid chemical agent. Washing with soap and water and bleach solutions is partially effective (see Appendix B).

BIOLOGICAL

2-9. Wash with soap and water. If water is not available, use SDKs in the same manner as described for chemical-agent decon.

RADIOLOGICAL

2-10. Locate radiological contamination with monitoring equipment and remove by brushing and shaking it off. Wash exposed areas of your skin and pay particular attention to your hair and fingernails. Avoid breathing the dust particles you shake off by wearing your protective mask or a piece of cloth over your nose and mouth. If wet, conduct a MOPP-gear exchange as soon as possible because brushing or shaking will not remove the contamination. Wipe off your equipment with warm, soapy water using rags or damp paper towels.

OPERATOR'S SPRAY DOWN

2-11. Decon other mission-essential portions of your equipment before continuing your mission. The CAM and/or M8/M9 detector paper is used to determine what surfaces require decon.

2-12. Operator's spray down is most effective when done within 15 minutes of contamination.

CHEMICAL

2-13. Decon those surfaces that you must touch on the exterior of the vehicle or equipment you must use to do your job with the on-board portable decon apparatus (such as the M11 or M13) (see Figure 2-2). The IEDK may be used on equipment that DS2 may cause damage to by corrosive action (see Figure 2-3).

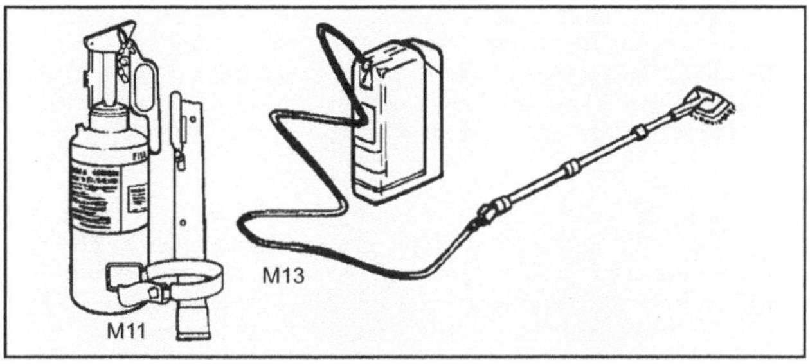

Figure 2-2. Portable Decon Apparatuses

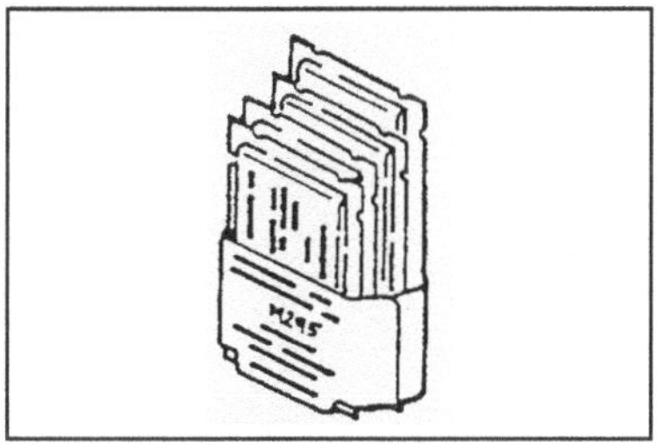

Figure 2-3. M295 IEDK

WARNING
Use extreme caution at all times when handling DS2. Do not mix DS2 and STB because it will cause a fire. Do not allow DS2 to be sprayed on personnel or protective clothing. DS2 is a combustible solution. Do not spray DS2 on hot engine blocks. Severe chemical burns can result if personnel fail to observe all safety precautions. DS2 can severely injure eyes and skin and, if inhaled, can cause illness. DS2 can damage the NBC protective overgarment. Long-term contact with DS2 can damage the NBC protective gloves, hood, and overboots.

2-14. Scrub the DS2 into the exterior surface with brushes. Wait 30 minutes, then wash off. If a decon apparatus is not available, use the field-expedient resources that are available to apply DS2 or STB from bulk containers. If necessary, use the nonstandard decontaminants that are discussed in Appendix B.

BIOLOGICAL

2-15. Use a bleach solution for biological contamination; however, if bleach is not available, use hot, soapy water. Apply with brushes and scrub the surface well. Rinse the surface after scrubbing (no wait is required). DS2 and STB are also effective against most known biological contamination, but because of their caustic nature, they are not preferred. Other nonstandard biological decontaminants are described in Appendix B.

RADIOLOGICAL

2-16. If you are contaminated by fallout, rain out, neutron-induced contamination, or any type of radiological agent, use your monitoring equipment to help locate it and decon as required. If detection equipment is not available and you suspect that you are contaminated, decon. Radiological contamination can usually be removed by brushing or scraping. Water is effective for flushing away radiological contamination; however, the runoff should be controlled by using drainage ditches that flow into a sump. Remember, you have not destroyed the contamination, it has just been moved. The runoff will still be hazardous. If you have time, brush or scoop away the top inch of soil from your fighting position to lower the amount of radiological contamination affecting you.

Chapter 3

Operational Decon

An operational decon limits the spread of contamination, allows temporary relief from MOPP4, and facilitates additional decon requirements. By speeding up the weathering process, the need for a thorough decon may be eliminated. This process requires about a 120-square-yard area and, depending on the number of contaminated vehicles, may not require a nearby water source. An operational decon consists of a vehicle washdown and a MOPP-gear exchange. The techniques that may be used to conduct an operational decon are decentralized company control, centralized battalion control, or centralized brigade control.

TECHNIQUES

3-1. Decentralized company control is provided by the unit commander. The commander requests PDDE support, selects the decon site, links up with the PDDE crew, and executes the operational decon.

3-2. Centralized battalion control is similar to the above except that the battalion chemical NCO controls the decon. The site is selected by the battalion commander, and the battalion chemical NCO directs site setup and security and provides command and control (C^2). The battalion chemical NCO travels with the PDDE and communicates via the radio.

3-3. Centralized brigade control is done when decon assets within the brigade have been consolidated. In this technique, the brigade chemical NCO performs those functions described for the battalion chemical NCO. The advantages and disadvantages of operational-decon techniques are listed in Table 3-1, page 3-2.

PHASES

3-4. The three phases for an operational decon are preparation, execution, and site clearance.

PREPARATION

3-5. The preparation phase starts with the decision to conduct an operational decon and ends with a site that is set up and ready for operation. (The decision to decon was described in Chapter 1.) The CAM is used to identify which vehicles need decontaminating, and PDDE support is requested. Coordination with higher headquarters (HQ) includes identifying the linkup point and the unit that will provide the PDDE support. Signal operating instructions (SOI) information is exchanged and the technique (decentralized company control, centralized battalion control, or centralized brigade control) is selected. The number and type of vehicles and the number of personnel for a MOPP-gear exchange are provided.

Table 3-1. Advantages and Disadvantages of Operational-Decon Techniques

	Decentralized Company Control	Centralized Battalion Control	Centralized Brigade Control
Advantages	Flexibility Dispersion	C^2 Flexibility Dispersion Planning Sustained operations	C^2 Synchronized support assets Rehearsal Time Planning Sustained operations
Disadvantages	C^2 Synchronized support assets Rehearsal Time Planning Sustained operations	Synchronized support assets Rehearsal Time	Flexibility Dispersion

3-6. Site selection is done by the controlling HQ. The site may have been designated in the operation order (OPORD) or selected based on the current METT-TC. For site characteristics, see Table 3-2.

3-7. Linkup at the designated point includes establishing security, positioning any necessary marshalling areas, camouflaging removal points, designating the direction of flow into and out of the site, and ensuring that unit supply has sufficient replacement MOPP gear.

3-8. Site setup requires the PDDE crew to position itself upwind from the entrance. Set up the MOPP-gear exchange about 164 feet upwind from the vehicle washdown at a 45° angle. Notify the contaminated unit when the site is ready.

EXECUTION

3-9. The execution phase starts with establishing positive control over the site and ends with processing the last contaminated vehicle and soldier through the site. Vehicles move from the predecon area to the site either on order or by watching the vehicle ahead. Contaminated soldiers leave the vehicle and proceed to the MOPP-gear-exchange point. The contaminated unit provides security.

3-10. Each vehicle receives a 2- to 3-minute washdown with hot, soapy water and moves to a point near the MOPP-gear exchange. Any remaining soldiers requiring MOPP gear exchange will do so then. When the squad or crew has finished, it remounts its vehicles and moves into an after-decon assembly area (AA) to await further instructions or moves to its next battle position.

SITE CLEARANCE

3-11. The site-clearance phase includes cleanup, marking, and reporting. METT-TC will dictate the cleanup requirements. The PDDE crew buries or double bags and removes contaminated MOPP gear and other expendables. Standard NBC contamination markers are posted, and the PDDE crew submits an NBC 5 report.

Table 3-2. Operational-Decon Phases

Preparation	
Decon assessment	Identify soldiers and equipment to be decontaminated.
Coordination	Request PDDE support. The chemical section conducts coordination with the contaminated unit on the linkup point. Decon operations should be done between 1 and 6 hours after becoming contaminated.
Site selection	Is selected by the controlling HQ. Is off the main route but has easy access. Is a large enough area (120 square yards per site for a squad-size element). Has good overhead concealment. Has a good water source (plan for 100 gallons per vehicle). Has good drainage.
Linkup	Ensure that the noncommissioned officer in charge (NCOIC) knows where to link up with the contaminated unit and the location for site setup. Radio communication is essential for operations.
Site setup	Ensure that the PDDE is positioned properly and ready to dispense hot, soapy water. Ensure that the contaminated unit sets up and operates the MOPP-gear exchange at the same time as the vehicle washdown. Consider contamination runoff when positioning the PDDE.
Execution	
Site control and security	Ensure that the drivers of the contaminated vehicles know when to move into position at the washdown location. Ensure that the contaminated unit has provided site security.
Processing	Ensure that the decon site NCOIC is processing vehicles at a rate of 2 to 3 minutes per vehicle. Also, ensure that soldiers are going through the MOPP-gear exchange at the rate of 30 minutes per squad/crew.
Site Clearance	
Cleanup	Decon site NCOIC ensures that the MOPP-gear-exchange area is cleaned up.
Marking and reporting	Decon site NCOIC ensures that his team properly marks the decon site and sends the NBC 5 report forward.

VEHICLE WASHDOWN

3-12. A vehicle washdown may be conducted with or without PDDE and in either a one- or two-lane configuration. An unsupported washdown requires the contaminated unit to have washing equipment that can produce 60 to 120 pounds per square inch (psi) of water pressure. The capacity to heat water and inject soap increases effectiveness. A supported washdown requires PDDE organic to the unit or from a supporting chemical decon or dual-purpose company. A two-lane washdown is simply a one-lane washdown parallel with another (see Figure 3-1, page 3-4).

3-13. Other configurations are limited only by the constraints of the METT-TC. Figure 3-2, page 3-5, illustrates a dispersed operational-decon setup. Table 3-3, page 3-5, describes the vehicle-washdown process.

MOPP-GEAR EXCHANGE

3-14. A MOPP-gear exchange can be done with a buddy team, a triple buddy team, or an individual (emergency). For planning purposes, estimate about 60 minutes for a squad or a platoon to complete the exchange. Since individual performance is severely degraded after 6 hours at MOPP4, the MOPP-gear exchange should be scheduled before that time.

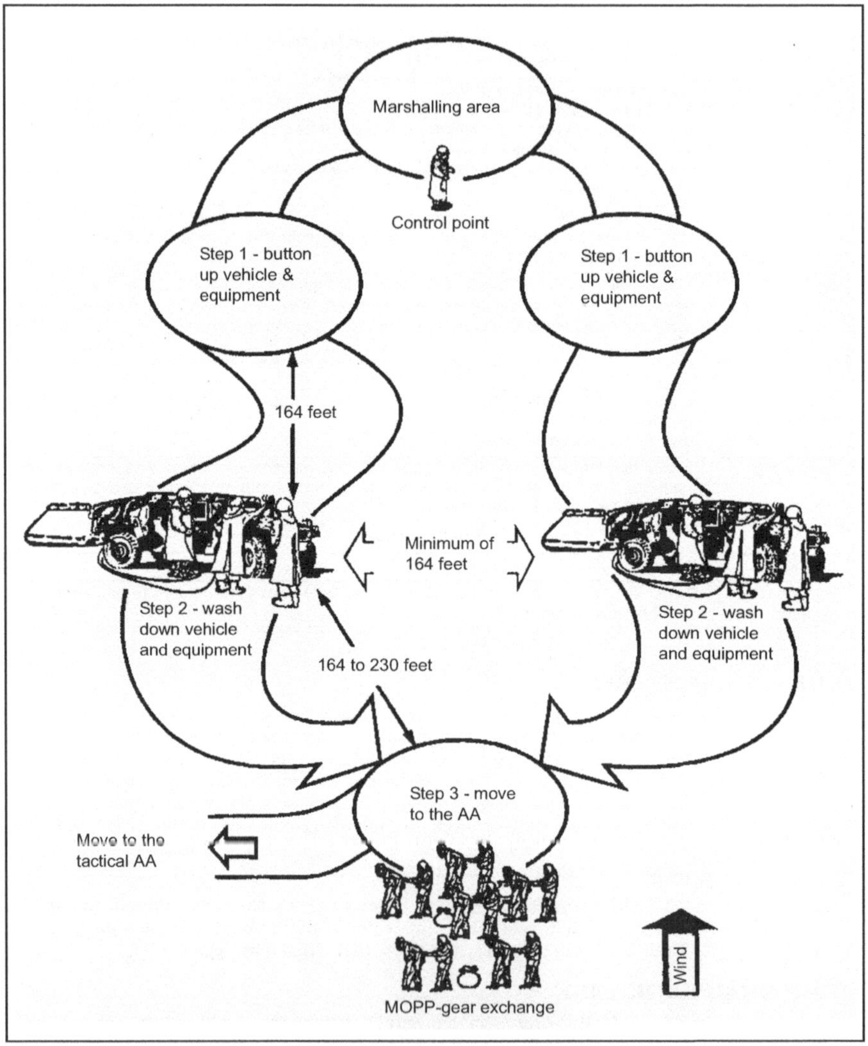

Figure 3-1. Two-Lane Washdown

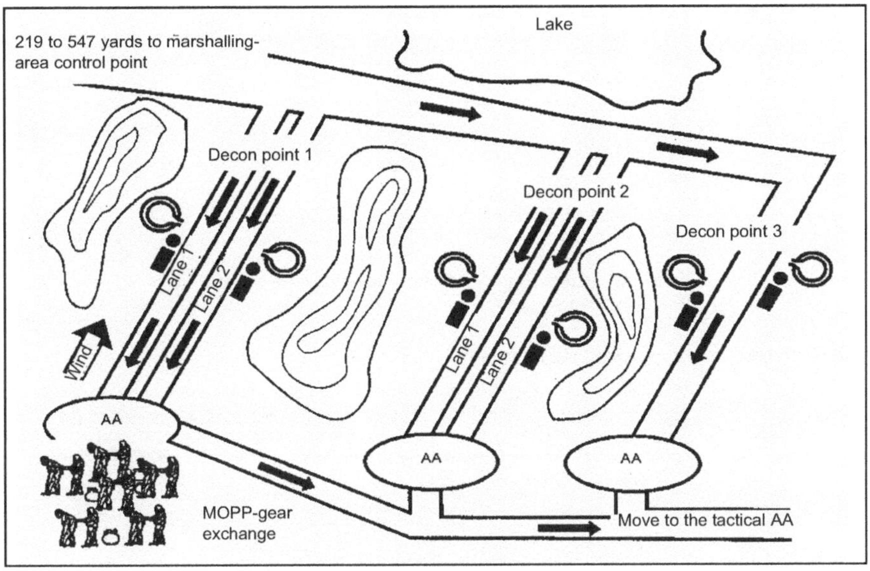

Figure 3-2. Dispersed Operational-Decon Setup

Table 3-3. Vehicle-Washdown Process

Steps	Equipment	Procedures
1. Assemble vehicles. Unit is tactically dispersed. Personnel at the control point direct movement.	Watch	Personnel at the control point supervise preparing vehicles and directing movement out of the AA.
2. Prepare vehicles.	None	Individual/crew closes all access doors, hatches, windows, and other openings. Remove camouflage and cover muzzles. If required, crews (less drivers) move to the MOPP-gear-exchange area. Move to the wash area on order.
3. Perform washdown.	Delivery system (such as the M12 and M17, 65-gallon-per-minute [GPM] pump, fire-fighting equipment, and so forth) that delivers hot, soapy water at 60 to 120 psi. Also sufficient water, fuel, and detergent for vehicles.	Soldiers/personnel wash equipment from top to bottom. Decon crew wears a toxicological apron, protective (TAP) or wet-weather gear over MOPP gear.
4. Move out to the next position.	None	Vehicles move to the MOPP-gear-exchange area (if required) or the next battle position.

BUDDY-TEAM METHOD

3-15. This method uses pairs of soldiers under the supervision of their squad/ team leader to conduct the buddy-team MOPP-gear exchange. See Table 3-4, pages 3-7 through 3-17, for detailed step-by-step procedures for this method. For units equipped with the combat vehicle crewman uniform system (CVCUS)/chemical protective undergarment (CPU), see Table 3-5, pages 3-18 through 3-21.

TRIPLE BUDDY-TEAM METHOD

3-16. This method is used by soldiers equipped with the M40A1, M42, or M43 mask with the quick-doff hood. A third soldier is needed to hold the filter canister and hose to prevent the transfer of contamination. See Table 3-6, pages 3-22 through 3-27, for detailed step-by-step procedures for this method. For units equipped with the CVCUS/CPU, see Table 3-7, pages 3-28 through 3-30.

INDIVIDUAL (EMERGENCY) METHOD

3-17. This method is used only when a soldier does not have buddy to help him and the risk of MOPP-gear failure demands that an exchange occur. See Table 3-8, pages 3-31 through 3-33, for detailed step-by-step procedures for this method. For units equipped with the CVCUS/CPU, see Table 3-9, pages 3-34 through 3-37.

Table 3-4. Steps for the Buddy-Team Method MOPP-Gear Exchange

Steps	Contamination Type	Required Equipment	Procedures
1. Decon gear—removes gross contamination from individual gear (weapons, helmet, load-bearing equipment, and mask carrier).	All contamination	Four long-handled brushes A large piece of plastic (poncho or similar material)	The soldier mixes three parts earth to two parts STB.
	Chemical/ biological	One 30-gallon container, STB dry mix, and one IEDK	The soldier removes and discards the chemical protective helmet cover if worn. He brushes or rubs STB onto his individual gear (helmet and mask carrier) and the hose of the M42 or M43 mask if worn. He gently shakes off any excess STB and sets aside his gear on an uncontaminated surface.
	Radiological	Hot, soapy water	The soldier brushes or wipes radiological contamination from his individual gear. He washes it with hot, soapy water (if available) then sets it aside to dry on an uncontaminated surface (plastic, poncho, or similar material).
2. Prepare to decon—facilitates later removal of overgarment trousers and overboots.	All contamination	Cutting tool	Buddy #1 unfastens the shoulder straps on Buddy #2's hood and pulls them over his shoulder and reattaches them to the Velcro® fasteners. He loosens the drawcord on Buddy #2's hood (see Figure 3-3, page 3-8). He removes the M9 detector paper from Buddy #2's overgarment. He unties the drawcords on the trouser legs of Buddy #2's overgarment. He unzips Buddy #2's trouser legs and rolls a cuff in each trouser leg, ensuring that the cuffs do not come above the tops of his overboots (see Figure 3-4, page 3-8). He unfastens or cuts the fasteners on Buddy #2's overboots. **NOTE: A soldier can do this step by himself or with the help of his buddy. When wearing overboots, unsnap both quick releases on the overboots. The M40 voice amplifier (M7) and the M42A2 detachable microphone cannot be decontaminated and will be disposed of as contaminated waste; however, ensure that these items are contaminated before disposing of them.**
3. Decon mask and hood—removes gross contamination.	Chemical/ biological	Two IEDKs per person	Buddy #1 uses an IEDK to wipe Buddy #2's eye-lens outserts from the top down. **(NOTE: Do not press so hard that you break Buddy #2's face-mask seal.)** He then wipes the rest of Buddy #2's hood from the top of the head to the bottom of the hood (see Figure 3-5, page 3-9). After he has finished wiping Buddy #2's mask, he must wipe his own gloves in preparation for rolling Buddy #2's hood. He starts from the rear and rolls Buddy #2's hood, using 2-inch tucks, until it reaches the center of his head (see Figure 3-6, page 3-9). He rolls the front of Buddy #2's hood tightly under the outlet valve and filter. He ensures that the hood is off of Buddy #2's garment (see Figure 3-7, page 3-10).

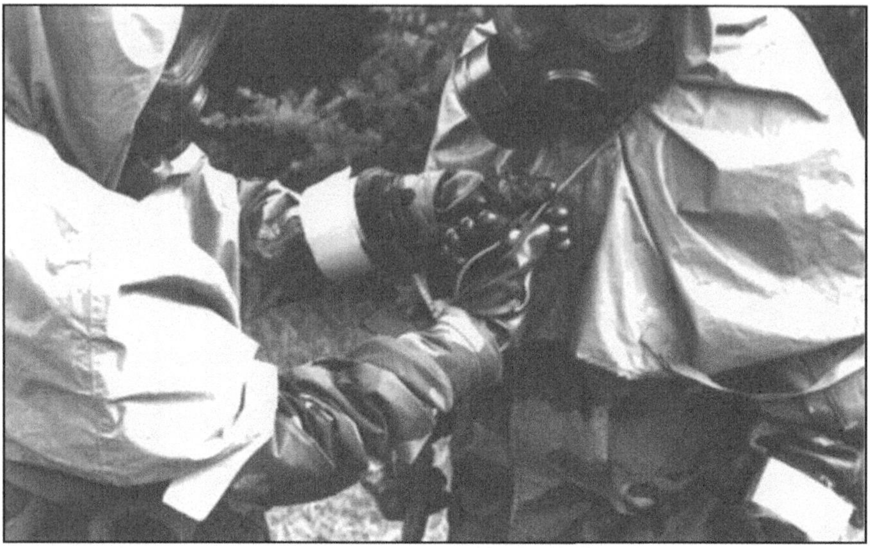

Figure 3-3. Loosening the Drawcord on the Hood

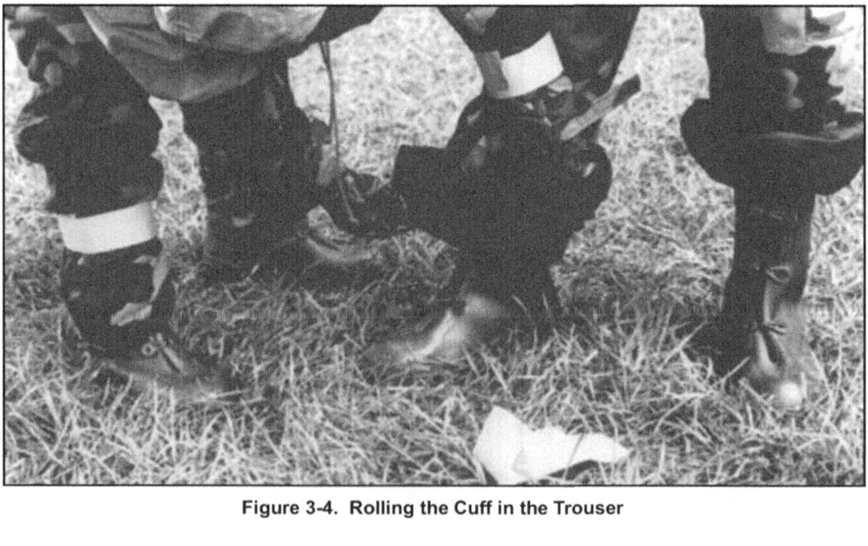

Figure 3-4. Rolling the Cuff in the Trouser

Figure 3-5. Wiping From the Head to the Bottom of the Hood

Figure 3-6. Rolling the Hood to the Center of the Head

Figure 3-7. Ensuring That the Hood is off the Garment

Table 3-4. Steps for the Buddy-Team Method MOPP-Gear Exchange (Continued)

Steps	Contamination Type	Required Equipment	Procedures
3. Decon mask and hood (continued).	Radiological	Three containers (about 3-gallon capacity) Two sponges Soapy water Rinse water Paper towels or similar drying material	Buddy #1 wipes Buddy #2's mask and hood with a sponge dipped in hot, soapy water and rinses them with a sponge dipped in clean water. He dries Buddy #2's mask and hood with paper towels or rags. Buddy #2 wipes his own gloves. **NOTE: Cool, soapy water is not as effective for removing contamination, but it can be used if you scrub longer.** If the water supply is limited, use drinking water from your canteen with a wet sponge or cloth. If water is not available, brush and dust off the radioactive dust particles. ***Do not reverse roles.*** Only Buddy #2's hood will be decontaminated and rolled at this time.
4. Remove overgarments and overboots—limits the spread of agents and helps prevent agents from penetrating through to the undergarments or the skin.	All contamination	Two discard containers (for example, plastic bags)	Buddy #1 grasps Buddy #2's overgarment jacket, unsnaps the snaps individually, and unties the drawcord at the bottom of the jacket (see Figure 3-8, page 3-12). He unfastens the Velcro at the wrist of Buddy #2's jacket and then refastens it. He unfastens the Velcro closure over the zippered front of Buddy #2's jacket and unzips the jacket. He grasps Buddy #2's jacket at the shoulders and instructs him to make a fist. He then pulls Buddy #2's jacket down and away from him, ensuring that the black part of the jacket is not touched (see Figure 3-9, page 3-12). He lays Buddy #2's overgarment jacket on the ground, black side up. (It will be used to stand on later.) He carefully unfastens and unzips Buddy #2's trousers. **NOTE: Do not loosen Buddy #2's waist tabs.** He instructs Buddy #2 to loosen his overboots by alternately stepping on each heel and pulling up on his foot (see Figure 3-10, page 3-13). He grasps Buddy #2's trousers and pulls them down to his knees. He instructs Buddy #2 to walk out of his trousers/overboots simultaneously and step onto the black side of the jacket (see Figure 3-11, page 3-13). Buddy #2 should step onto the jacket wearing his mask, battle-dress uniform (BDU), combat boots, and gloves.
5. Remove gloves.	All contamination	Two discard containers (from step 4)	The soldier holds the fingertips of his gloves and partially slides his hand out. When the fingers of both hands are free, he holds his arms away from his body and lets the gloves drop (see Figure 3-12, page 3-14).

Figure 3-8. Unsnapping the Trousers From the Jacket

Figure 3-9. Pulling the Jacket Down and Away

Figure 3-10. Loosening the Overboots

Figure 3-11. Stepping on the Black Lining of the Jacket

Figure 3-12. Removing the Gloves

Table 3-4. Steps for the Buddy-Team Method MOPP-Gear Exchange (Continued)

Steps	Contamination Type	Required Equipment	Procedures
6. Put on overgarment.	All contamination	One set of chemical protective overgarments per person	Buddy #1 opens the package containing the new overgarment without touching the inside of the package. Buddy #2 pulls out the overgarment without touching the outside of the package. He puts on the overgarment and fastens it, leaving the trouser legs open (see Figure 3-13, page 3-16). ***Do not reverse roles.*** Only Buddy #2 will put on the clean overgarment at this time.
7. Put on overboots and gloves.	All contamination	One set of chemical protective overboots per person One set of chemical protective gloves per person (correct size) M9 detector paper	Buddy #1 opens the package of clean overboots without touching the inside of the package. Buddy #2 removes the overboots from the package without touching the outside of the package, puts them on, and fastens his trouser legs. Buddy #1 opens the package of clean gloves without touching the inside of the package. Buddy #2 removes the gloves from the package without touching the outside of the package and puts them on. Buddy #2 puts on the M9 detector paper (see Figure 3-14, page 3-16). ***Do not reverse roles.*** Only Buddy #2 will put on the clean overboots and gloves at this time.

Figure 3-13. Putting on the New Overgarment

Figure 3-14. Putting on the M9 Detector Paper

Table 3-4. Steps for the Buddy-Team Method MOPP-Gear Exchange (Continued)

Steps	Contamination Type	Required Equipment	Procedures
8. Secure hood.	All contamination	One IEDK	Buddy #1 uses an IEDK to wipe Buddy #2's gloves. He unrolls Buddy #2's hood and attaches the straps and tightens the neck cord. Buddy #1 and Buddy #2 reverse roles and repeat steps 2 through 8.
9. Secure gear.	All contamination	One chemical protective helmet cover per person	The soldier secures his individual gear and puts it back on (see Figure 3-15). He puts on a new chemical protective helmet cover and moves to the AA. He uses the buddy system to check the fit of all secured gear.

Figure 3-15. Securing the Hood

Table 3-5. Steps for the CVCUS/CPU (Buddy-Team Method) MOPP-Gear Exchange

Steps	Contamination Type	Required Equipment	Procedures
1. Decon gear—removes gross contamination from individual gear (weapons, helmet, load-bearing equipment, and mask carrier).	All contamination	Four long-handled brushes A large piece of plastic (poncho or similar material)	The soldier mixes three parts earth to two parts STB.
	Chemical/ biological	One 30-gallon container, STB dry mix, and one IEDK	The soldier removes and discards the chemical protective helmet cover if worn. He brushes or rubs the STB onto his individual gear and the hose of the M42 mask if worn. He gently shakes off any excess STB and sets aside his gear on an uncontaminated surface.
	Radiological	Hot, soapy water	The soldier brushes or wipes radiological contamination from his individual gear. He washes it with hot, soapy water (if available) then sets it aside to dry on an uncontaminated surface (plastic, poncho, or similar material).
2. Prepare for decon—facilitates MOPP-gear removal.	All contamination	Cutting tool	Buddy #1 unfastens the shoulder straps on Buddy #2's hood and pulls them over his shoulder and reattaches them to the Velcro fastener. He loosens the drawcord on Buddy #2's hood. He removes the M9 detector paper from Buddy #2's CVCUS. He unzips the ankles on both legs of Buddy #2's CVCUS. He unfastens or cuts the fasteners on Buddy #2's overboots. **NOTE: A soldier can do this step by himself or with the help of his buddy.**
3. Decon mask and hood—removes gross contamination.	Chemical/ biological	Two IEDKs per person	Buddy #1 uses an IEDK to wipe Buddy #2's eye-lens outserts from the top down. **(NOTE: Do not press so hard that you break Buddy #2's face-mask seal.)** He then wipes the rest of Buddy #2's hood from the top of the head to the bottom of the hood (see Figure 3-5, page 3-9). After he has finished wiping Buddy #2's mask, he must wipe his own gloves in preparation for rolling Buddy #2's hood. He starts from the rear and rolls Buddy #2's hood, using 2-inch tucks, until it reaches the center of his head (see Figure 3-6, page 3-9). He rolls the front of Buddy #2's hood tightly under the outlet valve and filter. He ensures that the hood is off of Buddy #2's garment (see Figure 3-7, page 3-10).

Table 3-5. Steps for the CVCUS/CPU (Buddy-Team Method) MOPP-Gear Exchange (Continued)

Steps	Contamination Type	Required Equipment	Procedures
3. Decon mask and hood (continued).	Radiological	Three containers (about 3-gallon capacity) Two sponges Soapy water Rinse water Paper towels or similar drying material	Buddy #1 wipes Buddy #2's mask and hood with a sponge dipped in hot, soapy water and rinses them with a sponge dipped in clean water. He dries Buddy #2's mask and hood with paper towels or rags. Buddy #2 wipes down his own gloves. **NOTE: Cool, soapy water is not as effective for removing contamination, but it can be used if you scrub longer.** If the water supply is limited, use drinking water from your canteen with a wet sponge or cloth. Only Buddy #2's mask and hood will be decontaminated and rolled at this time.
4. Remove CVCUS/ overboots—limits the spread of contamination.	All contamination	Two discard containers (for example, plastic bags)	Buddy #1 opens all the zippers on Buddy #2's CVCUS in the following order: wrist, ankles, and front. While standing behind Buddy #2 and adjacent to the clean area, Buddy #1 grasps the shoulders of Buddy #2's CVCUS and instructs him to make a fist to ensure that his gloves do not come off. He then pulls Buddy #2's CVCUS down below his knees, turning it inside out. He has Buddy #2 to partially remove his overboots by alternately stepping on each heel and pulling up on his foot. While kneeling, Buddy #1 grasps one leg of Buddy #2's CVCUS and his overboot and instructs him to remove his leg and foot simultaneously (see Figure 3-16, page 3-20). As Buddy #2 removes each leg and foot, he steps directly onto the clean area.
5. Remove CPU jacket/gloves.	All contamination	Two discard containers	Buddy #2 holds the fingertips of his gloves and partially slides his hand out. When the fingers of both hands are free, he holds his arms away from his body and lets the gloves drop. Buddy #1 unzips and removes Buddy #2's CPU jacket by grasping it at the shoulders and pulling down, turning it inside out (see Figure 3-17, page 3-20). He places it on the clean area with the clean side up. He removes Buddy #2's protective glove inserts.

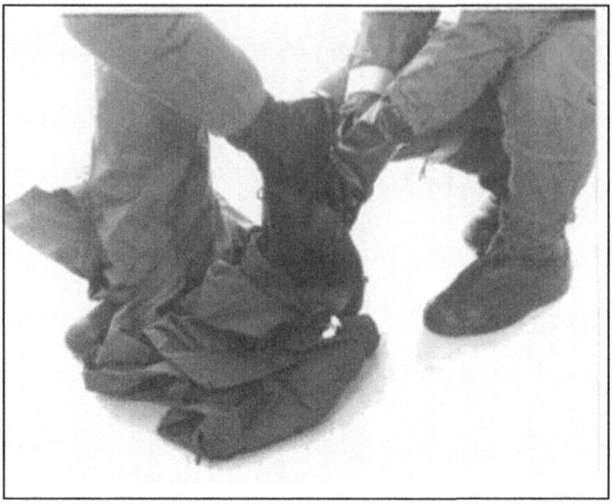

Figure 3-16. Removing Leg and Foot From the Garment

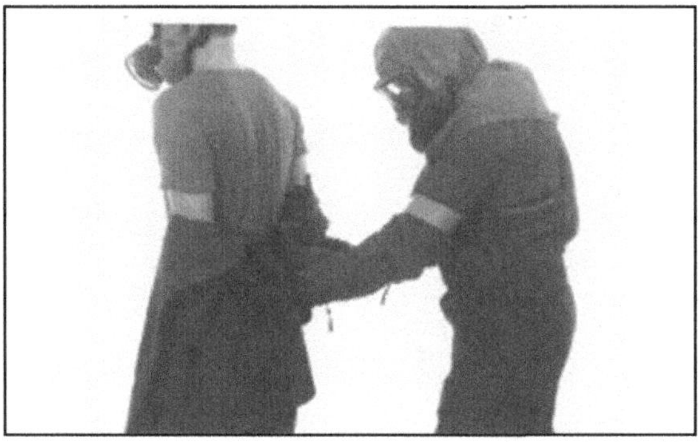

Figure 3-17. Removing the Garment and Turning it Inside Out

Table 3-5. Steps for the CVCUS/CPU (Buddy-Team Method) MOPP-Gear Exchange (Continued)

Steps	Contamination Type	Required Equipment	Procedures
6. Remove combat boots/ CPU pants.	All contamination	Two discard containers	Buddy #1 instructs Buddy #2 to stand on the clean area (but not on the CPU jacket) and unlace his combat boots. Buddy #2 partially removes his boots by alternately stepping on each heel and pulling up on his foot. Buddy #1 grasps one of Buddy #2's boots and instructs him to remove his foot. As Buddy #2 removes his foot, he will step directly onto the clean side of the CPU jacket. Repeat the procedure for the other foot. Next, Buddy #1 removes Buddy #2's CPU drawers by pulling down on the sides and turning them inside out. Buddy #2 is now stripped to his underwear and mask and is ready to put on new clothing. **NOTE: If Buddy #2's feet come in contact with the contaminated pants of the CPU, decon them.**
7. Put on CPU/ CVCUS.	All contamination	One CPU/CVCUS per soldier	Buddy #1 opens the package containing the new undergarment without touching the inside of the package. Buddy #2 pulls out the undergarment without touching the outside of the package. He puts on the new clothing in the following order: CPU drawers, CPU jacket, combat boots, and CVCUS (over the CPU drawers).
8. Put on overboots and gloves.	All contamination	One set of chemical protective overboots per soldier One set of chemical protective gloves per soldier M9 detector paper	Buddy #1 opens the package of clean overboots without touching the inside of the package. Buddy #2 removes the overboots without touching the outside of the package and puts them on. Buddy #1 opens the package of clean gloves without touching the inside. Buddy #2 removes the gloves without touching the outside of the package and puts them on. Buddy #2 puts on the M9 detector paper (see Figure 3-18, page 3-23).
9. Secure hood.	Chemical/ biological	One IEDK	Buddy #1 wipes his rubber gloves with an IEDK. He unrolls Buddy #2's hood and attaches the straps and tightens the neck cord. Buddy #1 and Buddy #2 reverse rolls and repeat steps 2 through 9.
10. Secure gear.	All contamination	One chemical protective helmet cover per person	The soldier secures his individual gear and puts it back on. He puts on a new chemical protective helmet cover and moves to the AA. He uses the buddy system to check the fit of all secured gear.

Table 3-6. Steps for the Triple-Buddy Method MOPP-Gear Exchange With the M40A1 and M42

Steps	Contamination Type	Required Equipment	Procedures
1. Decon gear—removes gross contamination from individual gear (weapons, helmet, load-bearing equipment, and mask carrier).	All contamination	Four long-handled brushes A large piece of plastic (poncho or similar material)	The soldier mixes three parts earth to two parts STB.
	Chemical/ biological	One 30-gallon container, STB dry mix, and one IEDK	The soldier removes and discards the chemical protective helmet cover if worn. He brushes or rubs STB onto his individual gear (helmet and mask carrier) and the hose of the M42 or M43 mask if worn. He gently shakes off any excess STB and sets aside his gear on an uncontaminated surface.
	Radiological	Hot, soapy water	The soldier brushes or wipes radiological contamination from his individual gear. He washes it with hot, soapy water (if available) then sets it aside to dry on an uncontaminated surface (plastic, poncho, or similar material).
2. Prepare to decon—facilitates later removal of overgarment trousers and overboots.	All contamination	Cutting tool	Buddy #1 unfastens the shoulder straps on Buddy #2's hood and pulls them over his shoulder and reattaches them to the Velcro fastener. He loosens the drawcord on Buddy #2's hood and ties off the microphone cord to the hose of his mask (see Figure 3-19). He removes the M9 detector paper from Buddy #2's overgarment. He unzips Buddy #2's trouser legs and rolls a cuff in each trouser leg, ensuring that the cuffs do not come above the tops of his overboots. He unfastens or cuts the fasteners on Buddy #2's overboots. **NOTE: A soldier can do this step by himself or with the help of his buddy.**

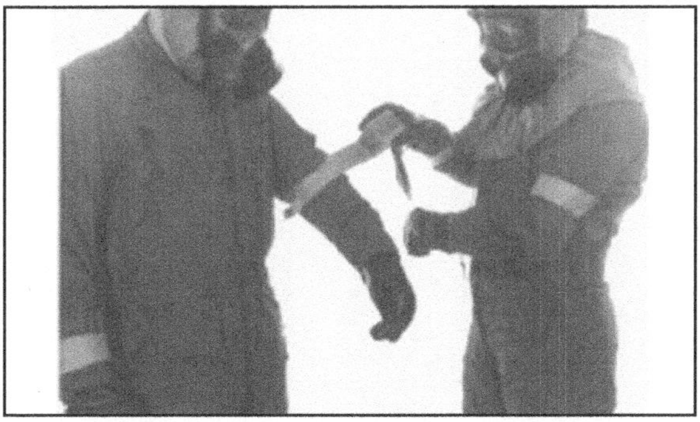

Figure 3-18. Reapplying the M9 Detector Paper

Figure 3-19. Tying the Microphone Cord to the Hose of the Mask

Table 3-6. Steps for the Triple-Buddy Method MOPP-Gear Exchange With the M40A1 and M42 (Continued)

Steps	Contamination Type	Required Equipment	Procedures
3. Decon mask and hood—removes gross contamination.	Chemical/ biological	Two IEDKs per person	Buddy #1 uses an IEDK to wipe Buddy #2's eye-lens outserts from the top down. **(NOTE: Do not press so hard that you break Buddy #2's face-mask seal.)** He then wipes the rest of Buddy #2's hood from the top of the head to the bottom of the hood (see Figure 3-5, page 3-9). After he has finished wiping Buddy #2's mask, he must wipe his own gloves in preparation for rolling Buddy #2's hood. He starts from the rear and rolls Buddy #2's hood, using 2-inch tucks, until it reaches the center of his head (see Figure 3-6, page 3-9). He rolls the front of Buddy #2's hood tightly under the outlet valve and filter. He ensures that the hood is off of Buddy #2's garment (see Figure 3-7, page 3-10).
	Radiological	Three containers (about 3-gallon capacity) Two sponges Soapy water Rinse water Paper towels or similar drying material	Buddy #1 wipes Buddy #2's mask and hood with a sponge dipped in hot, soapy water and rinses them with a sponge dipped in clean water. He dries Buddy #2's mask and hood with paper towels or rags. Buddy #2 wipes down his own gloves. **NOTE: Cool, soapy water is not as effective for removing contamination, but it can be used if you scrub longer.** If the water supply is limited, use drinking water from your canteen with a wet sponge or cloth. Only Buddy #2's mask and hood will be decontaminated and rolled at this time.
4. Remove overgarment overboots—limits the spread of agents and helps prevent agents from penetrating through to the undergarments or the skin.	All contamination	Two discard containers (for example, plastic bags)	Buddy #1 grasps Buddy #2's outside overgarment jacket, unsnaps the snaps individually, and unties the drawcord at the bottom of the jacket. He unfastens the Velcro at the wrist of Buddy #2's jacket and then refastens it. He unfastens the Velcro closure over the zippered front of Buddy #2's jacket and unzips the jacket. Buddy #1 grasps Buddy #2's jacket by the shoulders and instructs him to make a fist. He then pulls the jacket down and away from Buddy #2, ensuring that the black part of the jacket is not touched (see Figure 3-20, page 3-26). He lays Buddy #2's overgarment jacket on the ground, black side up. (It will be used to stand on later.) Buddy #1 carefully unfastens and unzips Buddy #2's trousers. **NOTE: Do not loosen Buddy #2's waist tabs.** He instructs Buddy #2 to break the seals on his overboots by alternately stepping on each heel and pulling up on his foot. He grasps Buddy #2's trousers and pulls them down to his knees. Buddy #1 instructs Buddy #2 to walk out of his trousers/overboots, taking care not to step on the contaminated side of the overgarment. If Buddy #2 is wearing overboots, Buddy #1 removes Buddy #2's trousers first then helps him step out of his overboots onto the black side of the jacket. Buddy #2 should step onto the jacket wearing his mask, BDU, combat boots, and gloves.

Table 3-6. Steps for the Triple-Buddy Method MOPP-Gear Exchange With the M40A1 and M42 (Continued)

Steps	Contamination Type	Required Equipment	Procedures
5. Remove gloves.	All contamination	Discard container (from Step 4)	The soldier holds the fingertips of his gloves and partially slides his hand out. When the fingers of both hands are free, he holds his arms away from his body and lets the gloves drop.
6. Put on overgarment.	All contamination	One set of chemical protective overgarments per person	Buddy #1 opens the package containing the new overgarment without touching the inside of the package. Buddy #2 pulls out the overgarment without touching the outside of the package. He puts on the overgarment and fastens it, leaving the trouser legs open. ***Do not reverse roles.*** Only Buddy #2 will put on the clean overgarments at this time.

Figure 3-20. Removing the Outer Garment

Table 3-6. Steps for the Triple-Buddy Method MOPP-Gear Exchange With the M40A1 and M42 (Continued)

Steps	Contamination Type	Required Equipment	Procedures
7. Put on overboots and gloves.	All contamination	One set of chemical protective overboots per person One set of chemical protective gloves per person M9 detector paper	Buddy #1 opens the package of clean overboots without touching the inside of the package. Buddy #2 removes the overboots without touching the outside of the package, puts them on, and fastens his trouser legs. Buddy #1 opens the package of clean gloves without touching the inside of the package. Buddy #2 removes the gloves without touching the outside of the package and puts them on. Buddy #2 puts on the M9 detector paper. ***Do not reverse roles.*** Only Buddy #2 will put on clean overboots and gloves at this time. **NOTE: The gloves and overboots will have a light powdery coating. This is normal. It is not necessary to remove it. It will not affect the protective qualities.**
8. Secure hood.	All contamination	One IEDK	Buddy #1 wipes his rubber gloves with an IEDK. He unrolls Buddy #2's hood and attaches the straps and tightens the neck cord. He checks the clips and neck cord on Buddy #2's hood. Buddy #1 and Buddy #2 reverse roles and repeat steps 2 through 8.
9. Secure gear.	All contamination	One chemical protective helmet cover per person	The soldier secures his individual gear and puts it back on. He puts on a new chemical protective helmet cover and moves to the AA. He uses the buddy system to check the fit of all secured gear.

Table 3-7. Steps for the CVCUS/CPU (Triple-Buddy Method) MOPP-Gear Exchange With the M40A1, M42, and M43 (Aviation and Armor)

Steps	Contamination Type	Required Equipment	Procedures
1. Decon gear—removes gross contamination from individual gear (weapons, helmet, load-bearing equipment, and mask carrier).	All contamination	Four long-handled brushes A large piece of plastic (poncho or similar material)	The soldier mixes three parts earth to two parts STB.
	Chemical/ biological	One 30-gallon container, STB dry mix, and one IEDK	The soldier removes and discards the chemical protective helmet cover if worn. He brushes or rubs STB onto his individual gear (helmet and mask carrier) and the hose of the M43 mask if worn. He gently shakes off any excess STB and sets aside his gear on an uncontaminated surface.
	Radiological	Hot, soapy water	The soldier brushes or wipes radiological contamination from his individual gear. He washes it with hot, soapy water (if available) and sets it aside to dry on an uncontaminated surface (plastic, poncho, or similar material).
2. Prepare to decon—facilitates later removal of overgarment trousers and overboots.	All contamination	Cutting tool	Buddy #1 unfastens the shoulder straps on Buddy #2's hood, pulls them over his shoulder, and reattaches them to the Velcro fastener. He loosens the drawcord on Buddy #2's hood and ties off the microphone cord to the hose of his mask. He removes the M9 detector paper from Buddy #2's overgarment. He unzips Buddy #2's trouser legs and rolls a cuff in each trouser leg, ensuring that the cuffs do not come above the top of his overboots. He unfastens or cuts the fasteners on Buddy #2's overboots. **NOTE: A soldier can do this step by himself or with the help of his buddy.**

Table 3-7. Steps for the CVCUS/CPU (Triple-Buddy Method) MOPP-Gear Exchange With the M40A1, M42, and M43 (Aviation and Armor) (Continued)

Steps	Contamination Type	Required Equipment	Procedures
3. Decon mask and hood—removes gross contamination.	Chemical/ biological	Two IEDKs per person	If wearing the M42 or M43 mask, use the buddy-team method (three buddies). Buddy #1 uses an IEDK to wipe Buddy #2's hood, mask, canister, and hose of the M42 or M43 mask. **NOTE: Do not press so hard that you break Buddy #2's face-mask seal.** He uses as many wipes as necessary. Buddy #3 wipes off his gloves and holds Buddy #1's hose and canister away from his contaminated overgarment. After Buddy #1 wipes Buddy #2's mask and then his own gloves, he rolls Buddy #2's hood. He starts from the rear and rolls Buddy #2's hood, using 2-inch tucks, until it reaches the center of his head (see Figure 3-6, page 3-9). He rolls the front of Buddy #2's hood tightly under the outlet valve and filter. He ensures that the hood is off of Buddy #2's garment (see Figure 3-7, page 3-10).
	Radiological	Three containers (about 3-gallon capacity) Two sponges Soapy water Rinse water Paper towels or similar drying material	Buddy #1 wipes Buddy #2's mask and hood with a sponge dipped in hot, soapy water and rinses them with a sponge dipped in clean water. He dries Buddy #2's mask and hood with paper towels or rags. Buddy #2 wipes his own gloves. **NOTE: Cool, soapy water is not as effective for removing contamination, but it can be used if you scrub longer.** If the water supply is limited, use drinking water from you canteen with a wet sponge or cloth. If water is not available, brush and dust off the radioactive dust particles. ***Do not reverse roles***. Only Buddy #2's hood will be decontaminated and rolled at this time.
4. Remove CVCUS overboots—limits the spread of contamination.	All contamination	Two discard containers (for example, plastic bags)	Buddy #1 opens all the zippers on Buddy #2's CVCUS in the following order: wrist, ankles, and front. While standing behind Buddy #2 and adjacent to the clean area, Buddy #1 grasps the shoulders of Buddy #2's CVCUS and instructs him to make a fist to ensure that his gloves do not come off. He then pulls Buddy #2's CVCUS down below his knees, turning it inside out. He has Buddy #2 to partially remove his overboots by alternately stepping on each heel and pulling up on his foot. While kneeling, Buddy #1 grasps one leg of Buddy #2's CVCUS and his overboot and instructs him to remove his leg and foot simultaneously. As Buddy #2 removes each leg and foot, he steps directly onto the clean area. Buddy #1 decons his gloves again.

Table 3-7. Steps for the CVCUS/CPU (Triple-Buddy Method) MOPP-Gear Exchange With the M40A1, M42, and M43 (Aviation and Armor) (Continued)

Steps	Contamination Type	Required Equipment	Procedures
5. Remove CPU jacket/ gloves.	All contamination	Two discard containers	Buddy #2 holds the fingertips of his gloves and partially slides his hand out. When the fingers of both hands are free, he holds his arms away from his body and lets the gloves drop. Buddy #1 unzips and removes Buddy #2's CPU jacket by grasping it at the shoulders and pulling down, turning it inside out. He places it on the clean area with the clean side up. He removes Buddy #2's protective glove inserts.
6. Remove combat boots/ CPU pants.	All contamination	Two discard containers	Buddy #1 instructs Buddy #2 to stand on the clean area (but not on the CPU jacket) and unlace his combat boots. Buddy #2 partially removes his boots by alternately stepping on each heel and pulling up on his foot. Buddy #1 grasps one of Buddy #2's boots and instructs him to remove his foot. As Buddy #2 removes his foot, he steps directly onto the clean side of the CPU jacket. Repeat the procedure on the other foot. Next, Buddy #1 removes Buddy #2's CPU drawers by pulling down on the sides and turning them inside out. Buddy #2 is now stripped to his underwear and mask and is ready to put on new clothing. **NOTE: If Buddy #2's feet come in contact with the contaminated pants of the CPU, decon them.**
7. Put on CPU/ CVCUS.	All contamination	One CPU/CVCUS per person	Buddy #1 opens the package containing the new undergarment without touching the inside. Buddy #2 pulls out the undergarment without touching the outside of the package. He puts on the new clothing in the following order: CPU drawers, CPU jacket, combat boots, and CVCUS (over the CPU drawers).
8. Put on overboots and gloves.	All contamination	One set of chemical protective over-boots per person One set of chemical protective gloves per person M9 detector paper	Buddy #1 opens the package of clean overboots without touching the inside of the package. Buddy #2 removes the overboots without touching the outside of the package and puts them on. Buddy #1 opens the package of clean gloves without touching the inside of the package. Buddy #2 removes the gloves without touching the outside of the package and puts them on. Buddy #2 puts on the M9 detector paper.
9. Secure hood.	Chemical/ biological	One IEDK	Buddy #1 wipes his rubber gloves with an IEDK. He unrolls Buddy #2 hood and attaches the straps and tightens the neck cord. He checks the snaps and neck cord on Buddy #2's hood to ensure that they are closed. Buddy #1 and Buddy #2 reverse rolls and repeat steps 2 through 9.
10. Secure gear.	All contamination	One chemical protective helmet cover per person	The soldier secures his individual gear and puts it back on. He puts on a new chemical protective helmet cover and moves to the AA. He uses the buddy system to check the fit of all secured gear.

Table 3-8. Steps for the Individual (Emergency Method) MOPP-Gear Exchange

Steps	Contamination Type	Required Equipment	Procedures
1. Decon gear—removes gross contamination from individual gear (weapon, helmet, load-bearing equipment, and mask carrier).	Chemical/ biological	One IEDK	The soldier uses M8 detector paper to determine the areas of gross contamination and field-expedient absorbents (sand, dirt, or rags) to remove gross liquid contamination. He should take special care to avoid touching these areas during overgarment removal. He uses an IEDK to decon his individual gear (helmet, load-carrying equipment (LCE), weapon, and mask carrier).
	Radiological	Hot, soapy water	The soldier brushes or wipes radiological contamination from his individual gear. He washes it with hot, soapy water (if available) then sets it aside to dry on an uncontaminated surface (plastic, poncho, or similar material).
2. Prepare to decon—facilities removal of overgarment trousers and overboots.	All contamination	Cutting tool	The soldier unfastens the shoulder straps on his hood, pulls them over his shoulders, and reattaches them. He loosens the drawcord on the hood of his protective mask. **NOTE: If wearing the M43 protective mask, tie off the microphone cord to the hose of the mask.** He removes the M9 detector paper from his overgarment and unties/cuts the drawcords on the trouser legs of his overgarment. He unzips the trouser legs and rolls a cuff in each trouser leg, ensuring that the cuffs do not come above the top of his overboots. He unfastens or cuts the fasteners on the overboots.
3. Decon mask and hood—removes gross contamination from the mask and hood.	Chemical/ biological	One IEDK	The soldier wipes the eye-lens outserts on his mask from the top down. He wipes his mask and gloves and rolls his hood. He grasps the straps of the hood and lifts the hood off his shoulders and partially over his head until most of the back of his head is exposed. He rolls the hood, starting at the chin, and works around the entire mask until the rolled hood will stay up and off of his shoulders. He tucks the straps and neck cord into the roll. He rolls the hood tightly against his mask without pulling the hood off the back of his head. He tucks the tail between the upper part of the canister and the mask. (Tie the tail over and under the hose for the M42 mask.) The soldier removes the applicator mitt from the package with his nondominant hand. Making a "V", he wipes down his hand, paying particular attention to the areas between his fingers. Once he has thoroughly wiped down his dominant hand, he must insert it into the applicator mitt and thoroughly wipe down his other hand. He gently pats the voicemitter with black powder until it is covered.

Table 3-8. Steps for the Individual (Emergency Method) MOPP-Gear Exchange (Continued)

Steps	Contamination Type	Required Equipment	Procedures
3. Decon mask and hood (continued).			The soldier starts at the top of his hood and wipes down and away, patting until the surface of the hood is covered by the black powder. He rewipes his gloves, starting with his nondominant hand. The soldier lifts the hood off of his shoulders by grasping the shoulder straps in one hand and placing the other hand on top of the head and pulling the hood over his head until the elastic band is over his knuckles and most of the back of his head is exposed. He should not expose his ears or pull the hood completely over his face or mask. He tucks his shoulder straps, underarm straps, and the rolled portion of his hood under the elastic band. Follow these procedures when using the M40A1 mask with the quick-doff hood: The soldier removes the underarm straps from the front of his hood and places them over his shoulders. He refastens them on the front of his hood. **NOTE: Anytime you decon your mask or roll your hood, place two fingers on the voicemitter of the mask to prevent accidental breakage of the seal.**
	Radiological	Three containers (about 3-gallon capacity) Two sponges Soapy water Rinse water Paper towels or similar drying material	The soldier wipes his mask and hood with a sponge dipped in hot, soapy water and rinses them with a sponge dipped in clean water. He dries the mask and hood with paper towels or rags. **NOTE: Cool, soapy water is not as effective for removing contamination, but it can be used if you scrub longer.** If water is not available, soldier should brush and dust off the radioactive dust particles.
4. Remove overgarment/ overboots—prevents agent from penetrating through to the undergarments or the skin.	All contamination	Two discard containers (recommend plastic bags). Chemical protective suit with protective gloves and overboots	The soldier grasps his overgarment jacket and unsnaps the snaps individually. He unties the drawcord at the bottom of the jacket. He unfastens the Velcro at the waist and then refastens it. He unfastens the Velcro closure over the zippered front of the jacket and unzips the jacket. He grasps the front of the jacket and pulls the jacket back until it is off his shoulders. He puts his arms behind his back and works his arms out of the sleeves. He should not let the outside of the jacket touch his body. When the jacket is off, he lays it on the ground with the black side up. He unfastens and unzips his trousers. He should not loosen his waist tabs. He loosens his overboots by alternately stepping on each heel and pulling up on his foot. He grasps his trousers and pushes them down to his knees. He walks out of his trousers/ overboots simultaneously and steps onto the black side of the jacket. He steps onto the jacket wearing his mask, BDU, combat boots, and gloves.

Table 3-8. Steps for the Individual (Emergency Method) MOPP-Gear Exchange (Continued)

Steps	Contamination Type	Required Equipment	Procedures
5. Remove gloves.	All contamination	Two discard containers (from step 4) One IEDK	The soldier wipes around the edges of the packages containing the new items (gloves, overgarments, and overboots) with an IEDK. He opens the new packages. He holds the fingertips of his gloves and partially slides his hands out. When the fingers of both hands are free, he holds his arms away from his body and lets his gloves drop.
6. Put on overgarment.	All contamination	One set of chemical protective overgarments per person One IEDK	The soldier removes the overgarment from its package without touching the outside. He puts on the overgarment and fastens it, leaving the trouser legs open until he puts on the new overboots.
7. Put on overboots and gloves.	All contamination	One set of chemical protective overboots per person One set of chemical protective gloves per person	The soldier removes the overboots (one at a time) from their package without touching the outside and puts them on. He removes the gloves from their package without touching the outside, puts them on, and fastens his trouser legs. **NOTE: The gloves and overboots will have a light powdery coating. This is normal. Do not remove it. It will not affect the protective qualities.**
8. Secure hood.	All contamination	One chemical protective helmet cover per person	The soldier secures his individual gear and puts it back on. He puts on a new chemical protective helmet cover and moves to the AA.

Table 3-9. Steps for the CVCUS/CPU, Individual (Emergency Method) MOPP-Gear Exchange

Steps	Contamination Type	Required Equipment	Procedures
1. Decon gear—removes gross contamination from individual gear (weapon, helmet, load-bearing equipment, and mask carrier).	Chemical/ biological	One IEDK	The soldier uses M8 detector paper to determine the areas of gross contamination and field-expedient absorbents (sand, dirt, or rags) to remove gross liquid contamination. He should take special care to avoid touching these areas during overgarment removal. He uses an IEDK to decon his individual gear (helmet LCE, weapon, and mask carrier).
	Radiological	Hot, soapy water	The soldier brushes or wipes radiological contamination from his individual gear. He washes it with hot, soapy water (if available) then sets it aside to dry on an uncontaminated surface (plastic, poncho, or similar material).
2. Prepare to decon—facilities removal of overgarment trousers and overboots.	All contamination	Cutting tool	The soldier unfastens the shoulder straps on his hood, pulls them over his shoulders, and reattaches them to the front of the hood. He loosens the drawcord on the hood of his protective mask. **NOTE: If wearing the M43 protective mask, tie off the microphone cord to the hose of the mask.** He removes the M9 detector paper from his overgarment and unties/cuts the drawcords on the trouser legs of his overgarment. He unzips the trouser legs and rolls a cuff in each trouser leg, ensuring that the cuffs do not come above the top of his overboots. He unfastens or cuts the fasteners on the overboots.
3. Decon mask and hood—removes gross contamination from the mask and hood.	Chemical/ biological	One IEDK	The soldier wipes the eye-lens outserts on his mask from the top down. He wipes his mask and gloves and rolls his hood. He grasps the straps of the hood and lifts the hood off his shoulders and partially over his head until most of the back of his head is exposed. He rolls the hood, starting at the chin, and works around the entire mask until the rolled hood will stay up and off of his shoulders. He tucks the straps and neck cord into the roll. He rolls the hood tightly against his mask without pulling the hood off the back of his head. He tucks the tail between the upper part of the canister and the mask. (Tie the tail over and under the hose for the M42 mask.) The soldier removes the applicator mitt from the package with his nondominant hand. Making a "V", he wipes down his hand, paying particular attention to the areas between his fingers. Once he has thoroughly wiped down his dominant hand, he must insert it into the applicator mitt and thoroughly wipe down his other hand.

Table 3-9. Steps for the CVCUS/CPU, Individual (Emergency Method) MOPP-Gear Exchange (Continued)

Steps	Contamination Type	Required Equipment	Procedures
3. Decon mask and hood (continued).			The soldier gently pats the voicemitter with black powder until it has been covered. He starts at the top of his hood and wipes down and away, patting until the surface of the hood is covered by the black powder. He rewipes his gloves, starting with his nondominant hand. The soldier lifts the hood off of his shoulders by grasping the shoulder straps in one hand and placing the other hand on top of the head and pulling the hood over his head until the elastic band is over his knuckles and most of the back of his head is exposed. He should not expose his ears or pull the hood completely over his face or mask. He tucks his shoulder straps, underarm straps, and the rolled portion of his hood under the elastic band. Follow these procedures when using the M40A1 mask with the quick-doff hood: The soldier removes the underarm straps from the front of his hood and places them over his shoulders. He refastens them on the front of his hood. **NOTE: Anytime you decon your mask or roll your hood, place two fingers on the voicemitter of the mask to prevent accidental breakage of the seal.**
	Radiological	Three containers (about 3-gallon capacity) Two sponges Soapy water Rinse water Paper towels or similar drying material	The soldier wipes his mask and hood with a sponge dipped in hot, soapy water and rinses them with a sponge dipped in clean water. He dries the mask and hood with paper towels or rags. **NOTE: Cool, soapy water is not as effective for removing contamination, but it can be used if you scrub longer.** If water is not available, the soldier should brush and dust off the radioactive dust particles.
4. Remove CVCUS/ overboots.	All contamination	Two discard containers (plastic bags)	The soldier opens all the zippers on his CVCUS in the following order: wrists, ankles, and front. While standing adjacent to the clean area, he grasps the front of his CVCUS and pulls it off of his shoulders. He makes a fist to ensure that his gloves do not come off when removing his CVCUS. He puts his arms behind his back and works his arms out of the sleeves. He loosens his overboots by alternately stepping on each heel and pulling up on his foot. He pulls his CVCUS down below his knees, turning it inside out. He should not let the outside of his CVCUS touch his body. He walks out of his CVCUS/overboots simultaneously and steps onto the clean area.

Table 3-9 Steps for the CVCUS/CPU, Individual (Emergency Method) MOPP-Gear Exchange (Continued)

Steps	Contamination Type	Required Equipment	Procedures
5. Remove jacket/ gloves.	All contamination	Two discard containers	The soldier wipes his gloves and around the edges of the packages containing the new items (gloves, undergarments, and overboots) with an IEDK. He opens the new packages. He holds the fingertips of the gloves and partially slides his hands out. When the fingers of both hands are free, he holds his arms away from his body and lets the gloves drop. The soldier opens all Velcro wrist fasteners and unzips his CPU jacket. He removes the CPU jacket by grasping it at the shoulders and pulling down, turning it inside out. He places the jacket on a clean area with the clean side up. He removes the protective glove inserts.
6. Remove combat boot/CPU pants.	All contamination	Two discard containers	The soldier unlaces his combat boots, removes them, and steps directly onto his CPU jacket. He takes off his CPU drawers by placing his hands beneath the waistband and removing them, turning them inside out. The soldier is now stripped to his underwear and mask and is ready to put on his new clothing. The soldier follows these procedures if wearing the BDU. He stands on the clean area (but not on the CPU jacket) and unlaces his combat boots. He partially removes his boots by alternately stepping on each heel and pulling up on his foot. He removes each trouser leg and combat boot simultaneously and steps directly onto the clean side of the CPU jacket.
7. Put on CPU.	All contamination	One set of chemical protective undergarments per soldier	The soldier removes the undergarment from the package without touching the outside. He puts on the undergarment in the following order: CPU drawers, CPU jacket, combat boots, and CVCUS (over the CPU drawers).
8. Put on overboots and gloves.	Chemical/ biological	One set of chemical protective overboots per soldier One set of chemical protective gloves per soldier M9 detector paper	The soldier removes the overboots from their package without touching the outside and puts them on. He removes the gloves from their package without touching the outside and puts them on. He puts on the M9 detector paper.

Table 3-9. Steps for the CVCUS/CPU, Individual (Emergency Method) MOPP-Gear Exchange (Continued)

Steps	Contamination Type	Required Equipment	Procedures
9. Secure hood.	Chemical/ biological	One IEDK	The soldier wipes his rubber gloves with an IEDK. He unrolls his hood, attaches the straps, and tightens the neck cord.
10. Secure gear.	All contamination	One chemical protective helmet cover per soldier	The soldier secures his individual gear and puts it back on. He puts on a new chemical protective helmet cover and moves to the AA.

Chapter 4

Thorough Decon

Thorough-decon operations reduce and sometimes eliminate contamination. They restore combat power by removing nearly all contamination from unit and individual equipment so that troops can operate it safely for extended periods at reduced MOPP levels. Since a small risk of contamination remains, periodic checks with the CAM, the M8/M9 detector paper, or the M256A1 detector kit must be performed after each operation. A thorough decon can be done as part of reconstitution, or it can be done to support operations throughout the battlefield. After a thorough decon, the unit moves out of the decon site into a tactical AA. The unit, while in this tactical assembly, may undergo reconstitution or prepare for future operations. Combat-service-support (CSS) elements replenish combat stocks, refit equipment, and replace personnel and equipment, as required. The contaminated unit, with some assistance from a chemical unit, does the DTD. A supporting chemical unit does the DED or DAD. The planning considerations that are required to conduct thorough-decon operations and the methods that various chemical units (decon and smoke/decon) use to conduct DTD/DED are discussed in this chapter. The exact layout of a thorough-decon site is determined by METT-TC.

PLANNING

4-1. A thorough decon is the most effective type of decon but is the most resource intensive. A combination of weather conditions, agent used, mission requirements, time, troops, and supplies available may make weathering the decon option of choice. Commanders decide which type of decon is required based on the recommendation of the chemical personnel.

4-2. Generally, thorough-decon operations are conducted beyond the range of enemy direct-fire systems. If a contaminated unit requires a thorough decon as part of reconstitution operations, the decon site is established near the reconstitution area or the brigade's rear area. Usually, company-sized units are reconstituted in their brigade's rear area while battalion-level units are reconstituted in their division's rear area. Organizations larger than a battalion will be reconstituted in the corps's rear area. Further information about reconstitution can be found in FMs 100-5 and 100-9.

4-3. All echelons prepare for thorough-decon operations as part of the overall planning process. From the commander's general guidance, the chemical staff can begin to develop the decon plan. Coordination with higher HQ is required to determine the availability of engineer support for site preparation and closure. Engineers can provide support for sump preparation, drainage ditches, and sumps. Coordinate with the civil-affairs office for host-nation support (personnel, equipment, and supplies) and for environmental requirements and restrictions.

4-4. The chemical staff selects possible decon sites throughout the unit's area of operation (AO) based on such factors as decon type, terrain, mission, threat, road network, and availability of water. After the decon sites are selected, linkup points are chosen to support each site. A site may have more than one linkup point.

4-5. Since decon assets are limited, the commander must establish a priority of decon support. The priority of support lists the units in the order they will receive decon support. This can change from phase to phase during an operation. The chemical staff develops the priority of support based on an understanding of the commander's intent.

4-6. To give priority of support to the lead task force during the assault phase may not be the best choice since any contaminated elements will most likely not stop for decon until after the assault is completed. The commander should establish a priority of work that specifies the order in which equipment will be decontaminated. For example, a priority of work may be in this order: engineer equipment, artillery pieces, main battle tank, and long-haul vehicles. A limiting factor is the availability of water. A typical vehicle requires 500 gallons of water during the DED. The actual amount of water required varies by the vehicle and its contamination level. The supported unit's chemical staff must develop a water resupply plan for thorough-decon operations.

4-7. A water resupply plan can be as simple as selecting a series of linkup points along a route where the chemical unit can link up with a bulk water truck. More complex water resupply plans include caching water throughout the AO, coordinating for the movement of water bladders by aircraft, and identifying water sources in the unit's AO. The use of nonpotable, salt, and brackish waters should be considered.

PREPARATION

4-8. Subordinate units review their higher HQ decon plan. If additional decon sites or assets are required, request them from the higher HQ. The supporting decon-platoon leader conducts a leader's recon of the designated decon sites. Since the decon sites are initially selected by map recon, the actual site may not be suitable. If the selected site cannot support decon operations, the chemical unit leader attempts to find another site close to the original site and notifies the supported unit and his higher HQ of the change.

4-9. The decon-platoon unit leader determines the support requirements that are needed to operate the decon site. Engineers are required to dig sumps, improve access to the site, and dig ditches for runoff. Air defense (AD) and military police (MP) are required to provide security. Medical support is also required. Depending on the operational situation and the threat, some DED sites may be prepared in advance. Routes must be established and published in the OPORD, leading the contaminated unit to the linkup point.

NOTE: Units must be informed not to proceed directly to the decon site.

EXECUTION

4-10. The actual decon operation begins once a unit requests decon support. The unit leader must assess his situation and accurately report it to his HQ. This situation report should include the type and extent of contamination, the current location, his assessment of the unit's ability to perform the current mission while contaminated, and the time he expects the unit to become combat ineffective. The contaminated unit should perform immediate-decon techniques to increase its survivability and to limit the spread of contamination. An operational decon should also be considered and performed, as appropriate. The request for decon support must contain several essential elements of information to assist the chemical staff and the decon-platoon leader in coordinating the decon operation. These essential elements of information are the—

- Designation of the contaminated unit.
- Location of the contaminated unit.
- Time the unit became contaminated.
- Earliest possible time the unit can move/begin decon.
- Type of contamination.
- Number and type of vehicles contaminated.
- Special requirements (patient-decon station, recovery assets, unit decon team, and so forth).
- Supported unit's frequency and call sign to facilitate initial linkup.

4-11. The supporting chemical unit is given a warning order (WO) to conduct a decon operation. Subsequent WOs provide more detailed information. After requesting decon support, the supported unit issues an OPORD/fragmentary order (FRAGO) to the chemical unit.

4-12. The supported unit must coordinate the movement of the contaminated unit to the linkup point. If the contaminated unit is out of the supported unit's AO, the move must be coordinated with the higher HQ.

4-13. Once the chemical unit is set up, the decon-platoon leader moves to the linkup point. Other elements may provide assets to support the decon operations, such as medical, engineer, AD, and MP. These elements linkup with the chemical unit before the arrival of the contaminated unit (see Table 4-1, page 4-4).

4-14. The supported unit is responsible for site security and overall control. The decon-platoon leader is responsible for operating the predecon area and processing vehicles.

4-15. After linkup is achieved and all support assets are in position, the actual decon operation begins. The decon-platoon leader, with help from the commander of the contaminated unit or the supported unit's command and control cell, supervises the decon operation. All equipment not necessary for decon operations should be located in a clear area to limit the spread of contamination.

Table 4-1. Thorough-Decon Support Matrix

Tasks	Division CMO	Division	Brigade	DISCOM	FSB	Chemical Unit	Contaminated Unit	Battalion
Preparation-Phase Tasks								
Request	S						P	S
Coordination	S		S					P
Site selection		S OR S				P		
Advance-party linkup						S	P	S
Site setup				S	S	P		
Execution-Phase Tasks								
Site control/security						S	P	
Predecon actions						P	S	
Processing						P	S	
Site-Clearance-Phase Tasks								
Cleanup						P	P	
Marking and reporting						P		
S - supporting responsibility P - primary responsibility								

4-16. A thorough-decon site consists of the following areas: predecon staging area, DED and DTD areas, and postdecon AA (see Figure 4-1). The chemical unit leader selects these areas based on operational guidance, road network, available cover and concealment, and water supply. The contaminated unit uses the predecon staging area to ready itself. The postdecon AA is the location where the vehicles and soldiers exiting the DED and DTD areas are linked up before moving from the decon site.

PREDECON STAGING AREA

4-17. The contaminated unit moves to the predecon staging area about 274 to 547 yards downwind of the thorough-decon site. The contaminated unit performs predecon actions to include segregating vehicles by checking for contamination.

4-18. For chemical contamination, use the CAM and the M8/M9 detector paper. The contamination level can be estimated by comparing the M8/M9 detector paper to Figure 4-2. When monitoring vehicles for contamination, there should be about 33 to 49 feet between each vehicle to prevent false positive readings with the CAM. For radiological contamination, use the AN/PDR-77 or AN/VDR-2 radiac detector. If the vehicle only has isolated areas of contamination, use the M11 or M13 to decon those areas.

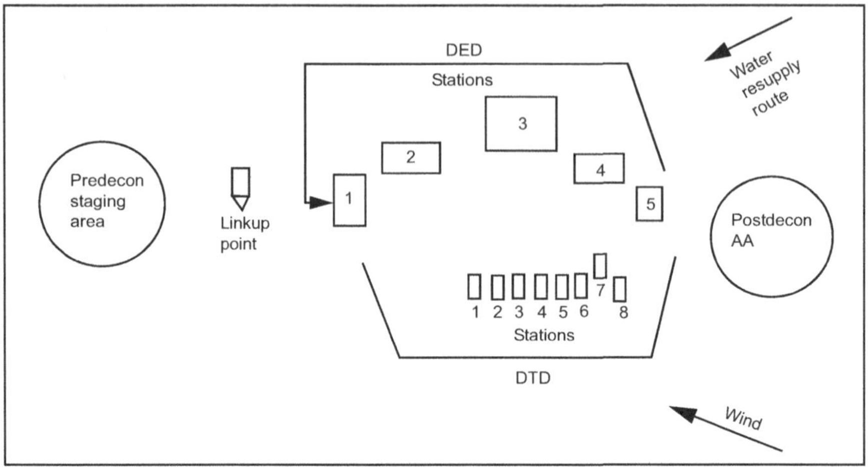

Figure 4-1. Thorough-Decon Site

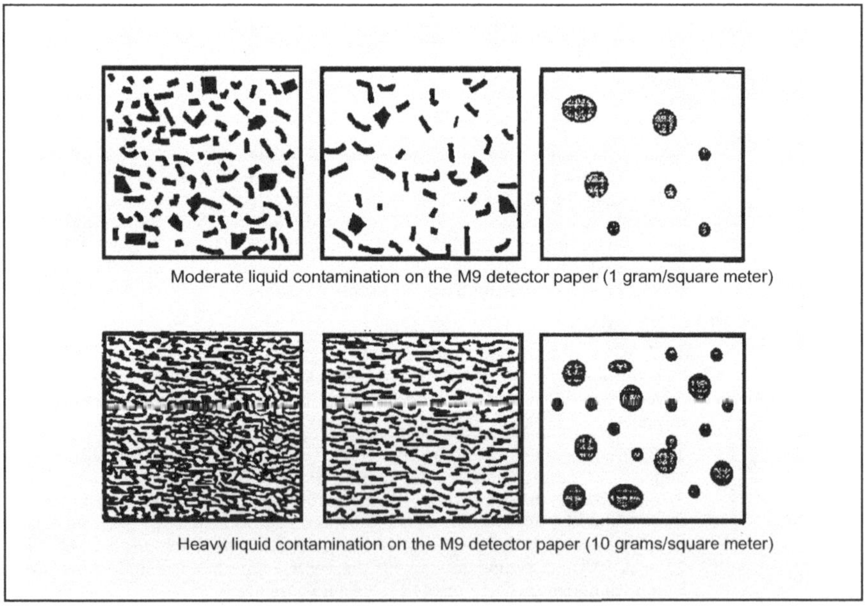

Figure 4-2. Contamination Levels

Vehicle Crews

4-19. The vehicle crews, except for the drivers, dismount. As the crews dismount, they remove all contaminated equipment, to include electro-optical systems from the top and sides of the vehicles. Once the crews have exited the vehicle, they will not reenter. This prevents contamination from being spread into the vehicle interior.

Vehicle Preparation

4-20. Using the pioneer tools from the contaminated vehicles, the crews remove all heavy mud and debris. They concentrate on the vehicle's undercarriage, which is the most likely place for contamination to collect and the hardest place to decon. Once the crews are finished with the pioneer tools, they are placed back on the vehicle. Initial removal of the mud and debris makes it more likely that the decon wash will remove any remaining contamination.

4-21. Seat covers (when applicable), canvas items, camouflage netting, and any other materials that can absorb liquid contamination are removed. These items create a potential transfer hazard and are not easily decontaminated. Left untreated, absorbed chemical agents will desorb after being decontaminated and will create a vapor hazard. Remove items that cannot be decontaminated by the standard methods used in the DTD, and place them at the collection point. Chemical-unit personnel provide advice concerning the decon or disposal of these items.

4-22. Design vehicle-loading plans to minimize the amount of equipment carried on the outside of the vehicle that cannot be readily decontaminated. Whenever possible, NBC covers should be used when a chemical attack is expected (see FM 3-4). All NBC covers are removed and disposed of as contaminated waste during the predecon actions.

DED AND DTD AREAS

4-23. In coordination with the chemical unit leader operating the decon site, the contaminated unit begins sending contaminated vehicles by priority for processing. Communication is maintained between the predecon staging area and the chemical unit leaders. All assistant vehicle drivers are the first individuals sent through the DTD to ensure that there is a driver exchange at station 3 (see paragraph 4-94).

POSTDECON AA

4-24. The chemical unit leader selects the general location for the postdecon AA. It must be big enough to hold the entire unit undergoing the thorough decon with proper cover and concealment. The postdecon AA is located about 0.62-mile upwind from the DED and DTD areas. The unit assembles in the postdecon AA after completing DTD and DED operations. The unit occupies the postdecon AA before moving to a reconstitution location or a tactical AA to prepare for future operations.

DETAILED TROOP DECON

4-25. The contaminated unit or its higher HQ is responsible for setting up, operating, manning, and closing the DTD area at the thorough-decon site. The chemical unit leader determines the general location of the DTD within the decon site and provides technical advice on setting up, operating, and closing the DTD area. The supervisor of the DTD must establish a work/rest cycle.

4-26. There are eight stations for a DTD. Spacing between the stations is 10 to 16 feet (see Figure 4-3, page 4-8). Recapitulation of personnel and equipment for a DTD is provided at Table 4-2, page 4-9.

STATION 1 - INDIVIDUAL-GEAR DECON

4-27. At this station, contamination is removed from individual gear (LCE, mask carrier, helmet, and weapon) to a negligible risk level (see Chapter 1 of this manual).

Preparation

4-28. At this station, the following equipment and supplies are needed:

- Four 30-gallon containers for each two decon lanes used.
- 2 long-handled brushes for each two decon lanes used.
- Sufficient STB slurry mix for chemical contamination.
- Hot, soapy water and a sponge for radiological contamination.
- Rinse water.
- 2 ponchos or plastic sheets.
- Engineer tape.

4-29. Dig a sump that is 6 feet long, 6 feet wide, and 4 feet deep (minimum). Place three 30-gallon containers near the sump for ease of changing. Fill two containers with an STB slurry mix. Fill the other two containers with clean water for rinsing and place it about 3 feet forward of the STB cans. Place two long-handled scrub brushes at each can of STB slurry.

4-30. To prepare the slurry, mix 100 pounds of STB with 20 gallons of hot water. The chemical unit provides the hot water. Change the mixture after 20 soldiers have decontaminated their gear. The rinse water should be changed after every 10 soldiers or when it appears dirty (place the waste mixture in the sump).

4-31. Place a poncho or a plastic sheet on the ground at the checkpoint. Divide the poncho or the sheet in half using engineer tape. This is the contamination control line. The checkpoint will be a minimum of 10 feet from all other stations in order to get a true reading on the detection equipment. Place the following chemical detection/identification equipment at the checkpoint for an average company-sized unit:

- 1 CAM.
- 8 books of M8 detector paper.
- 4 M256A1 detector kits.
- 100 plastic trash bags.
- 1 AN/VDR-2 for radiological contamination.

4-32. Three soldiers are required to operate this station. One soldier supervises the decon of the individual gear and takes the decontaminated equipment to the checkpoint. He also prepares a new slurry mixture as necessary. One soldier remains at the checkpoint and checks all the gear using the detection equipment to ensure that it is decontaminated. One soldier transports the decontaminated gear to the reissue point.

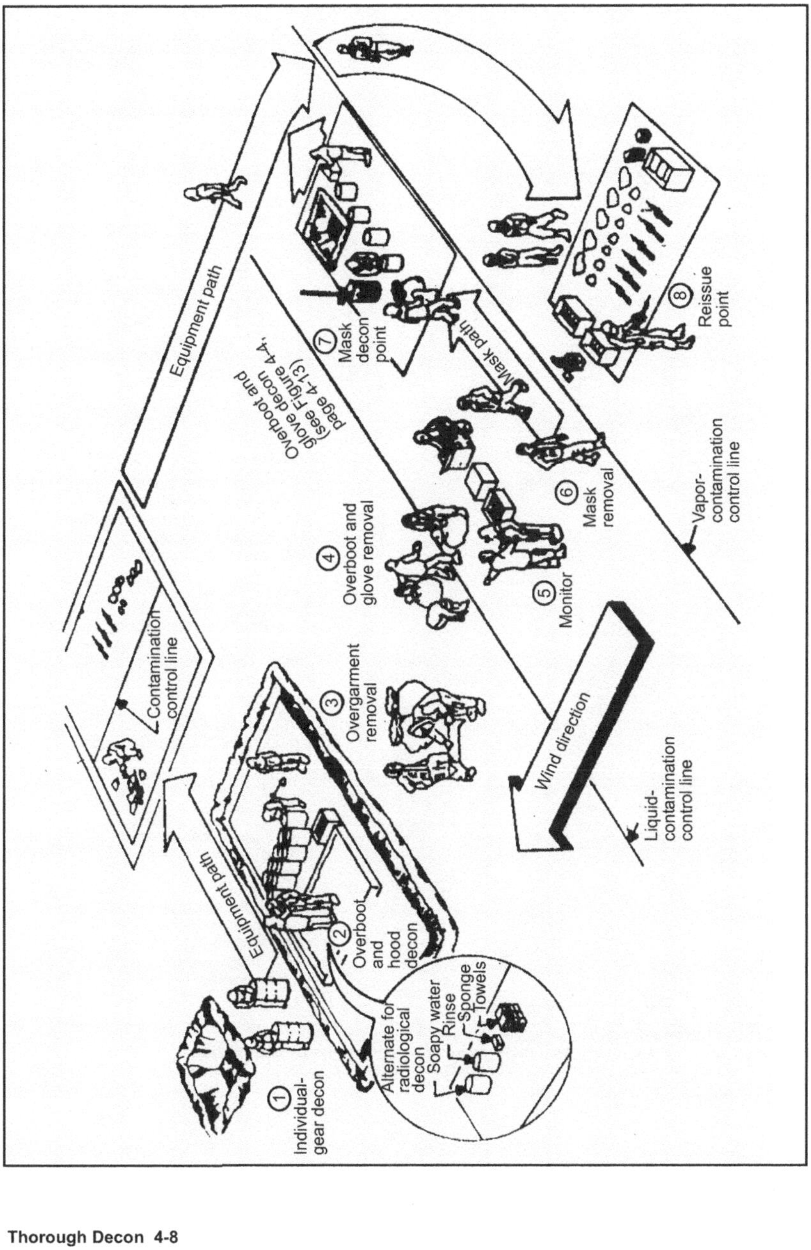

Figure 4-3. DTD Layout

Table 4-2. DTD Personnel and Equipment Recapitulation

Stations	Personnel	Equipment/Supplies
Station 1 - individual-gear decon	2 attendants 1 monitor (CAM operator)	4 30-gallon containers 2 long-handled brushes 2 ponchos or plastic sheets 1 CAM 8 books of M8 detector paper 4 M256A1 detector kits 100 plastic trash bags Sufficient STB slurry mix Rinse water *Hot, soapy water *AN/VDR-2 *Sponge
Station 2 - overboots and hood decon	1 attendant	Cutting tool One SKD or IEDK per person Large plastic sheet Plastic trash bags (as required) 10 drums of STB 1 shovel
Station 3 - overgarment removal	1 attendant	2 30-gallon containers 100 plastic trash bags 10 boxes of SDKs
Station 4 - overboot and glove removal	3 attendants	8 30-gallon containers 2 long-handled brushes M8 detector paper (as required) 1 immersion heater 100 plastic trash bags 1 CAM Engineer tape 1 cutting tool 2 ponchos or large plastic sheets 10 percent STB/HTH solution Hot, soapy water Cold rinse water
Station 5 - monitor	1 attendant (CAM operator) 1 aidman (or combat lifesaver)	1 CAM First aid supplies 5 books of M8 detector papers 1 case of SDKs
Station 6 - mask removal	2 attendants	1 M8A1 or M22 ACAA Engineer tape
Station 7 - mask decon point	2 attendants 1 monitor	1 30-gallon container 4 3-gallon containers 1 CAM 2 sponges One case of paper towels 1 immersion heater Mask sanitizing solution Hot, soapy water Rinse water
Station 8 - reissue point	Unit NBC NCO/supply sergeant	Mask PLL
* Denotes the equipment needed for radiological contamination. **NOTE: Assume that an operational decon was done before the thorough decon.**		

Execution

4-33. The soldier decontaminates his gear by washing and scrubbing it for 6 minutes in a decontaminant container with hot, soapy water or an STB slurry mix. If he is wearing the M42 mask, he should use hot, soapy water and a sponge or an STB slurry mix to decon the hose and canister.

4-34. The soldier dips his gear into the clean water and rinses it for 4 minutes then hands it to the attendant and proceeds to the next station. The attendant takes the gear to the equipment checkpoint and places the decontaminated gear on the "dirty" side of the contamination control line and returns to the containers to pick up more gear. The monitor at the checkpoint checks the gear using the appropriate detection device and the procedures associated with that device. If the residual contamination exceeds negligible risks, recycle the gear and decon it again (see Chapter 1 of this manual). If the gear passes the check, place it on the clean side of the contamination control line. The attendant will carry the equipment to the reissue point.

4-35. Depending on the time available, more extensive washing and checking procedures may be used. The longer the gear is washed or left out in the air after washing, the lower the contamination level. The gear may be put in closed areas or plastic bags and checked for hazardous vapors with the M256A1 detector kit or the CAM. The CAM only detects the G-series nerve-agent vapors and the H-series blister-agent vapors.

Risk

4-36. If this step is not done properly, contamination may remain on the equipment. The resulting vapor hazard could cause casualties to unmasked personnel, particularly in closed areas (vehicle interiors) or heavily wooded areas where air circulation is poor.

STATION 2 - OVERBOOT AND HOOD DECON

4-37. At this station, gross contamination on overboots, trouser legs, mask, and hood is neutralized. If ample hoods are available at the reissue point, the hood should be cut away.

Preparation

4-38. Prepare a shuffle pit by digging a shallow pit about 3 feet long, 3 feet wide, and 6 inches deep. Fill the shuffle pit with an STB dry mix or an STB slurry, depending on water availability. Prepare the STB dry mix by mixing three parts of earth to two parts of STB. Prepare the STB slurry in the same manner as at station 1 (see paragraph 4-30). Add more STB to the mix after 10 soldiers have processed through the shuffle pit. The chemical unit will provide 10 drums of STB for every company-sized unit that goes through the station.

4-39. At this station, the following equipment and supplies are needed for the mask and the shuffle pit: 1 cutting tool for each decon lane used, 1 SDK or IEDK per person, plastic trash bags (as required), 10 drums of STB, and 1 shovel (to mix).

4-40. One soldier is required to operate this station. The attendant directs and "observes" the soldiers as they decon their overboots and hoods.

Execution

4-41. The soldier walks into the shuffle pit and spreads his legs apart (double-shoulder width), bends at the waist, and uses his hands to thoroughly rub the STB dry mix or the STB slurry on his overboots and lower trousers legs. He should take special care to rub the rear of his overboots. He should also remove any excess decontaminant from his gloves.

4-42. If a replacement hood is available, remove the hood as follows. Buddy #1 cuts the shoulder straps and drawcord on Buddy #2's hood. Buddy #1 pulls Buddy #2's hood inside out over the front of the mask, being careful not to touch the exposed neck or head. Buddy #1 gathers Buddy #2's hood in one hand and using a cutting tool, cuts away the hood as close as possible to the eye-lens outsert, voicemitter, and inlet-valve covers. Make sure nothing is left dangling below the bottom of the mask.

4-43. If a replacement hood is not available, Buddy #1 decontaminates and rolls Buddy #2's hood in the same manner as for a MOPP-gear exchange. When the task is completed, Buddy #1 and Buddy #2 reverse roles.

NOTE: Soldiers should check their overboots, rubber gloves, and overgarment for damage. Any rips, tears, or punctures in these items should be reported to the monitor at station 5. This allows the monitor at station 5 to check the soldiers for chemical-agent symptoms and their clothing for possible contamination.

Risk

4-44. If this step is not done properly, contamination can be transferred to the combat boots and the head and neck area from the hood.

STATION 3 - OVERGARMENT REMOVAL

4-45. At this station, contaminated overgarments are removed before the agent penetrates the overgarment material and touches the undergarments or the skin.

Preparation

4-46. At this station, the following equipment and supplies are needed:

- Two 30-gallon containers for each two decon lanes used.
- 100 plastic trash bags (or about one per soldier).
- 10 boxes of SDKs.

4-47. One soldier is required to operate this station. He directs and "monitors" the soldiers as they remove their overgarments in the same manner as a MOPP-gear exchange.

Execution

4-48. The attendant assists the soldier in removing his overgarment. The attendant cuts and removes the M9 detector paper from around the soldier's wrist. He unfastens the Velcro closure over the jacket zipper, waist cord, and wrist Velcro straps on the soldier's jacket. He unfastens the back snaps and instructs the soldier to make a fist. He then pulls the soldier's jacket down and away from him.

4-49. The attendant cuts and removes the M9 detector paper from the soldier's trousers. He unfastens the Velcro straps and zippers on the cuffs of the soldier's trousers. He also unfastens the front waist snaps and unzips the front zipper. He has the soldier lift one leg and point that foot down and bend slightly at the knees for stability. The attendant grasps the cuff of the elevated foot with a hand on each side and pulls the cuff in an alternating, jerking motion until the soldier can step out of the trouser leg. The process is repeated on the other leg.

NOTE: The attendant ensures that the soldier steps wide enough so as not to rub his clean leg against the contaminated boot and overgarment.

Risk

4-50. If this step is not done properly, the agent may be transferred to the undergarment or the skin.

STATION 4 - OVERBOOT AND GLOVE REMOVAL

4-51. At this station, contaminated overboots and gloves are removed to limit the spread of contamination. The overboots and gloves may also be decontaminated for reissue (if serviceable).

Preparation

4-52. At this station, the following equipment and supplies are needed for every two lanes used:

- Engineer tape.
- 1 cutting tool for each lane used.
- 100 plastic trash bags (or about one per soldier).

4-53. If the overboots and gloves are not being decontaminated, two 30-gallon containers are needed.

4-54. If the overboots and gloves are being decontaminated, the following are needed:

- Eight 30-gallon containers.
- 2 long-handled brushes.
- 2 ponchos or large plastic sheets.
- 1 CAM.
- 1 AN/VDR-2 for radiological contamination.
- 4 M256A1 detection kits.
- 1 immersion heater.
- 100 plastic trash bags.
- M8 detector paper, as required.
- 10 percent STB/HTH solution.
- Hot, soapy water.
- Cold rinse water.

4-55. If replacement overboots and gloves are available, set up the station as follows (see Figure 4-4). Establish a liquid-contamination control line and set

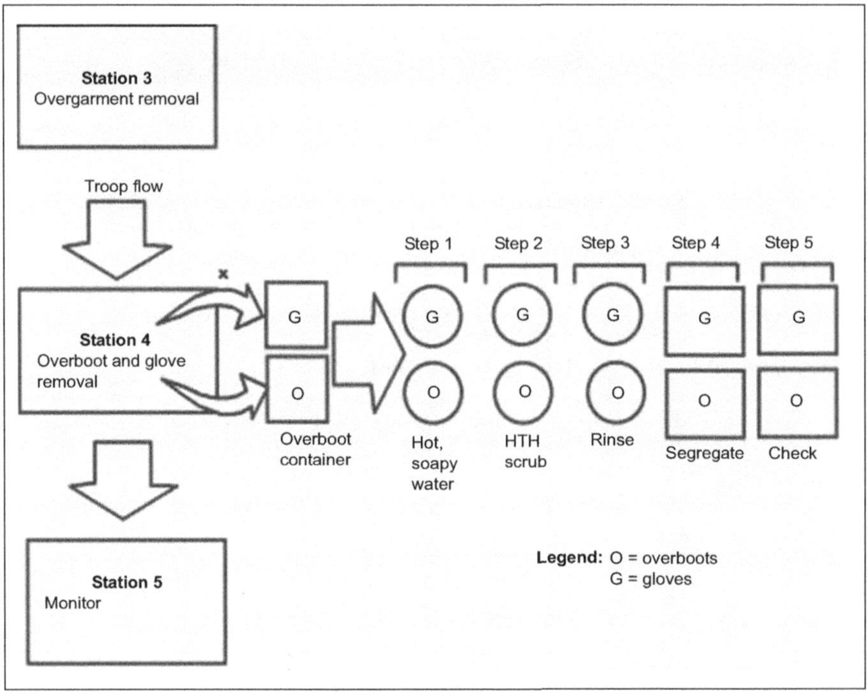

Figure 4-4. Station 4 Layout

two 30-gallon containers 1 foot back from the line. Soldiers support themselves using the containers and discard their overgarments into the containers. An attendant directs and monitors the soldiers as they remove their overboots and gloves in the same manner as a MOPP-gear exchange; however, the soldiers step over the control line instead of onto a jacket.

4-56. If replacement overboots and gloves are not available, set up the station as follows: Fill two 30-gallon containers with hot, soapy water and two 30-gallon containers with a 10 percent STB/HTH solution, placing two scrub brushes near the containers. Then fill two 30-gallon containers with cold rinse water. One attendant supervises and assists the soldiers wearing the M42 mask. The other two attendants decon the overboots and the gloves; one processes the overboots, while the other processes the gloves.

NOTE: Replace the water in the container once 20 items have been processed. When available, the decon platoon will assist with the water requirements.

4-57. For both situations, use engineer tape to mark the liquid-contamination control line on the ground. Place the cutting tools, two containers, and plastic bags on the "dirty" side of the liquid-contamination control line. The liquid-contamination control line separates the "dirty" and "clean" areas. No liquid agent should be tracked on the ground beyond the liquid-contamination control line.

Execution

4-58. The liquid-contamination control line separates the "dirty" and "clean" areas. The attendant unfastens or cuts the elastic closures on the soldier's overboots. The soldier faces the liquid-contamination control line and steps back from it about 12 inches. The attendant steps on the back of the soldier's overboot and instructs him to lift his heel and work his foot out of the overboot and step across the liquid-contamination control line. Repeat the process on the other foot. If the overboot cannot be removed by this process, the attendant cuts it off and discards it in the designated container. The soldier holds the fingertips of his gloves and partially slides his hands out. The attendant remove the soldier's gloves.

4-59. If the soldier is wearing the M42 protective mask, the attendant from station 6 carries the soldier's filter canister until it is removed.

NOTE: The attendant at station 4 performs his duty from the "dirty" side of the liquid-contamination control line.

4-60. The overboots and gloves are decontaminated using the following steps:

NOTE: Check all items for holes, tears, and punctures and discard any item with this defect. Do not decon any item that is unserviceable.

- Step 1. Submerge the gloves and overboots in their respective container of hot, soapy water. Some of the contamination is removed during this step. When the overboots and gloves are removed from the container, ensure that no water remains inside them.
- Step 2. Submerge the gloves and overboots in their respective containers of STB/HTH solution. Thoroughly scrub the items until no visible contamination remains. After scrubbing, submerge each item once more before moving to the rinse container.
- Step 3. Thoroughly rinse the scrubbed items, making sure that they are rinsed inside as well as out.
- Step 4. Place usable items on a poncho or a plastic sheet to air dry and weather.
- Step 5. Place usable items in plastic trash bags along with an M256A1 detector kit. If the detector kit shows contamination remaining, the attendants can recycle the items or discard them. However, if the kit show no contamination, the items can be reused.

Risk

4-61. If the overboots are not properly removed, the combat boots may become contaminated and contamination may be spread to clean areas. If the gloves are not properly removed, the undergarments and skin may become contaminated.

STATION 5 - MONITOR

4-62. At this station, contamination on personnel is identified, spot-decon capabilities are provided, and medical aid is provided, as required.

Preparation

4-63. At this station, the following equipment and supplies are needed:

- First aid supplies.
- 1 CAM.
- 1 AN/VDR-2 for radiological contamination.
- 5 books of M8 detector paper per 100 soldiers.
- 1 case of SDKs.

4-64. An aidman should be present to treat any soldiers suffering from chemical-agent symptoms. If an aidman is unavailable, a combat lifesaver should be present.

Execution

4-65. The attendant checks the soldiers for contamination using the CAM. Liquid agents can be detected with M8 detector paper. Small quantities of agent vapor can be detected with the CAM. Symptoms of agent poisoning are the most obvious indication of skin contamination. At this station, the medic checks each soldier for symptoms of agent poisoning and treats, as required. Soldiers should report any damage to their MOPP gear that was identified at stations 2, 3, and 4. The attendant can decon any areas identified as contaminated with an SDK. Soldiers are remonitored after decon. It is possible that all liquid-chemical contamination is absorbed into the clothing. If so, M8/M9 detector paper will indicate negative, even though there is a hazard.

Risk

4-66. If this station is omitted, the soldier could become a casualty. After this station, the soldier is not checked for contamination or decontamined again. Commanders can choose to conduct more extensive contamination checks here if medical assistance and time are available. This decreases the risk of casualties.

STATION 6 - MASK REMOVAL

4-67. At this station, the mask is removed without contaminating the soldier. The mask is taken to a mask decon point, limiting agent transfer at the station.

Preparation

4-68. At this station, the M8A1 or M22 automatic chemical-agent alarm (ACAA) and engineer tape are needed. Two soldiers are needed to operate this station. They remove and carry the masks to station 7.

Execution

4-69. If the hood is still attached to the mask, the attendant pulls the hood over the front of the mask, grabs the mask by the voicemitter cover, and pulls the mask off the soldier. The soldier holds his breath as the mask is removed. If the mask has optical inserts, the attendant holds the mask open so that the soldier can remove the inserts without touching the outside of the mask. The soldier walks upwind 16 feet, crosses the vapor-contamination control line, and then resumes breathing. The attendant brings the mask to station 7.

4-70. If the wind direction remains constant, no chemical vapor hazard is expected beyond the vapor-contamination control line. Position the M8A1 or the M22 ACAA upwind of the station to warn of vapor hazards. The soldier getting decontaminated moves straight ahead while his mask, which may still give off vapors, is held on the vapor dirty side of the line and taken to station 7 where it is decontaminated.

Risk

4-71. If the step is not done properly, the soldier could breathe toxic vapors. There is a high probability that the vapor hazard is still present on the mask and the hood. The soldier must not touch the outside of the mask because it could contaminate his bare hands. The soldier proceeds to station 8.

STATION 7 - MASK DECON POINT

4-72. At this station, all the contamination is removed from the mask.

NOTE: Once 20 items have been processed in the wash containers, replace the water. Once 10 item have been processed in the rinse water, replace it. Place the contaminated waste into the sump.

Preparation

4-73. At this station, the following materials and equipment are needed:

- 4 containers (about a 3-gallon capacity).
- 1 CAM (for chemical only).
- Hot, soapy water.
- Rinse water.
- Mask sanitizing solution.
- Immersion heater.
- One 30-gallon container.
- 2 sponges.
- 1 case of paper towels per company.

4-74. Dig a sump that is 4 feet long, 4 feet wide, and 4 feet deep in which to discard used filters and canisters. Place the equipment and materials as shown in Figure 4-3, page 4-8. Three soldiers are needed to operate this station. Two soldiers strip, wash, rinse, sanitize, and dry masks. The other soldier checks the masks and carriers them to the reissue point.

Execution

4-75. Remove the eye-lens outserts and the hood if the hood was not cut off at station 2. Remove and discard the filters or canisters. Put the items into the properly marked containers. Wash the mask, hood, eye-lens outserts, and the hoses on the M42 and M43 masks in hot, soapy water. Rinse these items in clean water, dip them into the sanitizing solution, agitate them for 5 minutes, and then rinse them again in clean water. Add one tube of mask sanitizing solution (calcium hypochlorite) to each quart of water. Wipe the masks with rags until they are almost dry. Discard each container of mask sanitizing solution into a sump after every 10 masks. The attendant checks the masks for contamination with a CAM. If the masks are still contaminated, the attendant recycles them for more decon and then decons his rubber gloves. If the masks are not contaminated, the attendant takes the masks to the reissue point. The attendant must take care not to contaminate the reissue point or himself.

Risk

4-76. If this step is not done properly, the soldiers may become contaminated when the masks are reissued to them at the reissue point. Even though the step is done correctly, there is still a possible danger when many masks are stacked together. Small amounts of residual vapor from each mask can become potentially dangerous.

NOTE: Runners between stations 7 and 8 are in MOPP2 and are prepared to go into MOPP4.

STATION 8 - REISSUE POINT

4-77. At this station, the mask with its components is provided to the soldier for reassembly.

Preparation

4-78. At this station, the protective-mask prescribed load-list (PLL) parts are needed. The unit NBC NCO/supply sergeant sets up the reissue point to provide the soldiers with replacement parts for all types of protective masks and assist in mask maintenance.

Execution

4-79. The mask with its components is reissued to the soldier who assembles it in the AA. The unit chemical NCO affixes canisters to the cleaned M42 and M43 hoses. The soldier picks up individual gear and moves to the postdecon AA.

Risk

4-80. If this step is not done properly, the soldier may be inadequately equipped for future operations.

NOTE: If time is not available, the NBC NCO will have replacement chemical suits, overboots, and gloves at this station for reissue. If time is available, personnel will receive this equipment at the postdecon AA.

RESUPPLY SUPPORT RESPONSIBILITIES

4-81. The supporting chemical unit (decon or smoke/decon) must supply the majority of the equipment and supplies that are required to operate the DTD. This does not relieve the supported unit of the requirement to maintain adequate supplies to conduct DTD procedures.

4-82. A chemical unit's operational control continues to receive support from its parent organization. The supported unit provides the chemical unit with replacements, supplies, and material at the end of the DED. This shortens the time the chemical unit is nonmission capable following the DED. The supported unit must coordinate for the supplies and equipment necessary to operate the DTD.

4-83. Typical nonchemical units do not have more than two CAMs per company-sized unit. However, the DTD requires a minimum of three CAMs. The supported HQ may need to provide additional CAMs. The division support command (DISCOM) sustains the division's chemical units. A chemical unit attached to a maneuver force identifies the resources that are needed to accomplish the mission and passes these requirements to the maneuver staff's Supply Officer (S4)/Logistics Staff at Corps and Division (G4). The S4/G4 then coordinates with the supporting CSS element to fill the requirement. The supply units that operate the Class II points issue the protective masks and overgarments. They normally package protective clothing as complete sets of MOPP equipment to support a predetermined number of soldiers. Decon supplies may be issued in preconfigured "push" packages.

DETAILED EQUIPMENT DECON

4-84. Chemical units (decon and smoke/decon platoons) are responsible for setting up, operating, and closing the DED portion of the thorough-decon operation. The DED for chemical and biological contamination is comprised of five stations. For radiological contamination, the DED uses all but station 2. Stations are normally 98 to 164 feet apart; however, spacing is METT-TC dependent.

DECON STATIONS

4-85. The actions that are taken at each of the stations are as follows:

Station 1 - Primary Wash

4-86. At this station, the gross contamination and dirt are removed from the vehicle. The vehicle is sprayed for 2 to 3 minutes with hot, soapy water. The vehicle is then scrubbed to help remove caked-on dirt. The mechanical action of scrubbing also helps remove thickened chemical agents. Although the undersurfaces are difficult to reach, try to remove as much dirt as possible. After scrubbing the vehicle, spray it again for 2 to 3 minutes to remove loosened dirt and contamination. This station uses about 250 gallons of water per vehicle. Larger vehicles with large quantities of dirt use more water. The runoff from this station is contaminated and must be treated as hazardous. The contaminated runoff must be controlled; if available, engineer support may be used. This station requires high water-pressure systems (M12A1 power-driven decon apparatus [PDDA] and M17 Lightweight Decontaminating System [LDS]) rather than low water-volume systems (65-GPM pumps).

NOTE: Thirty-five cubic feet of space per 250 gallons of liquid runoff should be used when calculating the size for the drainage pump.

4-87. The effectiveness of the wash depends on the type of wash (hot, soapy water; hot water; cold water; or steam). The relative effectiveness rankings for selected surfaces are listed in Table 4-3, page 4-20. Hot, soapy water is water heated to about 120° to 140° F to which a detergent has been added to reduce its surface tension. The detergent removes the agent by emulsification, which is followed by the mechanical displacement of the suspension. Hot water alone is less effective than hot, soapy water. Because of the high temperature, some agents are best removed by steam through vaporization. Finally, for some chemical agents, cold water exhibits better solvent characteristics.

Station 2 - DS2 Application

4-88. At this station, a decontaminant is applied to the entire vehicle. The vehicle is divided into four parts, and a member of the scrubbing team is assigned that part of the vehicle. This limits the workload of each member of the scrubbing team and avoids duplication of work. DS2 is applied starting at the top of the vehicle and working towards the undercarriage. Every effort is made to apply DS2 to the undercarriage, especially if the vehicle has crossed a contaminated area. The mop is the least tiring method of applying DS2. Using a mop to apply DS2 creates a large amount of spillage. However, continual use of the M13 decontaminating apparatus, portable (DAP) requires the scrubbing team to exert more energy than with using the mop. The M13 DAP can be used to apply DS2 to hard-to-reach areas. Planning factors for DS2 application are provided at Table 4-4, page 4-20.

4-89. Before starting the decon operation, the scrubbing team pours 5-gallon cans of DS2 into 30-gallon containers if mops are going to be used instead of M13 DAPs. Each member of the scrubbing team wears a TAP apron or wet-weather gear to protect himself from the DS2.

4-90. Water adversely affects the DS2's ability to react with chemical agents. If there is excess water remaining on the vehicle from station 1, the scrubbing team can—

- Wait for the majority of the water to evaporate.
- Remove the excess water.
- Increase the amount of DS2 applied.

4-91. There must be sufficient DS2 on the item being decontaminated for complete neutralization to occur. The DS2-to-agent ratio needs to be 55 to 1 for H agents and 25 to 1 for G agents. For a vehicle the size of an M1A1 tank, this corresponds to 15 and 7 gallons, respectively.

Station 3 - Contact Time/Interior Decon

4-92. At this station, the DS2 is allowed to completely neutralize the chemical agent and the interior of the vehicle is decontaminated. Vehicles are moved to a concealed position. Vehicles will remain in station 3 for no less than 30 minutes. DS2 reacts with most chemical agents within 5 minutes. However, by allowing the DS2 to remain on the contaminated surface for 30 minutes, the amount of agent that will later desorb (off gas) will be significantly reduced. When there is a 30-minute contact time, there will be no desorption after decon operations for most chemical agents. However, studies indicate that distilled-mustard (HD) vapors desorb after decon even if DS2 is allowed to remain for 30 minutes.

4-93. While the vehicle is held at this station for the DS2 to completely react, the driver inspects the interior of the vehicle for liquid contamination. He is given M8 detector paper to check for chemical contamination. If he identifies chemical contamination, he is given decon supplies to decon the interior of the vehicle. The best decon solution for use in the interior of vehicles is a 5 percent solution of HTH or STB. The driver wipes all reasonably accessible

Table 4-3. Effectiveness of Types of Wash

Agents/Surfaces	Relative Effectiveness[1]
TGD on alkyd-painted metal	Hot water and/or steam; hot, soapy water; cold, soapy water
TGD on CARC-painted metal	Hot, soapy water; hot water and/or steam; cold water
THD on alkyd-painted metal	Hot water and/or steam; hot, soapy water; cold water
THD on CARC-painted metal	Hot, soapy water; hot water and/or steam; cold water
HD on alkyd-painted metal	Hot, soapy water; hot water and or steam; cold water
HD on CARC-painted metal	Hot, soapy water and/or steam; cold water; hot water
VX on alkyd-painted metal	Steam; cold water; hot water and/or hot, soapy water
TGD on canvas/webbing	Steam; hot water and/or hot, soapy water; cold water
GD on canvas/webbing	Steam; hot, soapy water and/or hot water; cold water
THD on canvas/webbing	Steam; hot, soapy water; cold water; hot water
HD on canvas/webbing	Steam; hot water; hot, soapy water; cold water
VX on canvas/webbing	Steam; hot water and/or hot, soapy water; cold water
Overall	Steam; hot water and/or hot, soapy water; cold water

[1]The types of wash are listed in the order of effectiveness, starting with the most effective to the least effective.

Table 4-4. Planning Factors for DS2 Application

Planning Categories	Tank	Truck	APC
Gallons of DS2 applied	15	8	7
Minutes to apply solution	35	29	18

surfaces with a rag or sponge soaked in the HTH or STB solution. He should not attempt to decon areas where there is little likelihood of contamination (electrical assemblies, area beneath the turret floor, and so forth).

CAUTION

Do not mix HTH or STB with DS2. If mixed, a violent reaction will occur.

4-94. Once the interior decon is completed, the driver places a cover over the seat and floor of the vehicle. Then he dismounts from the vehicle and proceeds to the start of the DTD. The assistant driver, having completed the DTD, mounts the vehicle and moves it to the next station. Drivers must exercise caution when entering or exiting the vehicle. A DS2-coated surface is slippery, and the DS2 may react with chemical protective footwear. Personnel should avoid stepping in DS2 and tracking it into the vehicle.

4-95. For radiological contamination, use an AN/PDR-77 or AN/VDR-2 radiac detector to determine the extent and location of contamination inside the vehicle. If there is contamination, determine the intensity of the contamination inside the vehicle. If the contamination has an intensity greater than 0.33cGy, the interior of the vehicle must be decontaminated. Use a sponge to wipe the interior of the vehicle.

Station 4 - Rinse

4-96. At this station, the DS2 is removed from the vehicle. The vehicle is sprayed with water from top to bottom. Take care to rinse the undercarriage. This station uses about 200 gallons of water per vehicle. Failure to remove all DS2 from the vehicle may cause a false-positive reading at station 5. If high water-pressure systems (M12A1 PDDA and M17 LDS) are not available, large-volume water pumps (65-GPM pumps) should be used at this station. The driver removes plastic or other material (if present) covering the seats and floor and disposes of it as hazardous waste. See Table 4-5 for the planning factors for the rinse station.

Table 4-5. Planning Factors for the Rinse Station

Planning Categories	M1 Tank	M2 BFV	M113 APC	M109A PALADIN	HEMTT	5-Ton Truck	HMMWV
M12A1 PDDA rinse (gallons applied)	325	325	203	325	180	158	90
M12A1 PDDA rinse (minutes applied)	12	12	9	12	8	7	4
M17 LDS rinse[1] (gallons applied)	57	57	38	57	30	42	23
M17 LDS rinse[1] (minutes applied)	14	14	10	14	12	11	6
[1] Rinse is done with a spray wand.							

Station 5 - Check

4-97. At this station, the vehicle is checked to see if it has a negligible contamination level or if it still has significant contamination remaining. Detection procedures will vary depending on the type of contamination. If significant contamination is found on the vehicle, the vehicle will be recycled to station 2 for chemical contamination or station 1 for radiological contamination. An assistant driver takes the vehicle to the AA.

4-98. **Chemical.** The CAM is used to check for the presence of vapor from residual liquid contamination. A one bar or lower reading on the CAM indicates a negligible contamination level. Once the CAM indicates the presence of vapor contamination, M8 detector paper is used to verify the presence of liquid contamination. If it is suspected that both the CAM and M8 detector paper are producing a false positive, use an M256A1 detector kit to confirm or deny the presence of contamination. See Table 4-6, page 4-22, for a list of common interferences that can cause false-positive readings on the CAM. If the vehicle has significant contamination remaining, recycle it. The commander may modify the recycle criteria based on mission requirements.

4-99. There will be desorption of chemical agents from the surfaces after decon. The desorption of vapors on surfaces painted with the CARC will stop sooner then those surfaces painted with alkyd. Consider this when checking decontaminated items for overall decon effectiveness.

Table 4-6. Common Interferences of the CAM

Interferences	G-Bar Response	H-Bar Response
DS2	Low	
Insect repellent	Low to very high	
Brake fluid	High to very high	Very high
General-purpose cleaner	High	
Burning kerosene		High
Breath mints	High	
Gasoline vapors	Low	Low
Burning grass	Low to high	Low
Burning gas	Low	
Green smoke	Low	Low to high
Break-free oil	Low	
Ammonia	Very high	

4-100. **Radiological.** Use the AN/PDR-77 or AN/VDR-2 to determine if any contamination remains. If there is contamination remaining, determine the intensity of the contamination inside and outside of the vehicle. If the contamination has an intensity greater than 0.33cGy/hr, the vehicle is recycled to station 1.

RECYCLE CRITERIA

4-101. The commander, with the chemical unit leader's help, establishes the recycle criteria before starting decon operations. The recycle criteria determines which vehicles will return to station 1 after contamination is detected at station 5. If the unit has sufficient time and resources, any vehicle having more contamination than the acceptable level should be recycled. However, time and resources are usually limited, and not all vehicles can be recycled. The recycle criteria is based on the weathering effects.

RECONSTITUTION CRITERIA

4-102. The operational- and thorough-decon operations that are performed on vehicles or major equipment will be recorded on DA Form 2404 (Army) or NAVMC 10245 (Marine). This information will become a permanent record for these vehicles or major equipment. The entries that are required on these forms are the type of decon performed, decontaminant used, date/time group completed, location of decon site, and type of monitoring equipment used to verify decon completeness. These forms will remain a permanent part of the unit's, vehicle's, and equipment's records. This information will assist with reconstitution operations once the conflict or war has ended.

DED LAYOUTS

4-103. Dual-purpose and decon platoons establish thorough decon sites differently because of organization and equipment differences. The optimum layout for each type of platoon is described in the following paragraphs (see

Figures 4-5 through 4-8, pages 4-24 through 4-27). This layout provides the maximum output for platoons at 100 percent personnel and equipment. The equipment and personnel requirements for the optimum layout are identified for both the chemical unit (decon platoon) and supported unit (augmenter) (see Tables 4-7 through 4-10, pages 4-28 through 4-31).

4-104. Since it is likely that all platoons will not be at 100 percent strength, alternate layouts are discussed. For each alternate layout, chemical-unit equipment requirements are identified. However, personnel requirements are not identified by chemical unit and supported unit. The total number of personnel that are needed is identified, with a minimum number of chemical unit personnel.

ALTERNATE-LAYOUT PLANNING CONSIDERATIONS

4-105. Chemical unit leaders use METT-TC to determine the best possible DED layout to execute their mission. When determining alternate DED layouts, use the following guidelines:

- The ability to spray hot, soapy water or steam under pressure must be retained at station 1.
- The largest number of people is required at station 2.
- Experienced and qualified CAM operators are required at station 5.
- Water does not have to be hot to rinse off the DS2; however, the lower the water pressure the more water that is required for the rinse.

DECON PLATOON

4-106. The optimum DED layout for an M12A1 PDDA-equipped platoon requires the use of all authorized equipment and personnel (see Figure 4-5, page 4-24). While this DED layout is manpower and equipment intensive, it provides for the rapid decon of vehicles and equipment (eight vehicles processed per hour). This layout uses dual lanes at stations 1, 4, and 5 to process two vehicles at one time. Since the most time- and labor-intensive work takes place at station 3, this station is designed to process three vehicles at a time. The processing rate of this layout will be affected by any work/rest cycle (see Table 4-11, page 4-32).

SMOKE/DECON PLATOON

4-107. Smoke/decon platoons equipped with the M17 LDS will set up the DED differently then the decon units equipped with the M12A1 PDDA. While the optimum DED layout is manpower and equipment intensive, it provides for the rapid decon of vehicles and equipment (eight vehicles processed per hour). The layout uses a dual lane at stations 1, 4, and 5 to process two vehicles at a time. Since the most time- and labor-intensive work takes place at station 3, this station is designed to process three vehicles at a time. The processing rate of this layout will be affected by any work/rest cycle (see Table 4-11, page 4-32).

4-108. It may not be possible, for a variety of reasons, for an M17 LDS-equipped decon platoon to use the optimum DED layout. Figure 4-7, page 4-26, shows an example of an optimum DED layout. Limited personnel or equipment will affect the DED layout. Figure 4-8, page 4-27, shows an example of an alternate DED layout.

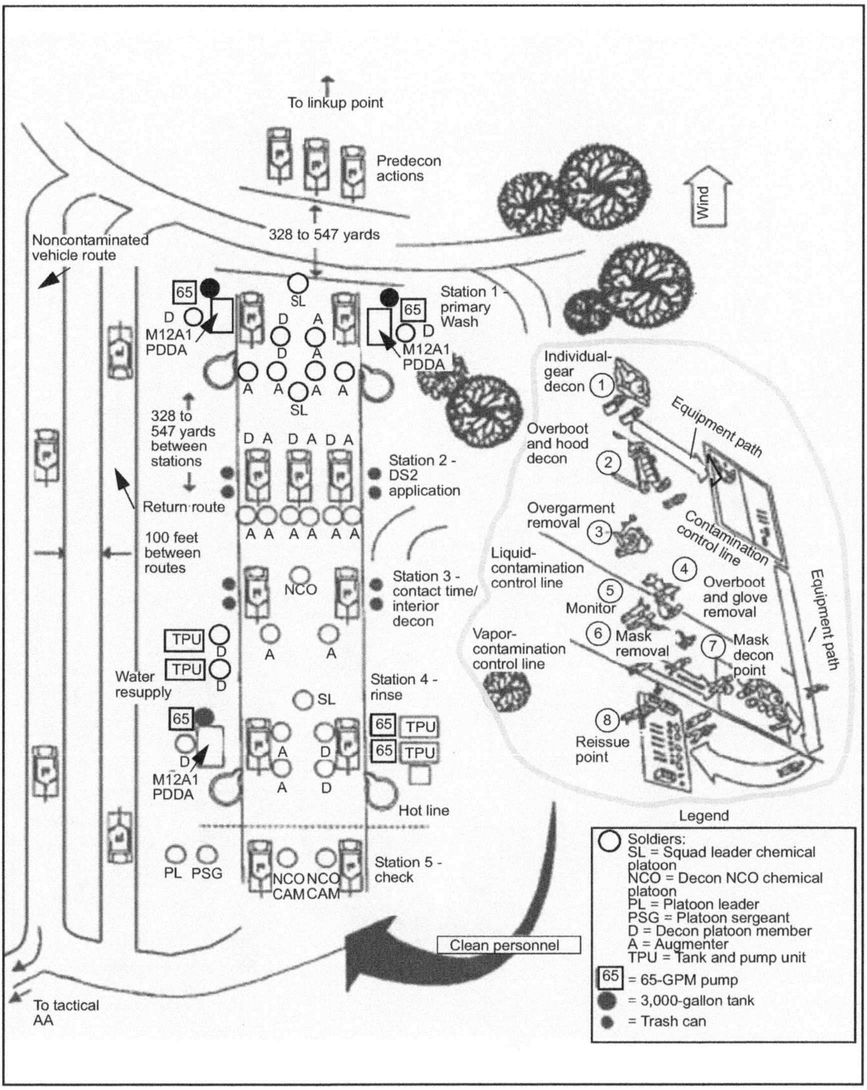

Figure 4-5. Optimum DED Layout for the M12A1 PDDA-Equipped Unit

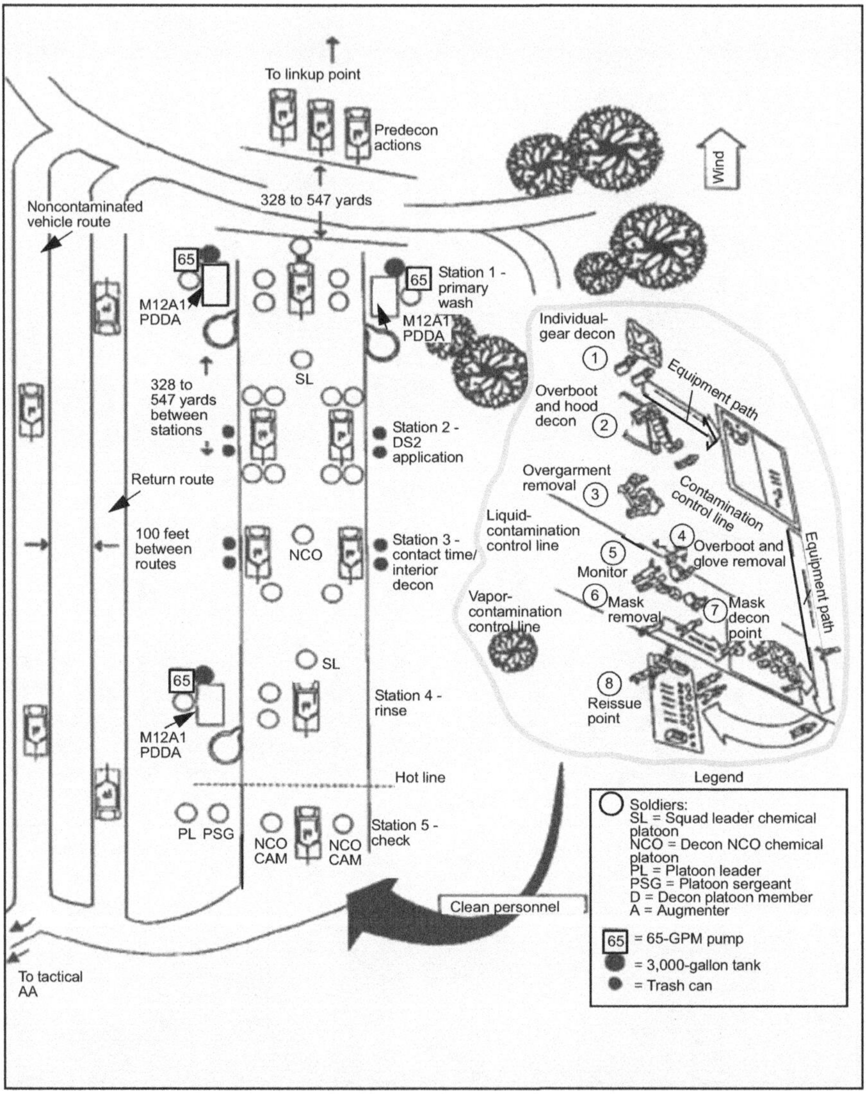

Figure 4-6. Alternate DED Layout for the M12A1 PDDA-Equipped Unit

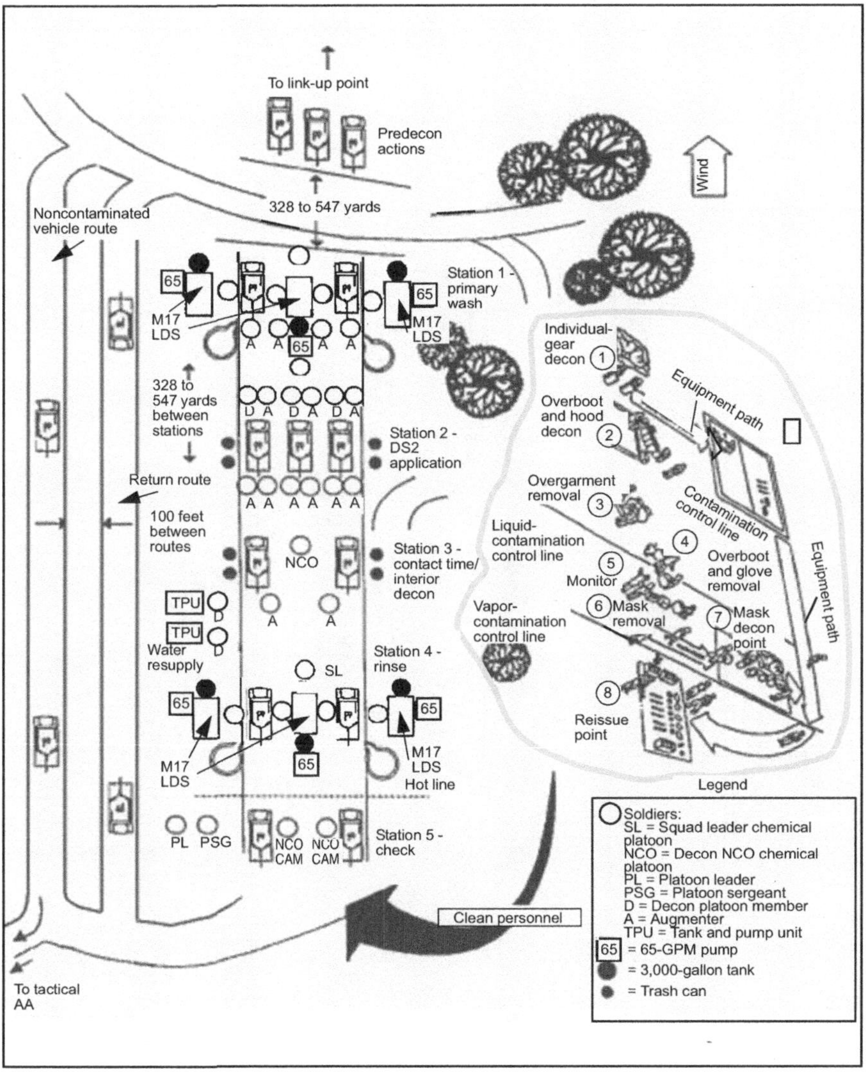

Figure 4-7. Optimum DED Layout for the M17 LDS-Equipped Unit

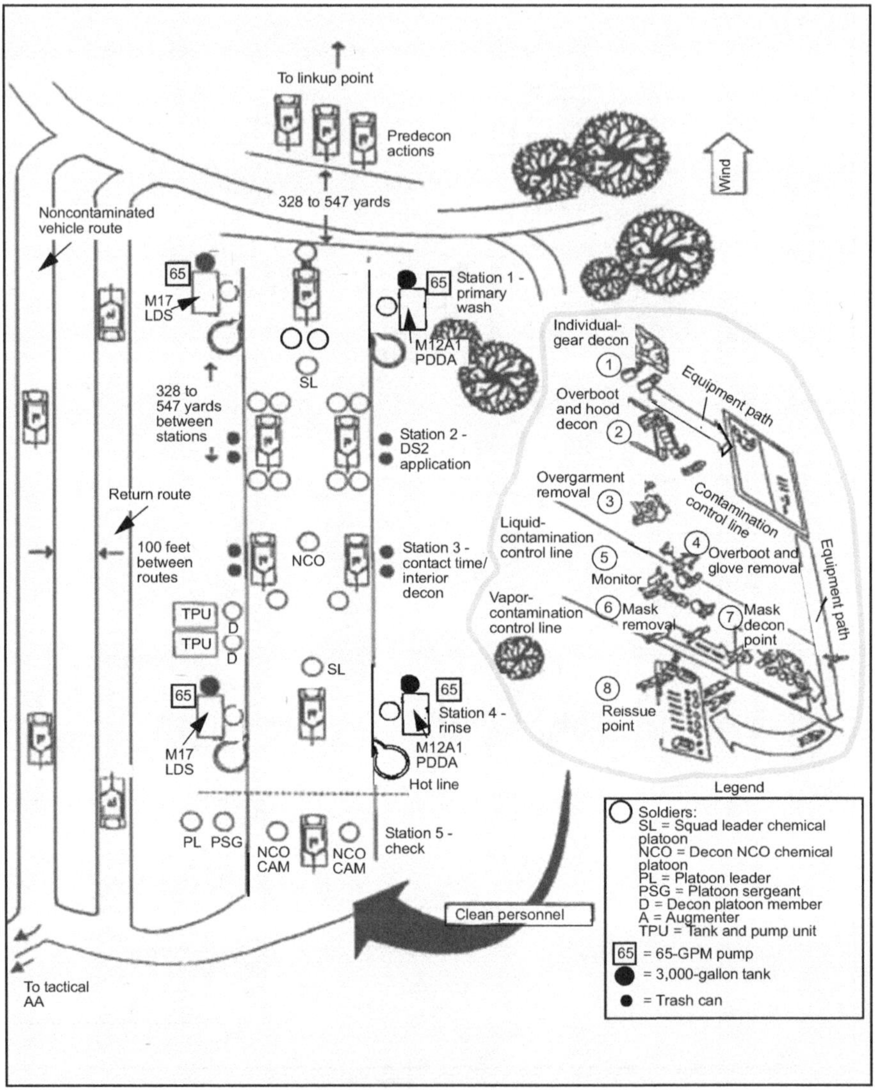

Figure 4-8. Alternate DED Layout for the M17 LDS-Equipped Unit

Table 4-7. Personnel and Equipment Requirements for the Optimum DED Layout of the M12A1 PDDA-Equipped Unit

Stations	Personnel		Equipment/Supplies
	Decon Platoon	Augmenter	
Station 1 - primary wash	1 squad leader 2 PDDE operators 4 sprayers	4 scrubbers	2 M12A1 PDDAs 2 3,000-gallon tanks 2 65-GPM pumps 6 long-handled brushes 8 TAP aprons Liquid detergent
Station 2 - DS2 application	1 squad leader 3 appliers	9 appliers	18 long-handled brushes 9 mops with extra mop heads 3 30-gallon containers 9 M13 DAPs Sufficient DS2
Station 3 - contact time/ interior decon	1 NCO	2 interior decon assis- tants	2 AN/VDR-2 3 TAP aprons 6 30-gallon containers 10 books of M8 detector paper 30 sponges 8 M256A1 detector kits 50 trash bags 1 clipboard with a pen 1 stopwatch
Station 4 - rinse	1 squad leader 1 PDDE operator 2 pump operators	2 sprayers	1 M12A1 PDDA 1 3,000-gallon tank 3 65-GPM pumps 2 TPUs 2 TAP aprons
Station 5 - check	2 NCO/CAM opera- tors		2 CAMs 10 M256A1 detector kits 20 books of M8 detector paper 2 AN/VDR-2 2 M8A1 or M22 ACAAs
C^2	1 PL 1 PSG		1 HMMWV with a radio 3 NBC marking kits
Total personnel	20	17	

WARNING

Chemical-unit leaders must consider the impact of the work/rest cycle on their unit's ability to process vehicles through the DED. Failure to initiate a work/rest cycle could result in heat casualties and mission failure.

4-109. It may not be possible, for a variety of reasons, for an M12A1 PDDA-equipped decon platoon to use the optimum DED layout. Limited personnel or equipment will affect the DED layout. Figure 4-6, page 4-25, shows an example of an alternate DED layout.

Table 4-8. Personnel and Equipment Requirements for the Alternate DED Layout of the M12A1 PDDA-Equipped Unit

Stations	Personnel	Equipment/Supplies
Station 1 - primary wash	1 NCOIC* 2 PDDE operators* 4 sprayers 2 scrubbers	2 M12A1 PDDAs 2 3,000-gallon tanks 2 65-GPM pumps 4 long-handled brushes 6 TAP aprons Liquid detergent
Station 2 - DS2 application	1 NCOIC* 8 appliers	14 long-handled brushes 8 mops with extra mop heads 3 30-gallon containers 8 M13 DAPs Sufficient DS2
Station 3 - contact time/ interior decon	1 NCO* 2 interior decon assistants	2 AN/VDR-2 3 TAP aprons 6 30-gallon containers 10 books of M8 detector paper 30 sponges 8 M256A1 detector kits 50 trash bags 1 clipboard with a pen 1 stopwatch
Station 4 - rinse	1 NCOIC* 1 PDDE operator* 2 sprayers	1 M12A1 PDDA 1 3,000-gallon tank 3 65-GPM pumps 2 TPUs 2 TAP aprons
Station 5 - check	2 NCO/CAM operators*	2 CAMs 10 M256A1 detector kits 20 books of M8 detector paper 2 AN/VDR-2 2 M8A1 or M22 ACAAs
C^2	1 PL* 1 PSG*	1 HMMWV with a radio 3 NBC marking kits
Total personnel	29 (Minimum of 11 soldiers from the decon platoon)	
*** These individuals should be from the decon platoon.**		

THOROUGH-DECON-SITE CLOSURE

4-110. Once all vehicles and soldiers from the contaminated unit have processed through the thorough-decon site, the site can be closed. Coordination with the supported unit is necessary before actually closing the site to ensure that all contaminated elements have been processed.

4-111. The chemical unit (decon or smoke/decon platoons) closes the DED first. Once the DED is closed, the chemical unit processes through the DTD. After the chemical unit has processed through the DTD, the DTD is closed. At that point, the chemical unit marks the area as a contaminated area and reports its exact location to the supported unit, using an NBC-5 report. The exact procedures for closing the DED and DTD are described as follows:

Table 4-9. Personnel and Equipment Requirements for the Optimum DED Layout of the M17 LDS-Equipped Unit

Stations	Personnel		Equipment/Supplies
	Decon Platoon	Augmenter	
Station 1 - primary wash	1 squad leader 4 sprayers 2 scrubbers	2 scrubbers	3 M12A1 PDDAs 3 1,500-gallon tanks 3 65-GPM pumps 6 long-handled brushes 8 TAP aprons Liquid detergent
Station 2 - DS2 application	1 squad leader 3 appliers	9 appliers	18 long-handled brushes 9 mops with extra mop heads 3 30-gallon containers 9 M13 DAPs Sufficient DS2
Station 3 - contact time/ interior decon	1 NCO 2 interior decon assis-tants		2 AN/VDR-2 3 TAP aprons 6 30-gallon containers 10 books of M8 detector paper 30 sponges 8 M256A1 detector kits 50 trash bags 1 clipboard with a pen 1 stopwatch
Station 4 - rinse	1 squad leader 4 sprayers		3 M17 LDSs 3 3,000-gallon tanks 3 65-GPM pumps 2 TPUs 4 TAP aprons
Station 5 - check	2 NCO/CAM opera-tors		2 CAMs 10 M256A1 detector kits 20 books of M8 detector paper 2 AN/VDR-2 2 M8A1 or M22 ACAAs
C^2	1 PL 1 PSG		1 HMMWV with a radio 3 NBC marking kits
Total personnel	23	11	

CLOSING THE DED AREA

4-112. The DED is closed in sequence, starting at station 1. All vehicles, equipment, and nonexpendable supplies are inspected for contamination. If contamination is found on an item, it is decontaminated. The actions at each station are described as follows:

Station 1 - Primary Wash

4-113. Spray all vehicles and equipment with hot, soapy water to remove any contamination that could have been transferred during the primary wash operations. Drain the water billets or fabric tanks of water. Inspect all equipment and vehicles for contamination using the appropriate detection equipment.

Table 4-10. Personnel and Equipment Requirements for the Alternate DED Layout of the M17 LDS-Equipped Unit

Stations	Personnel	Equipment/Supplies
Station 1 - primary wash	1 NCOIC* 4 sprayers 4 scrubbers	3 M17 LDSs 3 3,000-gallon tanks 3 65-GPM pumps 4 long-handled brushes 8 TAP aprons Liquid detergent
Station 2 - DS2 application	1 NCOIC* 8 appliers	14 long-handled brushes 8 mops with extra mop heads 3 30-gallon containers 8 M13 DAPs Sufficient DS2
Station 3 - contact time/ interior decon	1 NCO* 2 interior decon assistants	2 AN/VDR-2 3 TAP aprons 6 30-gallon containers 10 books of M8 detector paper 30 sponges 8 M256A1 detector kits 50 trash bags 1 clipboard with a pen 1 stopwatch
Station 4 - rinse	1 NCOIC* 4 sprayers	2 M17 LDS 2 3,000-gallon tanks 2 65-GPM pumps 4 TAP aprons
Station 5 - check	2 NCO/CAM operators*	2 CAMs 10 M256A1 detector kits 20 books of M8 detector paper 2 AN/VDR-2 2 M8A1 or M22 ACAAs
C^2	1 PL* 1 PSG*	1 HMMWV with a radio 3 NBC marking kits
Water resupply	2 drivers*	2 TPUs
Total personnel	32 (Minimum of 10 soldiers from the smoke/decon platoon)	
*** These individuals should be from the smoke/decon platoon.**		

Station 2 - DS2 Application

4-114. Throw the mops and brushes that were used in applying the DS2 into a sump or bury them. Load the unused cans of the DS2 and the M13 DAPs on a vehicle.

Station 3 - Contact Time/Interior Decon

4-115. Inspect any unused supplies and equipment for contamination. If no contamination is detected, load the equipment and supplies on a vehicle. Throw all contaminated supplies into the nearest sump. In calculating sump requirements, 35 cubic feet of space is required for 250 gallons of liquid drain off. Once a sump is 2/3 full, prepare a new sump.

Table 4-11. Work/Rest Table

Temperatures	Stations and Workload				
	Station 1 – Primary Wash	Station 2- DS2 Application	Station 3 – Contact Time/Interior Decon	Station 4 - Rinse	Station 5 - Check
	Physical Exertion				
	Moderate	Heavy	Moderate	Moderate	Light
Cool (<68°F)	60 work 15 rest	30 work 30 rest	60 work 15 rest	60 work 15 rest	60 work 15 rest
Warm (68° - 74°F)	45 work 45 rest	20 work 20 rest	45 work 45 rest	45 work 45 rest	50 work 50 rest
Hot (74° - 84°F)	30 work 60 rest	15 work 30 rest	30 work 60 rest	30 work 60 rest	40 work 80 rest
Very Hot (>84°F)	20 work 60 rest	10 work 30 rest	20 work 60 rest	20 work 60 rest	25 work 75 rest

NOTE: When operating in temperatures above 75°F, consider the ability of the soldiers to accomplish the mission. Once the soldiers have reached their maximum workload for heat stress, they cannot recover quickly enough to accomplish the decon mission. A viable option is to postpone the decon operation until a cooler part of the day or evening. This will reduce the heat-stress load on the soldiers and increase the probability of mission success. For further information, see FM 3-4.

Station 4 – Rinse

4-116. Spray all vehicles and equipment with hot, soapy water to remove any contamination that could have been transferred during the rinse operations. Drain the water billets or fabric tanks of water. Inspect all equipment and vehicles for contamination using the appropriate detection equipment. If no contamination is detected, load the equipment on the vehicles. Spread one can of STB dry mix into each of the sumps and then cover them. Ensure that DS2 is not present in the sumps. Post NBC hazard markers near the covered sumps. **(NOTE: DS2 destroys chemical agents; however, some of the by-products created from it are toxic.)**

Station 5 – Check

4-117. Ensure that all equipment has been checked for contamination. If it is not contaminated, load it on a vehicle. If it is contaminated, decon it according to the appropriate technical manual (TM). Throw any contaminated supplies into the nearest sump. Move all vehicles upwind of station 5 and inspect them again for contamination. If any contamination is detected, the crew will use either the M11 or M13 DAP to decon the identified areas. Once the vehicles are staged, all the soldiers will proceed to the DTD.

CLOSING THE DTD AREA

4-118. Once all the soldiers from the DED have processed through the DTD, it may be closed. After the last person has exited the DTD—

- Pick up all the used supplies from station 7 and put them in station 7's sump. Take up the contamination control line. If engineer tape was used, dispose of it in station 7's sump.
- Move all usable supplies and equipment from all the stations to station 1. Discard all unusable supplies from stations 3, 4, and 5 in station 1's sump.
- Decon all supplies and equipment collected at station 1 using the decon and rinse water at station 1. Empty the decon and rinse water from the station into the sump and decontaminant the containers.
- Mark the entire decon area. Remove your overgarment using the MOPP-gear-exchange technique and dispose of it in the sump at station 1.
- Move any equipment used to fill the sump upwind of the decon area. Decon your rubber gloves and move all equipment and supplies in station 1 upwind of the decon area. Keep this equipment and supplies separate from that used to fill the sump.

THOROUGH DECON UNDER UNUSUAL CONDITIONS

4-119. Decon operations are not restricted to daylight hours. The enemy may employ chemical and biological weapons at night, since weather conditions are usually more favorable for their employment at that time. Consequently, NBC contamination encountered at night may require decon before daylight.

4-120. Conducting decon operations at night is a challenge. "White light" cannot be used without possibly revealing your location. However, decon personnel must have illumination to perform such essential decon tasks as spraying water, applying decontaminants, using detection equipment, and taking off the MOPP gear.

4-121. Use M8/M9 detector paper to check for chemical contamination and to see how well the chemical decon was performed. If contamination is present, the paper color changes to red and cannot be seen if red-filtered light is used. The CAM and M256A1 detector kit can be used to identify and detect chemical agents but will require a light source in order to obtain a reading.

4-122. Several vehicles awaiting or undergoing decon present a significant infrared "signature" due to the hot engines. To enter, exit, and move within the decon site is difficult under blackout conditions.

4-123. Night-decon operations are difficult but can be accomplished. Whenever possible, conduct thorough-decon operations in built-up areas. Use a building for DTD operations and a warehouse for DED operations. This allows for the use of white light. Soldiers can see what they are decontaminating, see where they are going, and read the color changes of the M8/M9 detector paper.

4-124. There will be a need for additional soldiers to act as ground guides in the DED area. The NCOICs of stations 1, 2, and 4 will inspect each vehicle with a flashlight or chemical light before the vehicle proceeds to the next station. There should be at least two flashlights at each station in both the DED and DTD areas.

Chapter 5

Special Decon Operations

The decon of terrain or structures allows personnel to increase their stay time in an area and facilitates their passage through an area. Large-scale terrain decon requires extensive amounts of equipment, material, and time. Thus, terrain decon is limited to areas of critical importance. Contaminated areas and structures that can be avoided do not need to be decontaminated except by weathering. If they must be used, decon may be necessary. When decon is not practical, use protection to deal with the hazard. Most terrain and structural surfaces (excluding unpainted metal and glass) absorb the agent. Therefore, the agent may be transferred to clothing by physical contact. The agent may also be transferred through vaporization and subsequent inhalation or diffusion into the eyes and the clothing. The decon of equipment containing vulnerable components presents certain challenges. Electronics and optics are especially vulnerable to damage if not carefully decontaminated. Some materials, such as canvas, absorb chemical agents and may not be decontaminated and reused.

TERRAIN DECON

5-1. A terrain decon provides more long-term rather than short-term benefits. Generally, you cannot decon occupied terrain well enough to allow you to reduce your MOPP level right away. At best, all you can do is speed up the weathering process. This may allow you to reduce your MOPP level a little sooner, for example, in 2 days instead of 4 or 5. Terrain decon can be so expensive and so ineffective that you should consider relocating.

5-2. If you decide on a terrain decon, limit the amount you do. Build a causeway or breach a path through the contamination for entry/exit into a major facility. Restrict your efforts to areas where you must work or move. No matter what you do, you will probably have to remain in MOPP; but you can reduce the danger of picking up liquid contamination and spreading it to other areas.

5-3. The most effective techniques for a terrain decon are the natural methods (weathering or aging). Using standard decontaminants will quickly exhaust your resources (see Appendix B). Usually the most practical method is to scrap off contaminated surfaces. Fire is a terrain-decon method; however, consideration must be given to the potential downwind vapor hazard. If possible, a contaminated area should be evacuated until time and weather remove the hazard. An STB slurry can be used for vital areas. To determine the persistency of chemicals at various temperatures, refer to FM 3-9.

5-4. A terrain decon may be accomplished using several methods. These methods are listed in the order of decreasing effectiveness. Although they reduce the contact hazard, none will prevent hazardous vapors desorbing from contaminated surfaces. Therefore, appropriate protective measures must still be taken.

WEATHERING

5-5. Weathering is the simplest and easiest form of decon.

Chemical and Biological Contamination

5-6. Warm, windy weather can significantly reduce terrain contamination. In some cases, this occurs in a few hours or it may take a few days. Many variables affect the persistency of chemical and biological hazards, so it is impossible to accurately predict how long it takes such contamination to weather. Sunlight is especially effective against most biological agents.

Radiological Contamination

5-7. Although the term "weathering" is used to describe the decon process, weather has little effect on radiological hazards. Heavy rain and wind may remove some contamination, but only time will reduce the radiation emanating from the contamination.

REMOVING/COVERING

5-8. Removing or covering contamination does not destroy it, but it does keep the hazards away from you temporarily.

Chemical and Biological Contamination

5-9. Contamination can be removed from paths and unpaved roadways by scraping off a layer of earth using heavy earthmoving equipment (bulldozers and road graders). Contaminated paths and paved surfaces can be covered with a 4-inch layer of earth, roofing paper, plastic sheets, and wood mats. This is a temporary measure because the agents may penetrate the covering. When the contamination penetrates the covering or when the covering is removed, the hazards will reappear. The coverings may also extend the life of the contamination hazard by reducing its exposure to air and sunlight.

Radiological Contamination

5-10. Radiological contamination must be covered by thick layers of dense material, like earth. Three inches of earth will decrease radiation dose rates by half because of the shielding provided by the soil. However, 12 inches is more effective. The job will be easier with earthmoving equipment, but the equipment and operators probably will have to undergo decon.

NEUTRALIZING

5-11. The STB dry mix or STB slurry may be used effectively against chemical or biological contamination, but not against radiological contamination. Trained chemical troops and equipment are required when using the STB slurry, which is costly both in time and material.

Chemical and Biological Contamination

5-12. The STB dry mix is spread on solid surfaces and raked into soft surfaces like sand or earth.

5-13. The STB slurry is applied to terrain by using a spray hose that is attached to an M12A1 decon apparatus. Operators wear TAP aprons over their MOPP gear and sit on the front fenders of a vehicle to spray the STB slurry. The speed of the vehicle should be adjusted according to the terrain; this is normally 2 to 4 miles per hour. The operators spray a fan pattern 3 to 5 feet ahead of the vehicle. The pattern overlaps on the center and extends past both sides of the vehicle. The STB slurry forms a seal over the surface and must be renewed periodically due to deterioration from traffic or weathering. One load of STB slurry will decon a concrete surface about 328 feet long by 33 feet wide. Other surfaces may require more decontaminants. It takes about an hour to load and mix one load of SBT slurry.

Radiological Contamination

5-14. The STB dry mix and STB slurry have no effect on radiological contamination or its hazards.

BURNING

5-15. Burning works well against chemically or biologically contaminated vegetation; but it is of no value against radiological contamination.

Chemical and Biological Contamination

5-16. Fuel may be used to burn grass or short undergrowth. Burning also works on dirt surfaces. Soak the area with diesel fuel, kerosene, or fuel oil and ignite remotely. Do not use gasoline; it burns too quickly. Burning will cause vapor hazards downwind, and protective measures will have to be used by downwind units. Area commanders must warn these units of the vapor hazards.

Radiological Contamination

5-17. Radiologically contaminated surfaces should not be burned. Burning will not destroy radiological contamination or its hazards. It may spread contamination if radioactive particles become suspended in smoke spread by the wind.

EXPLODING

5-18. Explosives can be used to blow up areas that are chemically and biologically contaminated; however they are of no value against radiological contamination.

Chemical and Biological Contamination

5-19. Detonating cord, bangalore torpedoes, or mine-clearing line charges may be used with the STB dry mix to clear small paths through a contaminated area. The two methods that are used follow:

- First method. Remove the holding band and loosen the lid of the bleach drum. Set the drum with the lid in place upside down over about 3 feet of looped detonation cord. Remove the drum and leave the bleach piled on the lid as shown in Figure 5-1, A, page 5-4. Lids may be fired in a series with a single strand of detonation cord connecting the loops.

- Second method. Remove the holding band and loosen the lid of the bleach drum. Set the drum in a shallow hole as shown in Figure 5-1, B. Make sure the lid is loose. Use 5 feet of looped detonating cord (7 feet in soft ground) for each drum. The drums may be fired individually or in a series. For gross contamination, additional firings may be necessary.

5-20. For either emplacement method, the suggested spacing for charges is 33 feet apart for a 100-foot-front contaminated area. The charges should be placed along the upwind edge of the contaminated area (see Figure 5-1, C).

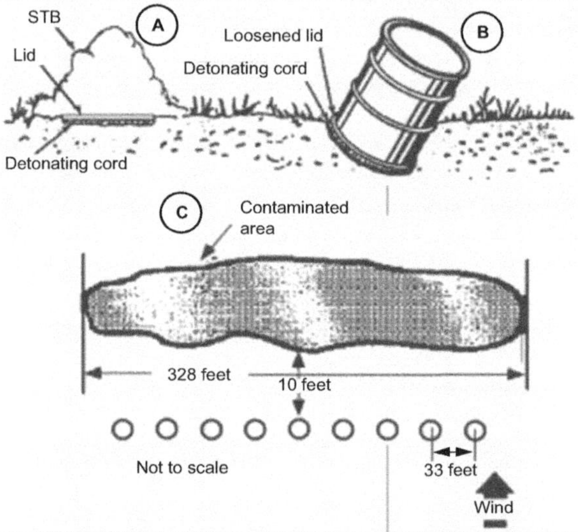

Figure 5-1. Two Methods of Emplacing Dry STB for Distribution by Detonation

Radiological Contamination

5-21. Explosives will not destroy radiological contamination or its hazards.

FLUSHING

5-22. Flushing with large quantities of water removes contamination. A pressurized stream of hot, soapy water delivered by the PDDA, M17 LDS, or fire trucks can remove significant amounts of an agent. Scrubbing removes even more.

Chemical and Biological Contamination

5-23. Flushing is not very effective against some kinds of agents, especially thickened agents. However, flushing may speed up the weathering process.

Radiological Contamination

5-24. Flushing is very effective for removing loose radioactive contamination such as fallout. Rain out, however, can coat surfaces with a film that resists flushing. Such films must be scrubbed. Any contamination removed by flushing and scrubbing will remain radioactive, so control the runoff.

CLEARING

5-25. To clear passageways, remove the contaminated layers covering the terrain.

Chemical and Biological Contamination

5-26. Most liquid chemical agents will not penetrate the soil more than 2 inches, and biological agents penetrate even less. Earthmoving equipment or hand tools can be used to scrap aside contamination. By scraping off 2 to 4 inches of earth, a passageway can be created that is free of transfer hazards. However, contamination on either side of the passageway will continue to present a vapor hazard.

Radiological Contamination

5-27. Fallout does not penetrate the top layer of soil unless it is followed by rain. It can be scraped aside. Move contaminated soil as far away as possible; the piles tend to concentrate radiation. Immediately below a nuclear blast area, the soil may be radioactive to a depth of 5 feet. In most cases, decon of this much earth is not practical.

FIXED-SITE DECON

5-28. Personnel and equipment need to be decontaminated to reduce the hazard and the spread of contamination, limit casualties, and minimize operational degradation. The goal is to restore the unit to an acceptable level of operational effectiveness. The inability of a fixed site to move away from contamination makes decon more critical. Decon is necessary when protective items cannot be worn in performing the mission or when long-term occupancy is anticipated.

5-29. Decon capabilities may be designed and built into a facility and used if the facility becomes contaminated. Mobile decon equipment may also be available at a fixed site to decon equipment, roads, and buildings. Interior decon may be accomplished with large commercial heaters. Loading docks, entries and exits, and building exteriors can be decontaminated with more conventional methods, such as using an STB dry mix or STB slurry, soap and water, or an M13 DAP. Commanders should identify all systems that are capable of contributing to the decon effort (for example, fire trucks, steam cleaners, and water pumps).

5-30. Decon equipment can be retrofitted to accommodate the process in each area. For instance, rest and relief shelters need rapid, personnel decon systems, and supply operations need a decon system that can handle moderate size pieces of equipment, but at a high volume. Light maintenance needs a system that can decon equipment sensitive to harsh decontaminants or water, and heavy maintenance needs a system that can clean large equipment.

5-31. The commander should designate and train teams that can perform decon for fixed-site operations. The following are fixed-site decon techniques that the teams could apply:

AIRFIELD

5-32. Decon the area required to sustain the mission (see Figure 5-2). Figure 5-3 illustrates the STB application on the entrance/exit of the airfield hangers. Note that the STB application includes 10 feet on each side of the runway between the entrance and the exit. A transfer hazard will remain a problem until all parts of the airfield have weathered. Buildings' or hangers' entrance/exit and doors must be decontaminated accordingly. Figure 5-3 illustrates a field-expedient method that could be employed at any fixed site to cover the major entry/exit toward the site.

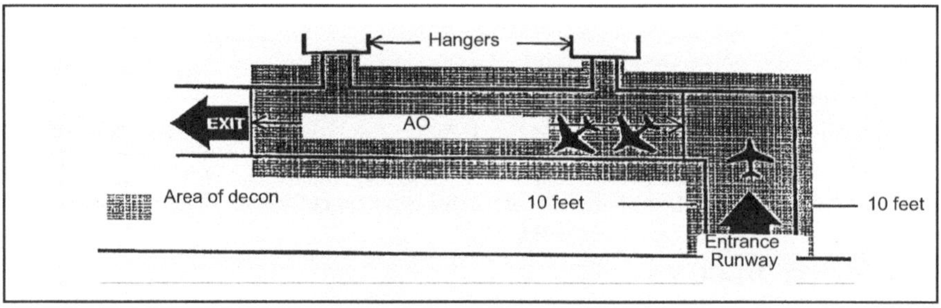

Figure 5-2. Airfield-Area Decon

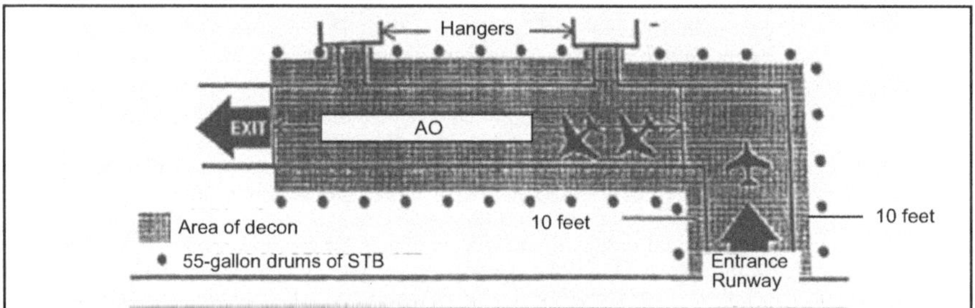

Figure 5-3. Field-Expedient Airfield Decon

HELIPAD

5-33. To decon a helipad site, spray the STB slurry on the entrance/exit approach or the path (see Figure 5-4). Since airborne STB particles can damage sensitive aircraft components, the decontaminated area should be washed with water before landing helicopters.

POMCUS/MOTOR PARK

5-34. At a prepositioning of material configured to unit sets (POMCUS)/motor-park site, decon entrance/exit approaches to limit contamination. Exposed vehicles in a POMCUS/motor park will become contaminated but most likely there will not be enough personnel to drive the vehicles to a decon station. Decon parked equipment by spraying it with an STB slurry or with hot, soapy water (see Figure 5-5). The M12A1 PDDA crew drives around the parked vehicles and performs vehicle washdown to accelerate the weathering process.

NOTE: For more detailed guidance on the decon procedures of fixed sites, ports, and airfields, refer to FM 3-4-1.

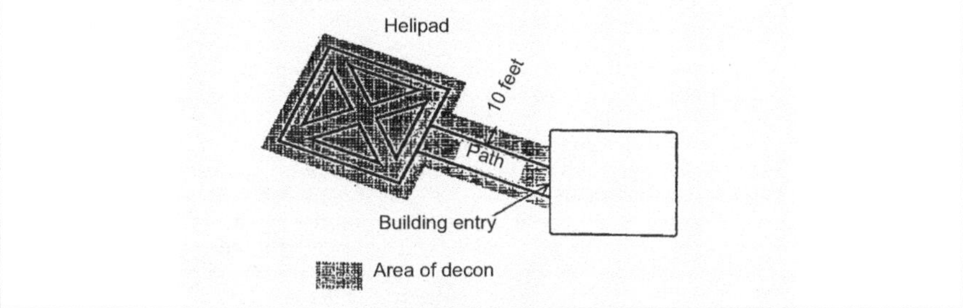

Figure 5-4. Helipad-Area Decon

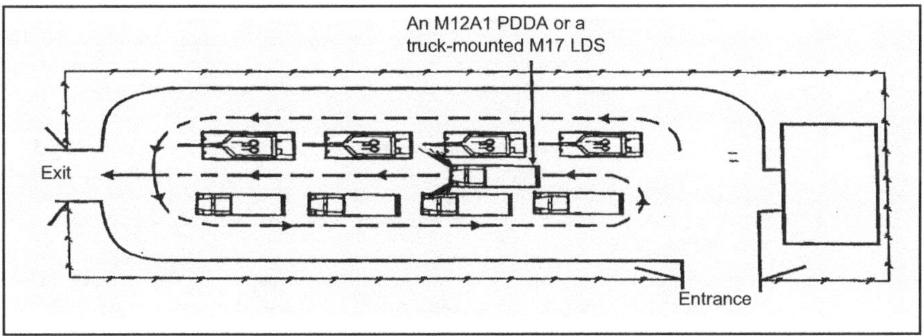

Figure 5-5. POMCUS/Motor-Park-Site Decon

VULNERABLE EQUIPMENT DECON

5-35. Most military equipment has not been critically assessed for its ability to withstand decon operations. As more testing is done, specific decon instructions will be included in the TM for all types of equipment. Some equipment is extremely vulnerable to damage when subjected to decon operations. Electronics and optic equipment are especially vulnerable. Some materials (such as canvas and rubber) tend to absorb chemical agents, and decontaminating absorbent surfaces is extremely difficult, if not impossible. Most decontaminants are highly corrosive and cannot be used on certain materials. They may also render ammunition unserviceable. Do not subject vulnerable equipment to unnecessary decon operations. Radiac meters may measure low levels of radiation that are not dangerous. The commander's operational exposure guidance (OEG) determines the maximum permissible radiation dose rate (usually 0.33 cGy per hour). If this dose rate is exceeded, you must decon.

ELECTRONICS

5-36. Nonsealed electronic-equipment circuitry can be damaged by corrosive decon materials. Most field electronic equipment is watertight for environmental protection. This also provides good protection against NBC contamination. Contamination probably will not penetrate gasket-equipped protective covers and sealed components on electronic equipment; but if exposed, the contaminants may be present on the outside of cases containing the electronic equipment. Wipe down the outside portions of the equipment case with a designated decontaminant. After decontaminating the outside, wipe down the equipment with either water or an approved solvent to remove traces of the decon solution. If equipment seals appear damaged or if penetration of NBC contamination into the inside of the equipment is suspected, then the unit should be treated as if it was nonsealed. Under no circumstances should electronic equipment be immersed in a decon solution or subjected to a high-pressure application of decon solutions.

5-37. For chemical contamination, wipe all metal electronic cases with DS2. If a chemical agent is identified as HD, allow 30 minutes contact time. Next, wipe the case down thoroughly with a damp cloth, dry it off, and wipe it down with a designated oil (for biological agents, wait 10 minutes). If DS2 is not available or the case is of some material other than metal, wipe the equipment's exterior with a cloth and hot, soapy water or use a nonstandard decontaminant (see Appendix B). If contamination is not extensive, use an IEDK.

5-38. Corrosive decontaminants, such as DS2, should never be used on nonsealed electronic equipment. This type of equipment is often found inside shelter assemblies and helicopters. Refer to the appropriate TMs or to Appendix C for acceptable decon procedures for nonsealed electronic equipment.

5-39. For radiological contamination, brush, wipe, or vacuum contamination from the equipment. Since contamination is not destroyed but just moved from one place to another, control the runoff and treat it is as a hazardous substance.

OPTICS

5-40. Optic systems are extremely vulnerable to decon materials that might scratch or adversely affect the lenses. Wipe optic systems with a soft, nonabrasive material such as lens-cleaning tissue, cotton wadding, or soft cloth dipped in hot, soapy water. Wipe the optic system with decontaminants. Do not immerse it.

5-41. Hot, soapy water is the preferred decontaminant for chemical and biological contamination. When DS2 is applied to coated lenses, it causes some damage. Likewise, exposing lenses to an STB slurry removes some of the magnesium from the coating. Small optic surfaces can be decontaminated with the IEDK.

5-42. Radiological contamination should be blown off with a stream of air or wiped off with hot, soapy water. Rinse by wiping with a sponge dipped in clean water.

AMMUNITION

5-43. Decon contaminated ammunition with cool, soapy water. Apply the soapy water with the PDDA or with brushes, mops, rags, or brooms. Cool, soapy water is the preferred decontaminant for all types of contamination.

WARNING

Do not use DS2 or an STB powder on ammunition. It removes critical markings from the ammunition. The STB powder may cause a fire when it comes in contact with a blister agent or DS2. It may also corrode and render the ammunition unserviceable. Do not use nonstandard decontaminants that are corrosive. They also may remove critical markings from the ammunition.

CANVAS ITEMS

5-44. These items include load-carrying equipment and web gear. It may be necessary to burn or bury them if they are heavily contaminated. Either an STB dry mix or an STB slurry may be used to decon them. The STB slurry is more effective. In many cases, weathering may be the preferred decon technique because scrubbing canvas frequently imbeds the contamination further and worsens the situation. If the item must be decontaminated, boil it for 1 hour in soapy water. This is the preferred method for chemical and biological decon. Radiological contamination can be removed by brushing it off, then washing. It may also be vacuumed off a surface.

5-45. Salvage of contaminated canvas requires a great amount of logistic and manpower requirements. Other options for disposing of contaminated canvas should be considered against METT-TC. NBC covers protect vulnerable items, but these covers must be buried or destroyed after use.

FOOD AND WATER

5-46. Most units cannot test the purity of food and water even if they can decon it. Food and water decon usually takes special handling by trained technicians. If your water becomes contaminated, contact medical personnel. Water-treatment specialists test and treat the water in an NBC environment. Direct contaminated-food questions to supporting veterinary personnel.

5-47. Food and water should be kept in sealed containers and protective packaging. Leave food in wrappers. Keep wrapped food in boxes, tins, or other containers. Store these containers in closed refrigerators or vans. See FM 3-3 for further discussion on avoiding contamination.

5-48. Contaminated food should be placed into three groups for handling.

Group I

5-49. Group I consists of canned and unopened items that have been exposed only to the vapors of an agent. Generally, the items in this group will be safe for issue to personnel after a brief period of outdoor airing to remove clinging vapors.

Group II

5-50. Group II consists of canned and unopened items that have been contaminated with a liquid chemical agent. Attempts to decon porous packaging materials, such as cardboard or wood, are likely to be unsuccessful and may result in spreading the contamination. To handle such items, strip off the outer contaminated coverings and examine the inner layer to determine if agent penetration has occurred. If it has, continue stripping off layers until an uncontaminated layer is reached. Items packaged for military operations are usually packed in boxes, so this procedure is feasible. When the inner uncontaminated package is reached, place it in group I. If the agent has penetrated to the food, place it in group III.

Group III

5-51. Group III consists of unpackaged or poorly packaged items that have been exposed to an agent in either a vapor or liquid form. Foodstuffs in this group should be decontaminated only when absolutely necessary.

5-52. If you do not know if the food is contaminated, do not use it. Do not decon items in group III without the help of veterinary personnel.

CHEMICAL MUNITIONS DISPOSAL AND DECON

5-53. In combat, chemical munitions may be encountered. Captured enemy chemical munitions may have sustained damage or have deteriorated to where they are leaking. Leaking munitions must be decontaminated, evacuated, and disposed of safely by specialized units with properly trained soldiers performing this task. All soldiers must use protective clothing and masks when there is a possibility of exposure to a toxic chemical agent. The MOPP level will be specified by the unit commander. Use detection equipment (CAM, M256A1 detector kit, M8A1 or M22 ACAA, and M8/M9 detector paper) to monitor chemical munitions.

5-54. Unexploded leaking munitions should be treated in the same manner as conventional ordnance. A spot report must be submitted to an explosive ordnance disposal (EOD) unit.

5-55. When disposing of and decontaminating chemical munitions, see Army regulation (AR) 385-10 and TM 10-277 for additional safety guidance.

HANDLING

5-56. Chemical projectiles should not be handled. A good rule of thumb is to treat them as you would a fused high-explosive (HE) round. Security and protective measures must match those given to conventional munitions. Notify an EOD unit for disposal and decon of chemical munitions. EOD units have the responsibility for sealing, packaging, and disposing of chemical munitions.

CONTAMINATION CONTROL

5-57. Whenever a chemical projectile is damaged and starts to leak, the round presents the same hazards to personnel in the area as an enemy chemical attack. Only munitions that have not completely detonated or have been severely damaged may exude liquid agent. It is the vapor hazard created when the chemical leakage occurs that poses the greatest threat to personnel. When a leaking chemical munition is discovered, react as if you are under chemical attack.

DEPLETED URANIUM (DU) DECON

5-58. DU is an extremely dense metal used in munitions to penetrate heavy armor or as a protective shielding (armor packages). DU is also used as equipment components. All components containing radioactive isotopes are listed in TB 43-0116.

5-59. DU exposure and incidents may occur anytime there is damage to the DU armor package, a vehicle is hit with DU munitions, DU munitions are damaged, or equipment components containing DU are damaged. The DU armor package can be damaged during vehicle maneuvers, on-board fires, maintenance activities, or ballistic impacts. DU munition problems may occur during storage, transportation, combat, testing, or manufacturing. DU contamination may be present on the ground in areas where equipment was destroyed or damaged.

5-60. DU contamination may include DU oxides (dust), contaminated shrapnel, munition components, or armor components. DU emits primarily alpha particles; however, beta, gamma, and x-ray ionizing radiation are also emitted. DU contamination can be inhaled, ingested, or injected. DU contamination does not pose an immediate health risk. Consequently, contamination should be removed from personnel or vehicle surfaces when directed by the unit commander based on METT-TC.

5-61. Visual signs that DU contamination is present include heavy, dull-black dust or small round holes. DU contamination can only be verified with a radiac meter. An AN/VDR-2 or AN/PDR-77 with an alpha or beta probe (flat-pancake surface) is used to detect and measure DU contamination. These probes are within the radiological protection officer (RPO) kit.

5-62. When working on or within DU contaminated equipment, soldiers should wear their protective mask. Depending on temperature, protective-clothing availability, DU contamination levels, and tasks to be performed, soldiers can wear the BDO or coveralls or they can roll down their sleeves and blouse their trousers as directed by the unit's chemical or medical personnel.

5-63. General decon procedures follow:

- Use a radiac meter to determine if DU contamination is present.
- Provide protection, including appropriate clothing, for workers as directed by the unit's chemical or medical personnel.
- Identify what is to be decontaminated.
- Obtain necessary equipment and materials (Appendix D).

- Brush, wash, or wipe off contamination with a damp cloth. Use a high-efficiency particulate air filter (HEPA) vacuum cleaner, if available.
- Work from the outside of the contaminated area to the inside.
- Cover fixed contamination with tape, paint, paper, plastic, or other disposable material.
- Use the standard double-bag-and-tag process for hazardous waste. The only contaminated waste generated by DU will be the vacuum-cleaner bags after use on multiple vehicles.
- Continue the decontaminating process until contamination levels are reduced to the appropriate level (see AR 11-9).
- Decon personnel and equipment according to Chapter 4 of this manual.

CONTAMINATED REMAINS DECON

5-64. Mortuary affairs personnel establish and operate the mortuary affairs decontamination collection point (MADCP). When an NBC event has occurred on the battlefield, there is a very high probability that many deceased personnel were exposed to contaminating agents. If the situation does not lend itself to the determination of a hazard on an individual basis, all remains within the affected area will be treated as if contaminated. If the theater surgeon or his staff determines that biological agents have been employed, all remains will be treated as if contaminated.

5-65. Whenever the threat of NBC warfare exists in a theater of operations, the unified commander will direct the Joint Mortuary Affairs Office (JMAO) to be prepared to handle contaminated remains. The JMAO will serve as the theater central point of coordination for the operation. Joint operations of contaminated remains are described in Joint Publication 4-06.

5-66. Personnel support is required after completing the evacuation mission to the MADCP. A DTD takes about 1 hour. The MADCP site will require a complete detailed decon by a chemical decon unit.

RADIOLOGICAL DECON

5-67. Radiological contamination may occur in the form of one element. In the following paragraphs, the decon of six specific, commonly found radioactive elements is discussed. The discussion is applicable not only to these elements but also to other elements having similar chemical properties.

CESIUM

5-68. The common radioisotope of cesium is cesium-137. It emits beta and gamma radiation, decaying to stable barium-137. Cesium-137 is widely used in gamma sources. It occurs in these sources as cesium-chloride pellets. Cesium chloride is a soluble salt. The contamination from a sealed-source leak absorbs water, becomes damp, and creeps. Contamination from a sealed cesium source is best decontaminated by wet procedures unless the contamination is on a porous surface, in which case wet procedures should be preceded by vacuuming. Cesium is known to adsorb from a solution onto glass surfaces. Decontaminating a cesium liquid-contaminated surface is best accomplished by wetting the surface, absorbing the solution with a rag or other absorbent material, and rinsing the area several times with water. If the contamination persists, use a detergent solution and scrub with a brush. A cesium-contaminated solution that has been standing for some time is best decontaminated by absorbing any remaining liquid, treating the surface several times with water (allowing the water to stand on the surface for about 1 minute each time), and then absorbing the liquid from the surface. If the contamination remains, further treatment depends on the surface. Metallic surfaces are treated with strong mineral or oxidizing acids. Waxed surfaces are removed. If contamination still persists, abrasives or other removal techniques are used.

COBALT

5-69. The common radioisotope of cobalt is cobalt-60, a beta gamma emitter. Metallic cobalt-60 is commonly used in sealed gamma sources. Particles of cobalt dust adhering to small articles are readily removed by ultrasonic cleaners or by dipping the article in a dilute solution of nitric, hydrochloric, or sulfuric acid. Cobalt-dust contamination that exists over a large area is best removed by vacuuming. Sealed cobalt sources may leak as a result of electrolytic action between the cobalt and the container. The result is often a soluble cobalt salt, which creeps and spreads. This is best decontaminated with a detergent or an ethylene diamine tetra acetic acid (EDTA) solution, followed by a treatment with mineral acids. Contamination from solutions containing cobalt may be treated with water.

PLUTONIUM

5-70. The most common isotope in which plutonium may be present as a contaminant is plutonium-239, an alpha emitter. This isotope is present in the AN/UDM-6 calibration source. Plutonium contamination may be the result of a nuclear weapon's accident, in which case, the plutonium will be scattered as a metal or oxide in a dust form. Both forms of plutonium are insoluble. Aging of plutonium-239 contamination is impractical since it has a 24,000-year half-life. Plutonium contamination that covers a small area is best decontaminated by vacuuming. If contamination remains, the area should be washed with a detergent solution. Any contamination that remains can be sealed in a protective coating of paint, varnish, or plastic. Plutonium oxide or metal dust spread over a large area, such as a field, is best decontaminated by removing the top layer of soil and disposing of it as radioactive waste. Personnel should wear respiratory protection when decontaminating or moving the soil.

STRONTIUM

5-71. The most common radioisotope of a strontium is strontium-90, a beta emitter. The daughter particle of strontium-90 is yttrium-90, which is also a beta emitter. Strontium-90-yttrium-90 is commonly used in sealed beta sources such as the M6 source. Generally, it is present as chlorine or carbonate. The chlorine is hygroscopic; it absorbs water and creeps out of the container. This contamination is best decontaminated by vacuuming, followed by a treatment with water, a complexing-agent solution, and a mineral acid, in that order. Contamination resulting from a dilution containing strontium is best decontaminated by absorbing the solution and washing the area with a detergent solution. If strontium contamination persists, the top layer of the surface should be removed by abrasives or other removal procedures and a sealing coat should be placed over the surface.

TRITIUM

5-72. Tritium is the radioisotope of hydrogen and is a weak beta emitter. If it is released to an area as a gas, the best decon method is to flush the area with air. Since inhalation of tritium can present an internal hazard, personnel entering an area containing tritium gas should wear an appropriate self-contained type of breathing apparatus. Objects in an area exposed to tritium for a length of time may absorb the gas and should be disposed of, if possible. They may be degassed, under a vacuum, by flushing with helium or hydrogen. A surface which monitors clean may become contaminated again in a matter of hours by percolation. There is no practical way of removing tritium oxide (T20) from water due to its similarity to natural water.

URANIUM

5-73. The most probable source of uranium contamination is a nuclear weapon accident in which the fissionable uranium is spread as a metal or oxide dust. The common isotopes of uranium contamination are uranium-235 and uranium-238. This metal or oxide is insoluble and is best removed from a contaminated surface by brushing or vacuuming, followed by a treatment with mineral acids or oxidizing acids and then sealed. Large-area uranium contamination is best decontaminated by removing the top layer of the surface or by sealing it.

Chapter 6

Effects of Environmental Conditions

US forces may find themselves anywhere in the world and subjected to the conditions in that region. Weather and terrain conditions will dictate how to fight and what equipment is most effective to support the fighting force under NBC conditions. The conditions in cold- and arctic-weather areas, warmed areas, urban areas, mountains, jungles, and deserts impact how decon operations are performed.

COLD- AND ARCTIC-WEATHER AREAS

6-1. These regions comprise about 45 percent of the North American continent and about 65 percent of the Eurasian land mass. They are characterized by extreme cold weather and deep snow during winter months. Spring breakup and mud restrict mobility. Whiteout and grayout cause loss of depth perception, making vehicle movement hazardous. Ice fog (clouds of ice crystals) covers troops, vehicles, bivouac areas, and permanent facilities, marking their location. Chemical agents become more persistent in low temperatures (see Table 6-1, page 6-2). As a result, some normally nonpersistent chemicals, such as sarin (GB), become persistent; this nerve agent could remain a transfer hazard for up to 30 days in arctic climates. Normally, persistent agents become more persistent. Although frozen agents do not present a significant problem in solid state, they become hazards when they warm up. Decon is a problem because low temperatures also reduce the effect of decontaminants.

6-2. The freezing points of decontaminants are as follows:

- Water, 32^oF.
- STB slurry with antiset, 0^oF.
- DS2, -25^oF.
- SKD, -50^oF.
- IEDK, -50^oF.

DECONTAMINANTS

6-3. When temperatures reach 32^oF, water can no longer be used as a decon. In such situations, use DS2. DS2 is effective down to -25^oF. For temperatures below -25^oF, use STB or HTH as a dry mix (two parts of STB to three parts of earth or snow) for decon. Apply the dry mix by shoveling it on contaminated surfaces or by filling sandbags with the mix and dusting it on the surfaces. Remove the dry mix by brushing, scraping, or using uncontaminated earth or snow to "wash" it off. Other decontaminating methods using nonstandard solvents and fuels may be used, but observe fire safety, protect personnel from corrosives, and take precautions against super-cooling effects. Because of their low freezing points, solvents such as aviation fuel (JP8), diesel fuel, and kerosene may be used to physically remove contamination. These solvents only flush the agent from the surfaces. They generally do not neutralize agents nor do they eliminate agents that soak into surfaces. Nonstandard solvents generally are very flammable and must be handled with care. See Appendix B for specific nonstandard solvents.

Table 6-1. Freezing Points of Selected Chemical Agents

Agents	Symbols	Contact Hazard	Vapor Hazard	Freezing Point
		Nerve		
Tabun	GA	Extreme	Low-moderate	+23°F
Sarin	GB	Extreme	Extreme	-69°F
Soman	GD	Extreme	Probable	-44°F
	GF	Extreme	Probable	-22°F
	VX	Extreme	Negligible	-60°F
	VR-55	Extreme	Probable	Unknown
	TGD	Extreme	Probable	Depends on the percent of thickener
		Blister		
Distilled mustard	HD	Extreme	Negligible	+57°F
Mustand-lewisite	HL	Extreme	Low	-14°F (Pure) -44°F (Plant purity)
Nitrogen mustard	HN-1	Extreme	Low	-29°F
Nitrogen mustard	HN-2	Extreme	Low	-76°F
Lewisite	L	Extreme	Negligible	0°F
Nitrogen mustard	HN-3	Extreme	Low	+25°F
Phosgene oxime	CX	Extreme	Low	+95°F
		Blood		
Hydrogen cyanide	AC	Low	Extreme	+8°F
Cyanogen chloride	CK	Low	Extreme	+20°F
Arsine	SA	Low	Extreme	-177°F
		Choking		
Phosgene	CG	Slight	Extreme	-198°F
Diphosgene	DP	Slight	Extreme	-71°F

EQUIPMENT-DECON DEVICES

6-4. Discussed in the following paragraphs are the different types of equipment decon devices:

M11 Decon Apparatus

6-5. This apparatus is designed to dispense DS2 at temperatures as low as -15°F. DS2 is effective down to -25°F, but it becomes thicker as the temperature drops. Normally, one nitrogen cylinder is used to charge the M11. In cold weather, a second nitrogen cylinder may be used if inadequate pressure is developed with one. Never use more than two nitrogen cylinders.

M13 Decon Apparatus

6-6. This apparatus is designed to dispense DS2 at temperatures as low as -15°F. As the temperature drops, the flow of decontaminants through the hose and wand becomes degraded.

M17 Lightweight Decon System and M12A1 Decon Apparatus

6-7. Both apparatuses have problems in the cold similar to any other water pump. Normal engine "cold-soaking" problems that cause starting difficulties and brittle metal parts have been observed. Also, internal pumps and lines will crack and break when water freezes within these systems. To avoid freeze-up problems, use the shutdown procedures described in TMs 3-4230-228-10 and 3-4230-209-20&P.

DETECTION AND WARNING DEVICES

6-8. In the following paragraphs, the different types of detection and warning devices are discussed:

M8/M9 Detector Paper

6-9. M8/M9 detector paper is limited in the cold because only agents in a liquid form can be detected. If the substance is thickened or solidified, collect a sample with a stick or scraper and wipe it onto a sheet of M8/M9 detector paper. Place the sample on a heated surface, such as an operating vehicle or a power generator, to thaw the suspected agent. Because of off gassing, do not do this in a heated vehicle or tent. Put a box or can over the paper to concentrate any vapors. Put an M256A1 detector kit inside the box or can. Heat the suspected agent enough to enable detection and allow agent identification.

M256A1 Detector Kit

6-10. As temperatures drop toward 32°F, the M256A1 detector kit takes longer to give a positive or negative indication. When temperatures reach freezing, carry the liquid components in the breast pockets of the chemical protective clothing. It may also be necessary to put the test sampler back into the shipping bag after sampling and place it on a heated surface to hasten the reaction. The kit will not work at temperatures below -25°F.

Chemical-Agent Monitor

6-11. The CAM will have difficulty in detecting agents at low temperatures. Contaminated personnel entering a building or a collective protection shelter will desorb agents as they warm up. Use the CAM in a vestibule or an air lock where the temperature has been raised to that of the building's or shelter's interior. Likewise, check aircraft and vehicles requiring maintenance as they pass through a warmed-up site before entry into a hanger or maintenance bay.

M8A1 Automatic Chemical-Agent Alarm

6-12. When operating an M8A1 ACAA at temperatures below 20°F, use the M253 winterization kit. Warm-up time with this kit is 50 minutes.

EQUIPMENT-DECON STATIONS

6-13. Place decon sites in built-up areas and near road junctions and intersections of forest lanes to facilitate access. Divide the decon sites and approach routes into sectors, and assign units to keep them open in spite of mud or heavy snow. With present technology, equipment decon problems in the field are difficult to overcome in an arctic climate. Commanders should consider fighting dirty in arctic regions. Fresh units can be rotated into the field so that dirty units can be moved back to built-up areas to decon.

6-14. Decon apparatuses and water trucks may have to be deployed with empty tanks instead of full ones as in temperate climates. To prevent freezing, it may be necessary to preheat water when loading the water trucks and the tanks of the decon apparatuses and keep it heated until it is used. These vehicles may have to be enclosed and warmed so that the engines will start. Decon apparatuses and water trucks must be drained immediately after use to prevent damage from freezing. Commanders may choose to conduct minimum decon until they can get back to fixed facilities and conduct a more thorough decon.

6-15. Vehicles and personnel covered with contaminated snow must be decontaminated before the snow has a chance to melt and freeze. Such snow forms layers of ice that make contamination difficult to remove. Radioactive fallout that is mixed with snowfall must be removed as soon as possible. Use tree branches (if available) to remove contaminated snow.

6-16. Snow can be used to cover contamination; however, the snow can blow away or the contamination can resurface when tracked vehicles, troop movement, or digging disturbs it. Snow cover provides some protection if left undisturbed, but this protection is too uncertain to rely on.

WARMED AREAS

6-17. Preventing contamination from entering warmed areas is a challenging problem. For example, if soldiers get frozen agents on their clothing, it will be hard to detect because low temperatures have slowed its effects. However, if the temperature warms or if the soldiers enter a warmed area, the agents become dangerous. Because of this, it may be necessary to set up a warming station for each warmed area to isolate the agents. Otherwise, all occupants may be subjected to hazardous liquids and vapors.

6-18. Additional soldiers and equipment are needed for these warming stations. Therefore, in cold-weather operations, the decon and detection priority must be heated support facilities. FM 3-4 discusses individual and collective protection in detail.

URBAN AREAS

6-19. The need for decon operations in urban areas may also exist, and it may be easier to support. Water sources are generally available and commercial chemicals may be used as decontaminants. Site security may be simplified because of limited observation and poor fields of fire. Decon operations in urban areas will not differ significantly from similar operations in the field; however, keep the following considerations in mind:

STREETS AND STRUCTURES

6-20. When buildings are contaminated with persistent chemicals, their value for cover, concealment, and shelter is reduced. Wood and concrete tend to absorb liquid agents, and they may give off toxic vapors for days or weeks. Building decon is very difficult and requires large quantities of decontaminants. The hazard can be reduced by covering the contamination with plastic sheets, STB slurry, sodium silicate, or other substances that cover or absorb the agent. Even though a particular part of a building is not intended for occupation, it may still need to be decontaminated to prevent the contamination from spreading.

6-21. Streets and sidewalks also absorb liquid agents, then give off toxic vapors when heated by the sun. These surfaces may need to be decontaminated several times to reduce hazards. Streets, sidewalks, or other porous surfaces are best decontaminated by weathering if the time and the situation permit.

CIVILIANS

6-22. Try to locate decon operations away from civilians because they may injure themselves or interfere with the work. Civilians must not be forced to prepare decon sites or to help with decon operations. However, volunteers who have been cleared by the Assistant Chief of Staff, G5 (Civil Affairs) (G5) may be used.

SUPPLIES AND AMMUNITION

6-23. Store ammunition, as well as other supplies, under cover to protect it from NBC contamination. FM 3-4 explains contamination covers in detail.

SANITATION SYSTEMS

6-24. Urban areas may have sophisticated sanitation systems. When those systems are destroyed, sanitary conditions become far worse than those in areas where sanitation systems have never existed. Sanitation systems must be maintained to avoid epidemics and to avoid overloading biological-decon capabilities. Contaminated water and residue must be controlled so that they will not create a hazard to the civilian population or interfere with sanitation systems. Engineer support may be necessary to construct controlled runoff areas.

MOUNTAINS

6-25. Excluding the extremely high, alpine-type mountains, most mountain systems are characterized by heavy woods or jungle, compartments and ridge systems, limited routes, and highly variable weather conditions. All these factors will affect decon operations.

MOBILITY

6-26. The terrain and the disruption of existing routes may dictate that mountain operations be decentralized. Additional water-carrying vehicles may be needed to support these operations. Forces must be organized to be self-sufficient and should have decon elements attached to them initially. Decon platoons are attached to brigades and decon squads are attached to battalions. When decon elements operate independently, their parent chemical companies may not be able to provide support. Decon elements must be equipped with greater hauling capabilities, especially for water. Administrative and logistical support to the decon element will have to come from the support unit.

WIND

6-27. Mountains have changing weather and constant winds that promote weathering at a faster rate. In high mountains, calm winds in the ridges and passes are seldom; whereas, in protected valleys, strong winds are rare. Normally, wind velocity increases with altitude. Generally, downwind contamination will be greater because high winds are more frequent. However, variable winds will make predictions unreliable.

TEMPERATURE

6-28. Cool or cold temperatures have an adverse effect on decon operations. With every 328-foot gain in elevation, the temperature will normally drop 3^{0} to 5^{0}F. The drier the air, the more pronounced this drop will be. At high altitudes, there may be differences of 40^{0} to 50^{0}F between the temperature in the sun and that in the shade. Consequently, the temperature rises swiftly after sunrise and drops quickly after sunset. Low temperatures that normally exist at extremely high altitudes may demand decon procedures and precautions similar to those used in cold-weather operations. Daytime operations can be scheduled to avoid some low-temperature problems.

SUNLIGHT

6-29. Sunlight destroys most biological agents. Above the timberline, there is little protection from the sun, so the effectiveness of a biological attack is reduced.

JUNGLE

6-30. Jungle climates vary with the location. Close to the equator, all seasons are nearly alike, with rains throughout the year. Farther from the equator, especially in India and Southeast Asia, jungles have distinct wet (monsoon) and dry seasons. Both zones have high temperatures (averaging 78^{0} to 98^{0}F), heavy rainfall (as much as 400 inches annually), and high humidity (90 percent) throughout the year.

6-31. Transportation is extremely difficult due to natural obstacles; therefore, decon operations will be decentralized as much as possible. Because of high humidity and temperatures, MOPP gear should not be worn for long periods. Since decon crews will be more susceptible to heat stress, staffs should plan for frequent crew rotations and provide enough decon personnel to conduct decon operations.

6-32. Contamination will be retained temporarily in the jungle canopy, reducing the immediate hazard. Later, rains will wash these particles to the ground and concentrate them in low areas. These areas are likely to become contamination "hot spots." Decon sites should be put on high ground during the rainy season.

6-33. Tropical rainstorms will flood decon sites unless the sites are adequately drained. Ground that appears firm may become impassable when it rains. Rain and intense heat will speed the deterioration of many supplies. Do not break the seals on packing material until it is absolutely necessary.

6-34. Rain and high humidity (wetness) degrade the protective abilities of the overgarment. Take care to ensure the maximum length of time of the overgarment's usefulness. Additional quantities of overgarments must be maintained to replace those that have gotten wet. Rain helps the decon process by washing away contamination on exposed surfaces. Rain can also hydrolyze some agents. However, runoff may contaminate the soil.

TEMPERATURE AND HUMIDITY

6-35. When temperatures rise to the 85^{0} to 100^{0}F range, troops can only continue medium or heavy workloads by reducing their MOPP level. More soldiers will be needed to make up for extra rest periods and more frequent rotation periods during decon operations.

6-36. Frequently oil exposed metal parts and grease wheel bearings to protect them from moisture. Since petroleum, oils, and lubricants (POL) tend to absorb chemical agents, additional decon may be needed. Measures taken to protect electronic communications equipment from moisture will reduce the need for decon since these measures provide good NBC protection. Seal radios with silicone compound and pack them in waterproof containers when they are not in use (remove batteries). Protect handsets and microphones with plastic bags and check the bags frequently for moisture buildup.

6-37. The jungle's high humidity and temperature increase the need for maintenance. Contact teams must operate independently as they move from site to site doing repairs. There may be no decon specialists to help them.

6-38. Biological agents thrive in the heat, humidity, and shade, which are characteristics of the jungle. Therefore, weathering is not a practical means of biological decon.

6-39. Solid decontaminants, such as STB powder, tend to cake and decompose at a faster rate than in temperate climates. Caking is no problem, but the decomposition eventually makes STB powder ineffective.

TIME

6-40. Many tasks in the jungle take more time than in other environments. More time for movement and security must be allowed, which may leave less time for planning and preparation. In the jungle, more decon personnel will be required to perform decon operations.

SECURITY

6-41. Decon stations are lucrative targets; therefore, security must be a prime consideration. Relocating a decon site may require a small security force, thus additional personnel will be required.

RESUPPLY

6-42. Because of the terrain, aerial resupply usually will be common practice. Keep this in mind when determining the amounts of decontaminants to keep on hand since they will compete with ammunition, POL, medical supplies, food, and water for priority of delivery.

PERSISTENCY

6-43. Sunlight and wind are reduced within thick jungles due to the vegetation. As a result, chemical and biological agents are less likely to disperse and will remain a hazard for an extended time.

DESERT

6-44. Desert regions are characterized by extreme temperatures with ranges varying between 30^{0} to 130^{0}F over a 24-hour period. These regions have long drought periods, which are interrupted by sudden rains that bring flash floods. There are shortages of suitable groundwater. Large areas suitable for tracked-vehicle maneuvers may sometime have impassable ravines; wet, spongy grounds; and sand areas.

WATER

6-45. The principal problem for decon in the desert is the lack of water. The use of STB slurry and DS2 will burden the logistical system because of the water required for mixing and rinsing. Nonwater-based decontaminants (natural solvents) may be required (see Appendix B of this manual). Contamination avoidance becomes increasingly important in desert operations because of limited water sources. Therefore, maximum use should be made of field-expedient covers, especially at night when an attack by chemical or biological agents is most likely.

HEAT STRESS

6-46. Heat stress is a critical problem for soldiers working in the desert environment in full MOPP. Operating decon stations in daytime temperatures may require short periods of work followed by long periods of rest. Performing operations at night to avoid heat stress may create light discipline problems.

WEATHERING

6-47. Weathering is a viable decon option. High daytime temperatures can increase evaporation of liquid contamination. As a result, vapor concentrations will be high but may not last long. Refer to FM 3-4 for MOPP opening/unmasking guidance to estimate the chemical hazards resulting from enemy attacks. If liquid contamination soaks into soft, porous soil (such as loose sand), evaporation is not as quick. Strong winds also increase the evaporation rate. Low temperatures during the night have a reverse effect and tend to increase the persistency of chemical and biological contamination. The sandblasting effect of sandstorms may remove contamination from surfaces facing the storm. Desert sunlight and high temperatures will destroy many chemical and biological agents without additional decon measures.

BEARINGS AND OTHER CRITICAL MOVING PARTS

6-48. Bearings and other critical moving parts need extra lubrication in the desert. This complicates decon because lubricants tend to absorb chemical agents. After a sandstorm, maintenance must be conducted regardless of the last scheduled maintenance. Therefore, perform decon first and if time is critical, decon only those surfaces that will be touched during maintenance. This will not eliminate vapor hazards.

Chapter 7

Aviation

Contamination in the battle space causes a unique problem for aviation. Since an aircraft uses its high mobility to maneuver across the entire battle space, avoiding contamination can be very difficult. Once an aircraft is contaminated, it is very difficult to decon. Therefore, aviation units need to prepare effective standing operating procedures (SOPs) that emphasize contamination avoidance, protection, and methods to decon each type of aircraft using an immediate, an operational, and a thorough decon.

CONTAMINATION AVOIDANCE

7-1. Contamination avoidance is critical due to the difficulty in decon. Aircrews must be aware of NBC hazards. For instance, an aircraft that hovers or lands in a contaminated area could splash contamination onto itself. Likewise, contaminated passengers or ground crews pose a danger of transferring contaminants into the aircraft where they are difficult to remove. Aircraft could also possibly pick up some contamination by flying through an invisible vapor cloud. Hazards that will only incapacitate ground personnel for a short time can be fatal in the air.

7-2. Some things that can be done to limit contamination are as follows:

- Know what areas are contaminated and avoid these, if possible.
- Pick landing zones that will have a reduced splash effect if an aircraft must land in a contaminated area.
- Limit the spread of contamination into the aircraft from the outside. For example, ground crews at the forward arming and refueling point (FARP) could conduct arming and refueling without requiring the aircrew to exit the aircraft.
- Conduct inspections without touching or shaking items. The contaminated crews should perform this task. Many inspection points can be inspected visually.
- Increase the use of covers when not flying. Use engine covers, flyaway gear, and hatches. If possible, provide overhead cover for parked aircraft.
- Limit the number of aircraft that must operate in a contaminated area or use aircraft already contaminated.
- Line the troop compartment with plastic (a field expedient way to limit the spread of contamination) when carrying contaminated personnel or casualties. A plastic curtain can be fastened between the troop compartment and the flight compartment with tape or Velcro to limit contamination transfer. The aircraft's heater can be used with

the curtain to create an overpressure in the pilot's compartment. This will limit vapors from entering the compartment.

- Apply M9 detector paper to the landing gear of the aircraft. FARP personnel should always check the M9 detector paper before servicing the aircraft. Another piece of M9 detector paper can be placed on the windscreen where the aircrew can see it.

CAUTION

M9 detector paper will show false positives if exposed to petroleum products.

DECONTAMINANTS

7-3. Only approved cleaning compounds may be used to decon aircraft. Caustic decontaminants such as DS2, STB, bleaches, and sodium hypochlorite are not considered safe. STB corrodes metal components and the aircraft's skin, and DS2 corrodes rubber, plastic, and acrylic-plastic windows.

7-4. Soap and water, kerosene, JP8, and diesel fuels are approved as decontaminants on selected parts of the aircraft. JP8 is effective in removing some agents from the aircraft's skin and components; however, it does not neutralize the agents. Personnel must use care when handling JP8. When using a cloth soaked in JP8, personnel must avoid wiping internal components near the exhaust. If water is available, use it to rinse off the JP8. Many portions of the aircraft are delicate and cannot stand high-pressure water or extreme hot air.

7-5. Sodium carbonate is a fair decontaminant against G nerve agents, but it is not effective against V nerve and blister agents. The chemical reaction of the V nerve agent and sodium carbonate is extremely slow and produces a product that is very toxic. Most of the field-expedient decontaminants are corrosive and could cause damage to the aircraft's skin or acrylic-plastic windows.

AIRCRAFT DECON

7-6. When an aircraft becomes contaminated, the aircrew must fly in MOPP gear. However, wearing MOPP gear degrades the aircrew's ability to perform its jobs over time.

7-7. If air and ground crews are careful when operating in a contaminated environment, contamination can be contained on the exterior of the aircraft. The increased airflow over the aircraft's skin increases the rate of evaporation. Still, some agents will remain and continue to be a hazard. Thickened agents evaporate more slowly and may remain a hazard even after prolonged flights. If the interior is contaminated, flying the aircraft with the doors open can help reduce the hazard. The heat from the running engine decontaminates the interior of the engine compartment. However, more active methods of decon may be needed. Use the immediate-, operational-, or thorough-decon method with modifications to perform an aircrew spot decon, an aircraft washdown, and the DAD.

DECON TYPES

7-8. Table 7-1 shows the types of aircraft decon. The crew of a contaminated aircraft will use the techniques described in previous chapters to decon. Unit personnel conduct the aircrew spot decon and aircraft washdown, but a chemical decon unit usually conducts the DAD.

Aircrew Spot Decon

7-9. After personal wipe down, aviators may conduct a spot decon on their aircraft (see Table 7-1). The aircrew spot decon provides the same benefits as the operator's spray down. It limits the transfer and spread of contamination by decontaminating the surfaces that must be touched during operations. Spot decon applies to the first six functional activities. The goal is to limit the transfer and spread of contamination while sustaining flight operations.

Table 7-1. Aircraft Decon Types

Decon Types	When, Why, Who Performs Decon	Areas to Decon	Types of Decon Agents
Aircrew spot decon	The crew decontaminates after the immediate decon to allow continued operations	Essential operating surfaces on the aircraft	IEDK, JP8[1]
Aircraft washdown	The unit decon PDDE crew or decon unit performs decon within 6-24 hours to reduce contact hazards[2]	The entire exterior surface of the aircraft[3]	Hot, soapy water
DAD	The decon units decon when the mission allows to reduce the hazard to negligible risk levels	The entire exterior surface and selected interior surfaces of the aircraft	10 percent sodium-carbonate solution (10 pounds of sodium carbonate to 12 gallons of water)

[1]JP8 should not be used inside the aircraft.

[2]An aircraft washdown is most effective if conducted within 1 hour of contamination.

[3]An aircrew spot decon is performed to reduce contact hazards inside the aircraft. Do not spray water inside the aircraft.

7-10. Air or ground crews may perform the aircrew spot decon. Wash exterior surfaces with decontaminants to flush off contamination. Fuel, soap, and water are most commonly used. Exact procedures and areas to decon are identified in the aircraft's maintenance manual. See Table 7-2, pages 7-4 through 7-6, for guidance on the aircrew spot decon.

Aircraft Washdown

7-11. An aircraft washdown site consists of a single station; with other decon sites, there may be several stations. Before arriving at the site, the aircrew should fly its aircraft at high speeds to help evaporate exterior contamination. If the aircraft's interior has been contaminated, open the doors; however, before arriving at the washdown site, close the doors. At the washdown site, the aircraft will land and reduce its speed to an idle. If the aircrew is contaminated, it should exit the aircraft and perform a MOPP-gear exchange.

Table 7-2. Aircraft Spot Decon

Actions and Locations	Areas to Decon	Personnel Responsible	Decontaminants	Procedures	Remarks
Refueling at FARP	Fuel ports, hatches, and all areas that FARP personnel touch	POL handler	Diesel fuel; JP8; or hot, soapy water	Wipe fuel ports and hatches with a sponge dipped in the decontaminant. Do not allow the decontaminant to enter the fuel system. Control runoff because the agent will not be neutralized. This method simply flushes contamination from the surface.	FARP personnel should conduct an operational decon before servicing aircraft if they are contaminated.
Arming at FARP	Armament system	Ammo handler	JP8 or hot, soapy water	Wipe the armament system with a sponge dipped in the decontaminant. Control runoff because the agent will not be neutralized. This method simply flushes contamination from the surface.	**CAUTION** **Take care to prevent certain areas of the armament system from being exposed to the decontaminants.** The TM on the armament system should be checked to know what areas to avoid.
Entering and exiting aircraft anywhere	Door handles, steps, ladders, handholds, and all other areas that the aircraft crew is likely to touch	Crew members and FARP personnel	JP8 or hot, soapy water	Apply the decontaminant. Take care to prevent the spread of liquid contamination from the outside of the aircraft to the inside. This includes controlling the contaminated runoff. This method simply flushes contamination from the surface. Precautions might include having the crew chief do the decon, which reduces the chance of contamination spreading into the pilot and copilot compartment. All crew members should diligently practice NBC avoidance.	Procedures should be developed for each type of aircraft. Before entering the aircraft, use an IEDK on boots and gloves. FARP personnel can decon the area for exiting aircrew members.

Table 7-2. Aircraft Spot Decon (Continued)

Actions and Locations	Areas to Decon	Personnel Responsible	Decontaminants	Procedures	Remarks
Preflight and postflight checks anywhere. Maintenance inspections at aviation intermediate maintenance (AVIM) facilities	Areas that must be touched as part of the inspection	Aircrew	Diesel fuel; JP8; or hot, soapy water on exterior surfaces. Use hot air on interiors or areas that are not compatible with decontaminants or liquids.	Wipe areas that are required to be touched for preflight and postflight checks with a sponge dipped in the decontaminant. Wash gloves in a decontaminant before touching uncontaminated surfaces. Decon gloves with an IEDK after the inspection is completed. Control runoff because the agent will not be neutralized. This method simply flushes contamination from the surface.	Overheated air should not be used directly on instrumentation. Crews may want to wear wet-weather gear to keep most of the contamination off overgarments.
Repair and recovery anywhere	Parts or areas that will need to be decontaminated depending on the situation	Battle-damage repair team	Diesel fuel; JP8; or hot, soapy water on exterior surfaces. Use hot air on interiors or areas that are not compatible with liquid decontaminants.	Decon only those parts that need to be touched during repair.	Contaminated aircraft should not be brought into an uncontaminated area.
Cannibalization anywhere on the battle space	Parts or assemblies being cannibalized	Maintenance personnel (someone who knows what assembly or parts are needed)	DS2; diesel fuel; 5 percent chlorine solution; or hot, soapy water on exterior parts or surfaces. Use hot air on interior parts that are not compatible with liquids or corrosive decontaminants.	Decon only those parts or assemblies that need to be touched during cannibalization.	Caustic decontaminants should be used only on those areas that have been removed from the aircraft. Assemblies must be rinsed thoroughly before replacing.

Table 7-2. Aircraft Spot Decon (Continued)

Actions and Locations	Areas to Decon	Personnel Responsible	Decontaminants	Procedures	Remarks
Overhaul at AVIM facilities	All areas and equipment required to be worked on during the overhaul	Maintenance personnel and/or a chemical decon platoon from division or corps	DS2; diesel fuel; 5 percent chlorine solution; or hot, soapy water. Surfaces that are easily destroyed by liquid or corrosion should be decontaminated using hot air.	Wash with diesel fuel and then hot, soapy water. Then rinse. Check for contamination with the M8 detector paper or the CAM. If time permits, allow the equipment to weather to reduce chemical contamination to negligible risk levels.	Caustic decontaminants should be used only on those assemblies that have been removed from the aircraft. Assemblies must be rinsed thoroughly before replacing. This process has the same results as a deliberate decon.

To remove the BDU, follow the procedures for removing the protective overgarment as described in paragraphs 8-124 and 8-125. Do not remove the patient's identification tags.

Step 5 - Decon the Patient's Skin

8-130. **Spot decon.** Wash exposed skin surfaces with soap and warm water, to include areas around the neck, wrists, and lower parts of the face. Decon the patient's identification tags and chain, if necessary.

8-131. **Combat medic care.** The combat medic gently cuts away the bandage. He decontaminates the area around the wound and irrigates it with a 0.5 percent chlorine solution. If bleeding begins, he replaces the bandage with a clean one. He replaces the old tourniquet by placing a new one ½ to 1 inch above the old one. He then removes the old tourniquet and decontaminates the patient's skin with a 0.5 percent chlorine solution. He does not remove a splint. He decontaminates the splint by thoroughly rinsing it, to include the padding and cravats, with a 0.5 percent chlorine solution.

> **WARNING**
> **Do not use an SDK around any wounds. Do not remove splints. The splint will not be removed until the patient has been evacuated to a forward MTF or hospital.**

8-132. **Completeness of decon check.** Check the patient with M8 detector paper or the CAM for completeness of decon.

NOTE: Other monitoring devices may be used, if available.

8-133. **Contaminated-waste disposal.** Dispose of contaminated bandages and coverings by placing them in a contaminated-waste bag. Seal the bag and place it in the contaminated-waste dump.

Step 6 - Transfer the Patient to the Hot Line

8-134. Move the patient to the hot line. Two medics from the clean side of the hot line move the patient into the clean treatment area.

NOTE: Before decontaminating another patient, each decon team member drinks about half a quart of water. The exact amount of water consumed is increased or decreased according to the work level and temperature (see Table 8-2, page 8-11).

AMBULATORY PATIENT

8-135. The decon procedures for an ambulatory patient that is contaminated with radiation are discussed in the following paragraphs. Some procedures can be done with one soldier, while others require more than one.

8-136. All ambulatory patients requiring EMT or AMT in the clean area of the BAS will be decontaminated. Stable patients not requiring treatment at the BAS, but requiring evacuation to the medical company's clearing station or a corps hospital for treatment (for example, a patient with a broken arm), should be evacuated in their protective overgarments and masks by any

where it will pose a vapor hazard and be difficult, if not impossible, to decon later.

NOTE: Many (but not all) aircraft have been painted with CARC paint. Aircraft will be painted with CARC paint when sent to the appropriate maintenance shop. Contamination stays on top of CARC surfaces where it can be more easily neutralized or removed.

7-14. An aircraft washdown should be conducted within 6 hours and is most effective if conducted within 1 hour. Some amounts of chemical contamination may remain after the aircraft washdown. The aircrew should continue to wear, as a minimum, its protective masks and rubber gloves for protection until a deliberate decon can be conducted.

7-15. Presently, there is no timely detection method to check the effectiveness of decon for biological contamination. Medical personnel will monitor the aircrew's disposition if biological contamination is suspected.

Detailed Aircraft Decon

7-16. The goal of a DAD is to reduce contamination hazards to a negligible risk level so that air and ground crews can operate the aircraft without protective measures. The DTD can be conducted at the same time. The decon site should be set up in a similar fashion to the thorough-decon site (see Chapter 4). The decon site consists of the following areas: a predecon staging area, DAD and DTD areas, and a postdecon AA. For safety reasons, it is not possible to decon aircraft and ground vehicles at the same site. A separate DED site can be established in the vicinity to support the contaminated aviation unit's ground vehicles.

7-17. Aviation units must be supported by a chemical unit to set up and operate the DAD site. The chemical unit, assisted by the supported aviation unit, operates the DAD site while the aviation unit sets up and operates the DTD. See Chapter 4 for DTD operations.

7-18. Typically, a DAD is either performed as part of a reconstitution effort or before maintenance operations (after every 25 hours of flight time). When the aircraft is either overhauled or grounded for scheduled maintenance, contamination hazards should be reduced to negligible risk levels by going through a DAD. When components are removed from the aircraft for repair, care must be taken because of residual contamination. Decon these components either before cannibalization or overhaul.

7-19. Chemical units (decon and smoke/decon) are responsible for setting up, operating, and closing the DAD portion of the thorough-decon operation. The site setup is the same for all types of contamination (chemical, biological, and radiological). The DAD area is composed of the following five stations: primary wash, decon-solution application, contact time/interior decon, rinse, and check (see Figure 7-2).

DECON STATIONS

7-20. The actions that are taken at each of the stations are as follows:

Station 1 - Primary Wash

7-21. At this station, gross contamination is removed from the aircraft. Spray the aircraft for 2 to 3 minutes with hot, soapy water. Avoid hitting the

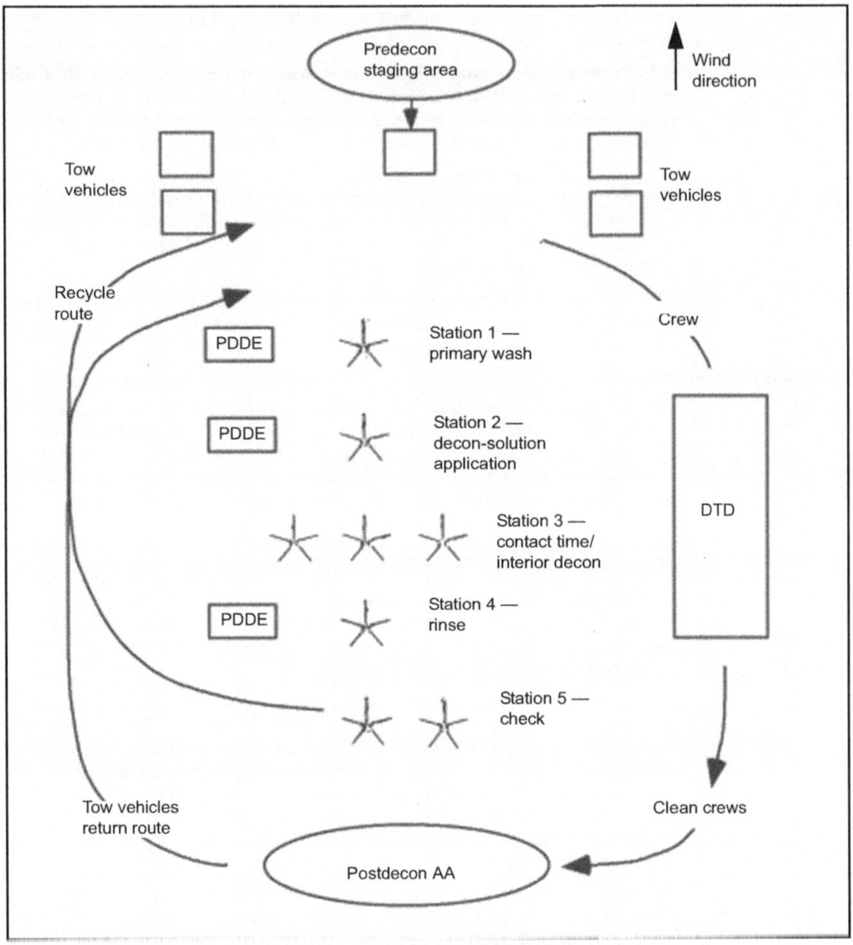

Figure 7-2. Optimum Layout for a DAD Station

aircraft's skin at a 90^0 angle to prevent damage. Also avoid the sensitive areas shown in Figures 7-3 through 7-6, pages 7-10 through 7-13.

Station 2 - Decon-Solution Application

7-22. At this station, a decon solution is applied using the M12A1 PDDE or the M17 LDS (in the siphinjector mode). After the solution is applied, a member of the scrubbing team scrubs the aircraft's skin. This allows the chemical agent

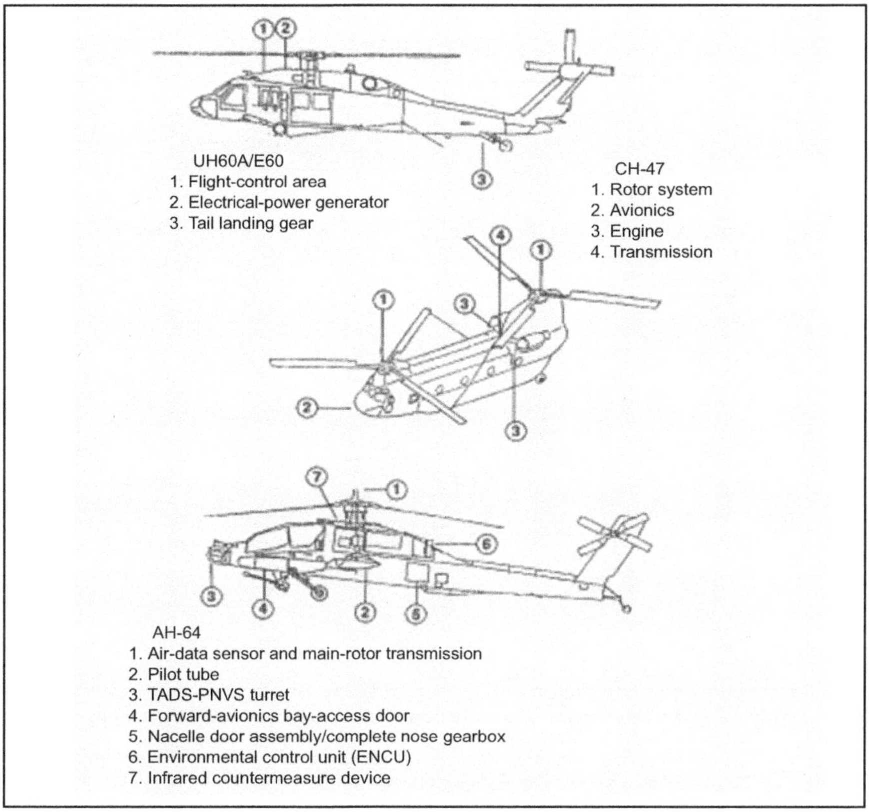

Figure 7-3. No Direct Water-Pressure Areas for the UH60A/E60, CH-47, and AH-64 Helicopters

to mix with the decon solution, especially when thickened chemical agents are present. For decontaminants, refer to Appendix B. STB and HTH should not be used because of their corrosive nature.

Station 3 - Contact Time/Interior Decon

7-23. At this station, the decon solution is allowed to completely neutralize the chemical agent; also, the interior of the aircraft is decontaminated, if necessary. The aircraft will remain in station 3 for no less than 30 minutes. The decon solution reacts with most chemical agents within 5 minutes. However, allowing the decon solution to remain on the contaminated surface for 30 minutes ensures a complete neutralization.

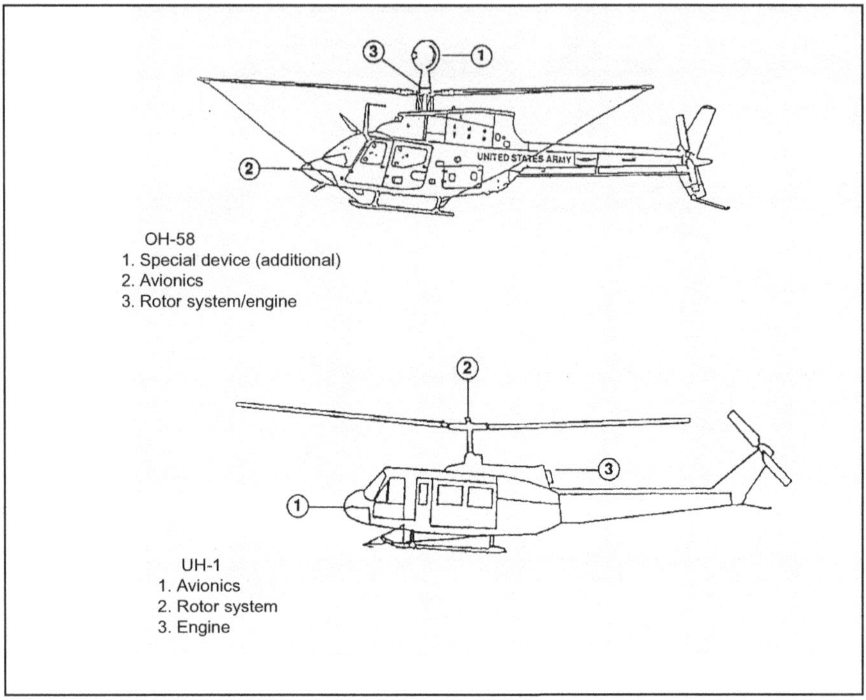

Figure 7-4. No Direct Water-Pressure Areas for the OH-58 and UH-1 Helicopters

7-24. While the aircraft is in this station, an attendant inspects the interior of the aircraft for liquid contamination using the CAM and M8 detector paper. If liquid-chemical contamination is identified, he will decon the interior of the aircraft.

WARNING
Personnel should not be in the aircraft during this operation.

7-25. Decon the aircraft's interior with hot, soapy water as long as maintenance personnel approve the use of it on certain pieces of equipment. Use IEDKs or damp, soapy washcloths to decon optical and electrical equipment and instruments.

NOTE: The IEDK leaves a charcoal residue on surfaces.

7-26. Field studies show the Herman-Nelson portable duct heaters to be effective in decontaminating chemically contaminated vehicle and aircraft

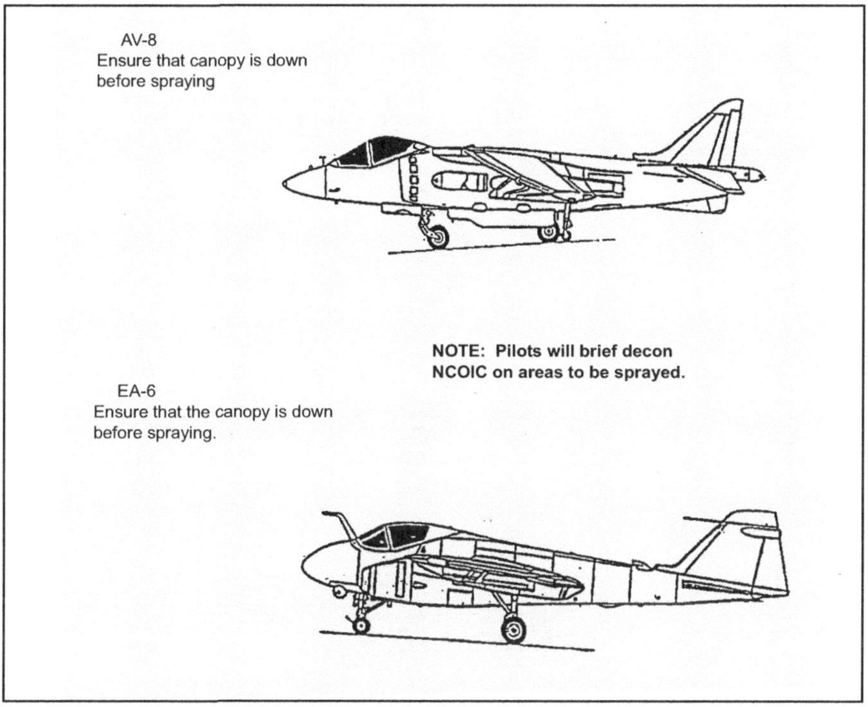

Figure 7-5. No Direct Water-Pressure Areas for the AV-8 and EA-6

interiors. However, using a hot-air decon has the following drawbacks: time required, heat damage, and downwind vapor hazards. A hot-air decon for a helicopter or light aircraft requires about 30 to 120 minutes to remove all of the contamination. Times are based on the amount of contamination and interior surface materials. Enough time must be provided to get the contaminated surface hot enough to vaporize the agent. The Herman-Nelson heaters discharge air at a temperature between 150° and 280°F. Pressurized containers, such as fire extinguishers, must be removed before starting a hot-air decon. These items will fail catastrophically and could cause severe damage. The heat will also damage sensitive electronic equipment. Additionally, chemical agents irreversibly bind to many materials, particularly gaskets, vinyl coatings, seat belts, and seat cushions. These items should be removed since they cannot be decontaminated. A hot-air decon does not destroy chemical agents. The air exiting the aircraft may contain hazardous levels of the agent. A downwind exclusion area of about 164 yards must be maintained to prevent unwanted casualties.

7-27. While a hot-air decon is an alternate to using hot, soapy water or an IEDK, the technique has many drawbacks. Maintenance personnel must be

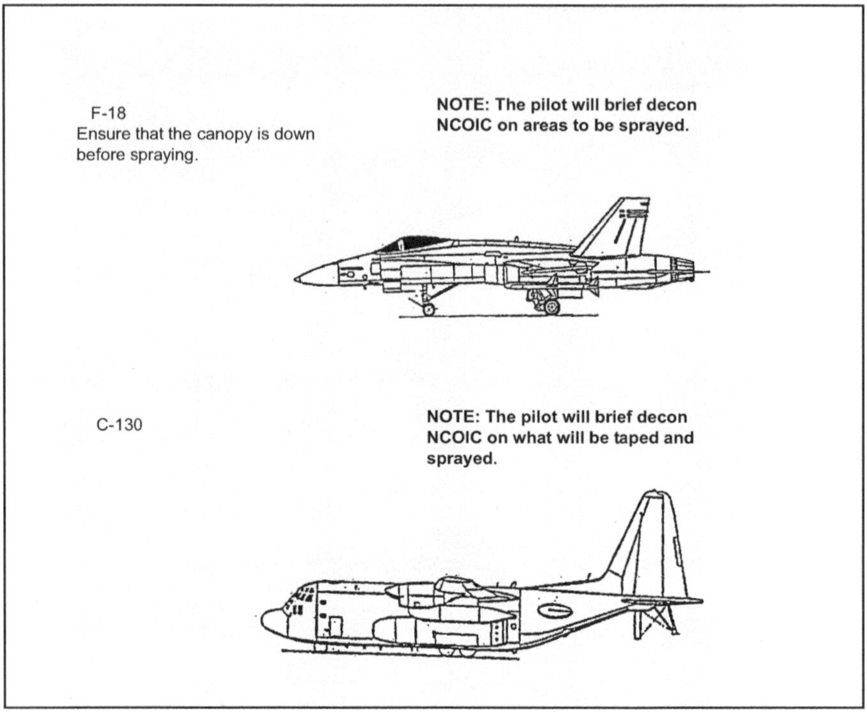

Figure 7-6. No Direct Water-Pressure Areas for the F-18 and C-130

consulted before using this technique to ensure that the aircraft is not damaged by the heat.

7-28. For radiological contamination, use a radiac meter to determine the extent and location of contamination inside the aircraft. If there is contamination, determine the intensity of the contamination. If the contamination has an intensity greater than 0.33 cGy, the interior of the aircraft must be decontaminated. Use hot, soapy water to wash the contaminated areas and a sponge to mop up the water and the contamination.

Station 4 - Rinse

7-29. At this station, the decon solution is removed from the aircraft. Spray the aircraft with water from top to bottom. Take care not to damage the aircraft's skin. This station will use about 250 gallons of water. Failure to remove all decon solution from the aircraft's skin may cause corrosion.

Station 5 - Check

7-30. At this station, the completeness of the decon is checked. Determine if the aircraft has a negligible risk or if it still has significant contamination remaining. Detection procedures vary depending on the type of contamination. If significant contamination is found on the aircraft, it will be recycled to station 2 for chemical contamination or station 1 for radiological contamination.

7-31. **Chemical.** Use the CAM to check for the presence of vapors. A one bar or lower reading on the CAM indicates a negligible risk. If the CAM indicates the presence of vapors, use M8 detector paper to check for liquid contamination. If it is suspected that these are producing a false positive, use an M256A1 detector kit to confirm or deny the presence of contamination. If the aircraft has significant contamination remaining, recycle it. Desorption of chemical agents may occur after the decon. On CARC-painted surfaces, the desorption of vapors will stop sooner than on alkyd-painted surfaces.

7-32. **Radiological.** Use the radiac meter to determine if any contamination remains. If so, determine the intensity of the contamination. If the contamination is greater than 0.33 cGy, recycle the aircraft to station 1.

RECYCLE CRITERIA

7-33. The commander, with the chemical unit leader's help, establishes the recycle criteria before starting a decon operation. If contamination is detected at station 5, the recycle criteria is used to determine which aircraft are returned to station 2 or, if radiologically contaminated, station 1. If the unit has sufficient time and resources available, any aircraft having more contamination than a negligible risk should be recycled. However, time and resources are usually limited and not all aircraft can be recycled.

AIRCRAFT DECON-SITE SELECTION

7-34. Operational- and thorough-decon sites for aircraft must be selected with care. The decon area must be big enough for the required number of aircraft, have a readily available water source, and allow for adequate drainage. The site should be relatively secure; however, it should be close enough to the AOs and FARP to allow for a reasonably quick turnaround of aircraft. The site must not have less than a 5 percent slope. Decon sites should be integrated into the tactical plan.

DECON-SITE LAYOUT

7-35. The equipment and personnel requirements for a DAD layout are identified in Table 7-3, page 7-16. Chemical units can establish other layouts based on METT-TC.

7-36. It may not be possible, for a variety of reasons, for a chemical platoon to use the optimum layout. Limited personnel or equipment will affect the layout of the thorough aircraft decon site. The platoon leader will use METT-TC and the concepts outlined in this chapter to establish a DAD station within his capabilities.

WARNING

Chemical unit leaders must consider the impact of the work/rest cycle on their operation to process aircraft. Failure to initiate a work/rest cycle could result in heat casualties and mission failure.

Table 7-3. Personnel and Equipment Requirements for an Optimum DAD Layout

Stations	Personnel		Equipment/Supplies
	Decon Platoon	Augmenter	
Station 1 - primary wash	1 squad leader 1 PDDE operator 2 sprayers	2 scrubbers	1 PDDE 1 3,000-gallon tank 2 65-GPM pumps 6 long-handled brushes 5 TAP aprons Liquid detergent
Station 2 - decon-solution application	1 squad leader 2 appliers 2 sprayers 1 PDDE operator	4 appliers	1 PDDE 18 long-handled brushes 9 mops with extra heads Liquid detergent
Station 3 - contact time/interior decon	1 NCO (CAM operator)	2 interior decon assis-tants	2 AN/VDR-2 or AN/PDR-77 3 TAP aprons 6 30-gallon containers 10 books of M8 detector paper 30 sponges 8 M256A1 detector kits 50 plastic trash bags 1 clipboard with a pen 1 stopwatch 1 CAM
Station 4 - rinse	1 squad leader 1 PDDE operator 2 sprayers		1 PDDE 1 3,000-gallon tank 3 65-GPM pumps 1 TPU 2 TAP aprons
Station 5 - check	1 NCO 2 CAM operators		1 CAM 10 M256A1 detector kits 20 books of M8 detector paper 2 AN/VDR-2 or AN/PDR-77 2 M8A1 or M22 ACAA
C^2	1 platoon leader 1 platoon sergeant		1 HMMWV 3 marking kits
Aircraft moving team		6 drivers 18 ground guides	
Total personnel	20	32	

Chapter 8

Patient Evacuation and Decon

Evacuating patients under NBC conditions forces the unit commander to consider to what extent he will commit evacuation assets to enter the contaminated area. Generally, if most or all of a supported force is operating in a contaminated area, most or all of the medical-evacuation assets will operate there also. If it is possible to keep some ambulances free from contamination, every effort should be made to do so.

EVACUATION CONSIDERATIONS

8-1. On the modern battle space, there are three basic modes of evacuating patients (personnel, ground vehicles, and aircraft). Personnel who physically carry the patients incur a great deal of inherent stress. Cumbersome MOPP gear, climate, increased workloads, and fatigue will greatly reduce the effectiveness of unit personnel.

8-2. If evacuation personnel are to be sent into an area that is contaminated with radiation, an OEG must be established. Radiation-exposure records must be maintained by the supported unit chemical NCO and made available to the commander, staff, and medical leader. Based on the OEG, the commander or medical leader decides which evacuation elements to send into the contaminated environment. Every effort is made to limit the number of evacuation assets that are contaminated; however, a number of these assets will become contaminated in the course of a battle. Therefore, optimize the use of resources, medical and nonmedical, which are already contaminated before employing uncontaminated resources.

8-3. Once a vehicle has entered a contaminated area, it is highly unlikely that it can be spared long enough to undergo a complete decon. This will depend on the contaminant, the tempo of the battle, and the resources available for casualty evacuation. Normally, contaminated vehicles (air and ground) will be confined to dirty environments.

8-4. Ground ambulances should be used instead of air ambulances in contaminated areas. They are more plentiful, are easier to decon, and can be replaced more easily. However, this does not preclude the use of aircraft, if required.

8-5. The patient's medical condition and the relative positions of the contaminated area, the forward line of troops (FLOT), and the threat's air-defense systems determine if and where air ambulances may be used in the evacuation process. One or more air ambulances may be restricted to the contaminated areas. To the greatest extent possible, use ground vehicles to cross the line separating the contaminated and clean areas.

8-6. After the patients are loaded, the medical-evacuation vehicles proceed to a medical-treatment facility (MTF) with a patient-decon station. (For the setup of a patient-decon station, see FM 8-10-7.) The patients are decontaminated and treated at this station. If further evacuation is required, transfer the patients to a clean ground or air ambulance. The routes that the ground vehicles use to cross between contaminated and clean areas should be considered dirty routes and should not be crossed by clean vehicles. The effects of the wind and time on the contaminants must be considered.

8-7. The rotor wash of the helicopters must always be kept in mind when evacuating patients. The intense winds will disturb the contaminants, increasing vapor hazards.

8-8. The helicopter should be allowed to land and reduce to a flat pitch before the patients are brought near. Additionally, the helicopter must not land too close to a decon station because any trace of contaminants in the rotor wash may be spread into the clean treatment area.

8-9. Evacuating patients must continue even under NBC conditions. The medical leader must recognize the constraints that NBC operations place on him, and he must plan and train to overcome these constraints.

PATIENT DECON

8-10. Patient decon presents special problems for units and combat-health-support (CHS) personnel. Under NBC conditions, contaminated wounded soldiers create increased hazards to rescuers and CHS personnel. In the following paragraph, decon procedures at unit level are discussed:

8-11. On the NBC battle space, two classifications of patients will be encountered—contaminated and uncontaminated. Those contaminated may suffer from the effects of an NBC agent, a conventional wound, or both. For more information on the treatment of NBC patients, see FM 8-9. Some may suffer battle fatigue or heat injuries induced by the stress of NBC conditions and the extended time spent in MOPP4. It is important to follow proper decon procedures to limit the spread of contamination. The most important decon is performed at the contamination site. Later decon may be too late to prevent injury. All agents should be promptly removed from the skin.

8-12. Patient decon must begin at the platoon and company level with the individual soldier. The soldier himself or members of his team must perform immediate decon and administer nerve-agent antidotes and convulsant antidotes, if required. Tag the patient with a DD Form 1380 or a field-expedient tag, noting the time and type of contamination (see Figure 8-1). If available, use the CAM or M8/M9 detector paper to determine the type and concentration of contamination. When the patient's condition and the battle permits, the patient may go through a MOPP-gear exchange (see Chapter 3 of this manual). The MOPP-gear exchange must not cause further injury to the patient.

PATIENT DECON AT AN MTF

8-13. In the following paragraphs, the types of MTFs are discussed:

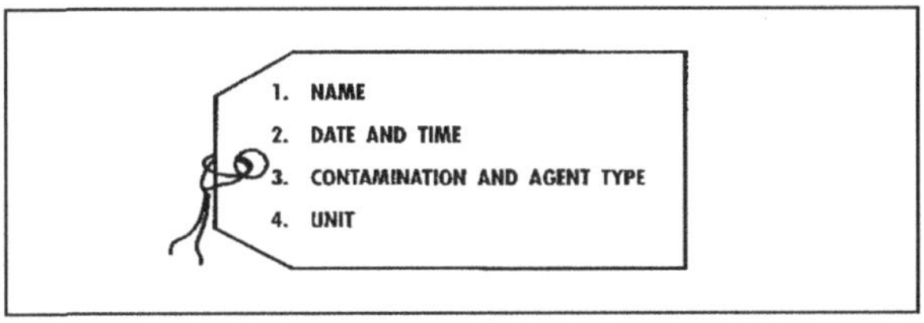

Figure 8-1. Field-Expedient NBC Patient Tag

BATTALION AID STATION (BAS)

8-14. When battle conditions prevent decon procedures forward, the patient may have to be evacuated to the BAS before decon. Contaminated patients arriving at the MTF must be decontaminated before admission into the clean treatment area.

8-15. Patient decon is the systematic removal of clothing and contaminants from a patient who is unable to decon himself. Patient decon is done by an 8-man, patient-decon team consisting of nonmedical personnel from the supported unit at the BAS. The team operates under the supervision of medical personnel to ensure that no further injury is caused to the patient by the decon process.

MEDICAL COMPANY'S CLEARING STATION

8-16. The medical company's clearing station may receive patients from the BAS or directly from other areas who may be contaminated. The clearing station must also have a patient-decon area. For more information on BAS operations under NBC conditions, see FMs 8-10-4 and 8-10-7. As with the BAS, the clearing station must have at least an 8-man, patient-decon team consisting of nonmedical personnel from the supported units to perform the decon. Procedures for patient decon at the clearing station are the same as for the BAS.

HOSPITAL

8-17. To the maximum extent possible, hospitals are located away from tactical or logistical targets. Patients evacuated from forward areas should already be decontaminated; however, contaminated patients may arrive from forward MTFs and units located within the geographical area of the hospital and require decon. Patient decon is done by 20 nonmedical personnel from units located in the geographical area/base cluster of the hospital.

8-18. If the hospital does not have CPSs and becomes contaminated with a persistent agent, patients are rerouted to other hospitals. All inpatients are evacuated, if possible, and the hospital is decontaminated. On completion of the decon, the hospital will return to normal operations.

8-19. A hospital with CPSs will decon the areas around the entry to these facilities, then continue receiving and caring for patients. Forward medical facilities and hospitals use the same patient-decon procedures. Several patient-decon stations can be operated at a hospital decon site. All patients arriving at the hospital will be considered contaminated. They must be decontaminated before being admitted into the clean areas. Perform decon as required.

CHLORINE-SOLUTION PREPARATION FOR PATIENT DECON

8-20. To decon a patient, use an SDK; however, if an SDK is not available, use a chlorine solution. If the chlorine solution is used, two concentrations of it are required. A 5 percent chlorine solution is required to decon gloves, aprons, litters, scissors, patient's hood, and other nonskin contact areas. A 0.5 (½) percent chlorine solution is required to decon the patient's mask, skin, and splints and to irrigate his wounds. To prepare the solutions, use calcium hypochlorite granules or sodium hypochlorite (household bleach) (see Table 8-1).

Table 8-1. Preparation of Chlorine Solution for Patient Decon

HTH Ounces	HTH Spoonfuls*	Household Bleach	Solution Percentage in 5 Gallons of Water
6	5	2 quarts	0.5
48	40	**	5

*These measurements are used when bulk HTH is used. To measure this preparation, use the plastic spoon supplied with your meal, ready-to-eat (MRE). Use a heaping spoonful of chlorine (all that the spoon will hold). Do not shake any granules off the spoon before adding to the water.

**Do not dilute in water because household bleach is about a 5 percent solution.

NOTE: HTH is supplied in a 6-ounce jar in the chemical-agent patient's treatment and decon medical-equipment set.

CAUTION

Only use a 0.5 percent chlorine solution on the patient's skin. Avoid vigorous scrubbing because it can force the agent into the skin.

CHEMICAL-AGENT PATIENT DECON PROCEDURES

8-21. Decon procedures for chemical-agent patients are discussed in the following paragraphs:

LITTER PATIENT

8-22. Before patients receive medical treatment in the clean treatment area, the 8-man patient-decon team decontaminates them. Figure 8-2 shows one way to establish the patient-decon station. Place the bandage scissors in a container of 5 percent chlorine solution between each use. The team members decon their gloves and aprons with a 5 percent chlorine solution.

NOTE: Litter patients requiring emergency medical treatment (EMT) or ambulatory medical treatment (AMT) in the clean area of the MTF will be completely decontaminated. However, a patient requiring immediate evacuation should have only his wound area and MOPP gear spot decontaminated to remove any gross contamination (for example, a stable patient with a partial amputation of a lower extremity). The patient should be evacuated in his MOPP gear.

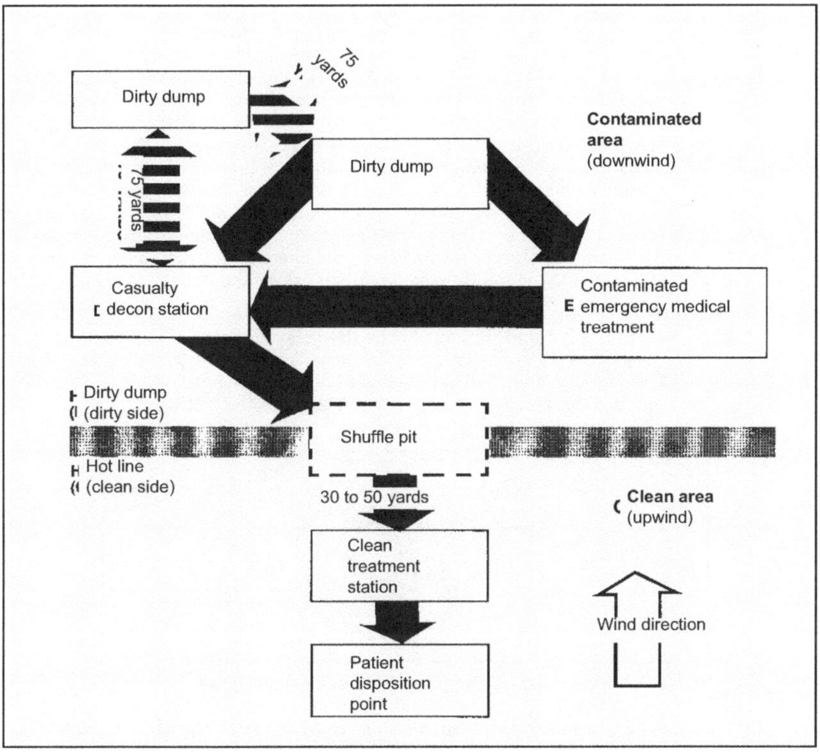

Figure 8-2. Layout for a Patient-Decon Station and a Clean Treatment Area Without a CPS

8-23. Decon the patient's skin, bandages, wounds, mask, identification tags with chain, and splints by wiping them with a 0.5 percent chlorine solution. For treatment procedures, refer to FMs 8-9, 8-33, and 8-285.

8-24. Some procedures in the following steps can be done with one soldier, while others require more than one soldier.

Step 1 - Decon the Patient's Mask and Hood

8-25. **Move the patient to the clothing removal station.** After the patient has been triaged and stabilized (if necessary) by the senior medic in the patient-decon area, move him to the litter stands at the clothing removal station.

8-26. **Decon the hood.** Use either an IEDK or a 5 percent chlorine solution (or household bleach) to wipe down the front, sides, and top of the hood.

8-27. **Remove the hood.** Remove the hood by cutting it with scissors or by loosening it from the mask attachment points for the quick-doff hood or other similar hoods (see Figure 8-3). Before cutting the hood, dip the scissors in a 5 percent chlorine solution. Cut the neck cord and the small string under the voicemitter. Release or cut the hood shoulder straps and unzip the hood zipper. Cut the hood upward to the top of the eye-lens outsert, staying close to the filter-inlet cover and eye-lens outsert, then across the forehead to the outer edge of the other eye-lens outsert. Proceed downward toward the patient's shoulder, staying close to the eye lens outsert and filter-inlet cover, then across the lower part of the voicemitter to the zipper. After dipping the scissors in the 5 percent chlorine solution, cut the hood from the center of the forehead over the top of the head. Fold the left and right sides of the hood to the side of the patient's head, laying the sides of the hood on the litter.

8-28. **Decon the protective mask and exposed skin.** Use an SDK or a 0.5 percent chlorine solution to wipe the external parts of the mask. Cover the mask's air inlets with gauze or your hand to keep the mask filter dry. Continue by wiping the exposed areas of the patient's face, to include the neck and behind the ears. Do not remove the protective mask.

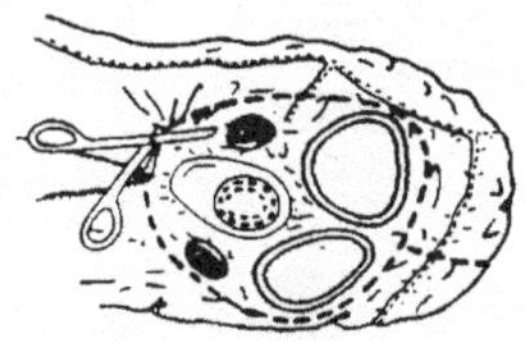

Figure 8-3. Cutting the Protective Mask Hood

8-29. **Remove the field medical card (FMC).** Cut the patient's FMC tie wire, allowing the FMC to fall into a plastic bag. Seal the plastic bag and rinse the outside of the bag with a 5 percent chlorine solution. Place the plastic bag under the back of the protective mask's head straps. The FMC will remain with the patient in the contaminated area and a clean copy will be made before the patient is moved to the clean area.

Step 2 - Remove Gross Contamination From the Patient's Protective Overgarment

8-30. Remove all visible contamination spots from the overgarment with an SDK or a 5 percent chlorine solution.

Step 3 - Remove the Patient's Personal Effects and Protective Overgarment

8-31. **Remove the patient's personal effects.** Remove the patient's personal effects from his protective overgarment and BDU pockets. Place the articles in a plastic bag, label the bag with the patient's identification, and seal the bag. If the articles are not contaminated, they are returned to the patient. If the articles are contaminated, place them in the contaminated holding area until they can be decontaminated, then return them to the patient.

CAUTION
Bandages may have been applied to control severe bleeding; therefore, they are treated like tourniquets. Only medical personnel will remove bandages, tourniquets, and splints.

NOTE: The overgarment jacket and trousers will be cut simultaneously. One soldier cuts the jacket, while another soldier cuts the trousers.

8-32. **Cut and remove the overgarment jacket.** Before cutting the overgarment jacket and trousers, dip the scissors in a 5 percent chlorine solution to prevent contamination of the patient's BDU or undergarments. Make two cuts, one up each sleeve from the wrist up to the shoulder and then across the shoulder through the collar (Figure 8-4). Cut around bandages, tourniquets, and splints, leaving them in place. Do not allow your gloves to touch the patient along the cut line. Keep the cuts close to the inside of the arms so that most of the sleeve material can be folded outward. Unzip the jacket and roll the chest sections to the respective sides, with the inner surface outward. Continue by tucking the clothing between the arm and chest. Roll the cut sleeves away from the arms, exposing the black liner.

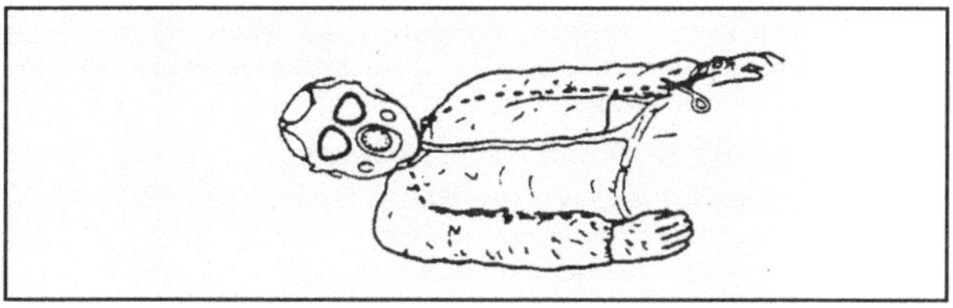

Figure 8-4. Cutting the Overgarment Jacket

8-33. **Cut and remove the overgarment trousers.** Cut both trousers legs starting at the ankle as shown in Figure 8-5, page 8-8. Keep the cuts near the inside of the legs, along the inseam, to the crotch. With the left leg, cut all the way to the waist, avoiding the pockets. With the right leg, cut across at the crotch to the left leg cut. Cut around bandages, tourniquets, and splints, leaving them in place. Place the scissors in a 5 percent chlorine solution. Fold the cut trouser halves away from the patient and allow the halves to drop to the litter with the contaminated (green) side down. Roll the inner leg portion under and between the legs.

8-34. **Remove the outer gloves.** Before touching the patient, the patient-decon team decontaminates its gloves with a 5 percent chlorine solution. Lift the patient's arms up and out of the cutaway sleeves unless detrimental to his condition. Grasp the fingers of the gloves, roll the cuffs over the fingers, and turn the gloves inside out. Do not remove the inner cotton gloves at this time. Carefully lower the patient's arms across the chest after the outer gloves have been removed (Figure 8-6, page 8-8). Do not allow the patient's arms to come into contact with the exterior of his overgarment. Drop his gloves into the

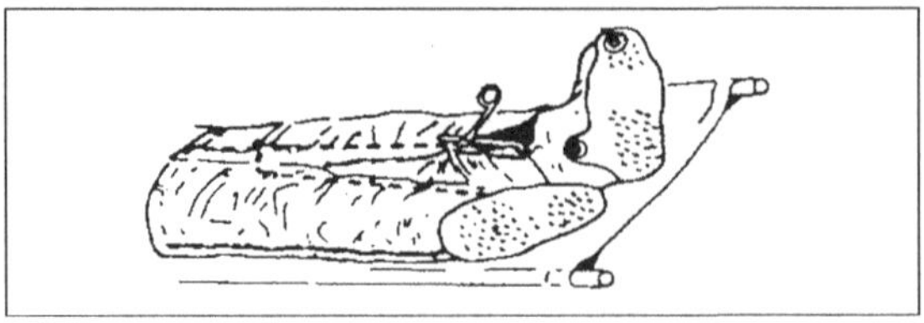

Figure 8-5. Remove the Overgarment Trousers

contaminated-waste bag. The team members decon their gloves with the 5 percent chlorine solution.

8-35. **Remove the overboots.** Cut the overboot laces and fold the lacing eyelets flat outward. While standing at the foot of the litter, hold the patient's heel with one hand. Pull the overboot downward, then toward you to remove it. Remove the two overboots simultaneously. This reduces the likelihood of contaminating one of the combat boots.

8-36. While holding the patient's heels off the litter, have a team member wipe the end of the litter with a 5 percent chlorine solution to neutralize any contamination that was transferred to the litter from the overboots. Lower the patient's heels onto the decontaminated litter. Place the overboots in the contaminated-waste bag. The team members decon their gloves with the 5 percent chlorine solution.

Step 4 - Remove the Patient's BDU and Undergarments

8-37. **Cut and remove the BDU.** To cut and remove the BDU jacket and trousers, follow the procedures for removing the protective overgarment as described in paragraph 8-32 and 8-33.

8-38. **Remove the combat boots.** Cut the bootlaces along the tongue. Remove the boots by pulling them toward you. Place the boots in the contaminated-waste bag. Do not touch the patient's skin with your contaminated gloves when removing his boots.

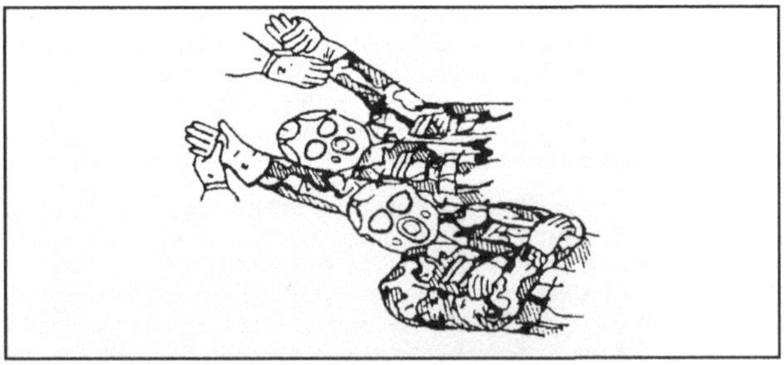

Figure 8-6. Removing the Outer Gloves and Positioning the Arms After the Glove Removal

8-39. **Cut and remove the undergarments.** Follow the procedures for cutting away the protective overgarment and rolling it away from the patient (see paragraph 8-32 and 8-33). If the patient is wearing a brassiere, cut it between the cups. Cut both shoulder straps where they attach to the cups and lay them back off of the shoulders. Remove the socks and cotton gloves. Do not remove the identification tags.

Step 5 - Transfer the Patient to a Decon Litter

8-40. After the patient's clothing has been cut away, transfer him to a decon litter or a canvas litter with a plastic-sheeting cover. Three decon team members decon their gloves and aprons with a 5 percent chlorine solution. One member places his hands under the patient's legs at the thighs and Achilles tendons, a second member places his arms under the patient's back and buttocks, and a third member places his arms under the patient's shoulders and supports the head and neck. They carefully lift the patient using their knees (not their backs) to minimize back strain. While the patient is elevated, another decon team member removes the litter from the litter stands and replaces it with a decontaminated (clean) litter. The team members carefully lower the patient onto the clean litter. The clothing and overgarments are placed in a contaminated-waste bag and moved to the contaminated-waste dump. The dirty litter is rinsed with the 5 percent chlorine solution and placed in the litter storage area.

Step 6 - Decon the Patient's Skin

8-41. **Spot decon.** With the patient in a supine position, spot decon the skin by using an SDK or a 0.5 percent chlorine solution. Decon areas of potential contamination, to include areas around the neck, wrists, and lower parts of the face. Decon the patient's identification tags and chain, if necessary.

NOTE: A complete body wash is not appropriate and may be harmful to the patient. During a complete body wash, the patient would have to be rolled over to reach all areas of the skin. This is not necessary for an adequate decon.

8-42. **Combat medic care.** The combat medic gently cuts away the bandage. He decontaminates the area around the wound and irrigates it with a 0.5 percent chlorine solution. If bleeding begins, he replaces the bandage with a clean one. He replaces the old tourniquet by placing a new one ½ to 1 inch above the old one. He then removes the old tourniquet and decontaminates the patient's skin with an SDK or a 0.5 percent chlorine solution. He does not remove a splint. He decontaminates the splint by thoroughly rinsing it, to include the padding and cravats, with a 0.5 percent chlorine solution.

> **WARNING**
> **Do not use an SKD around any wounds. Do not remove splints. Splints will not be removed until the patient has been evacuated to a forward MTF or hospital.**

8-43. **Completeness of decon check.** Check the patient with M8 detector paper or the CAM for completeness of decon.

NOTE: Other monitoring devices may be used, if available.

8-44. **Contaminated-waste disposal.** Dispose of contaminated bandages and coverings by placing them in a contaminated-waste bag. Seal the bag and place it in the contaminated-waste dump.

Step 7 - Transfer the Patient Across the Shuffle Pit

8-45. The patient's clothing has been cut away and his skin, bandages, and splints have been decontaminated. Transfer the patient to the shuffle pit and place the litter on the litter stands. The shuffle pit is wide enough to prevent the decon team members from straddling it while carrying the litter. A third team member will assist with transferring the patient to a clean treatment litter in the shuffle pit. Decon personnel rinse or wipe down their aprons and gloves with a 5 percent chlorine solution.

8-46. The three team members lift the patient off the decon litter (see step 5 for lifting procedures). While the patient is elevated, another team member removes the litter from the stands and returns it to the decon area. A medic from the clean side of the shuffle pit replaces the litter with a clean one. The patient is lowered onto the clean litter. Two medics from the clean side of the shuffle pit move the patient to the clean treatment area. The patient is treated in this area or awaits processing into the CPS. The litter is wiped down with a 5 percent chlorine solution in preparation for reuse. Once the patient is in the air lock of the CPS and the air lock has been purged, his protective mask is removed. Place the mask in a plastic bag and seal it.

NOTE: Before decontaminating another patient, each decon team member drinks about half a quart of water. The exact amount of water consumed is increased or decreased according to the work level and temperature (see Table 8-2).

AMBULATORY PATIENT

8-47. All ambulatory patients requiring EMT or AMT in the clean treatment area of the BAS will be decontaminated. Stable patients not requiring treatment at the BAS, but requiring evacuation to the medical company's clearing station or a corps hospital for treatment (for example, a patient with a broken arm), should be evacuated in their protective overgarments and masks by any available transportation. However, before evacuation, spot remove all thickened agents from their protective clothing.

NOTE: Place the bandage scissors that are used in this procedure in a container of 5 percent chlorine solution when not in use. Most ambulatory patients will be treated in the contaminated treatment area and returned to duty. Upon removal of an ambulatory patient's clothing, he becomes a litter patient. The BAS and clearing station do not have clothing to replace those that are cut off during the decon process. The patient must be placed in a patient protective wrap (PPW) for protection during evacuation (Figure 8-7, page 8-12).

8-48. The ambulatory patient is decontaminated and undressed as follows. Some procedures in the following steps can be done with one soldier, while others require more than one.

Table 8-2. Heat-Injury Prevention and Water Consumption

Criteria			Controls	
Heat		Water Intake Quart/Hour	Physical Activity for Soldiers/Trainees	
Condition	WBGT Index °F*		Acclimatized to Work/Rest**	Not Acclimatized to Work/Rest
White	78-81.9	At least ½	Continuous	
Green	82-84.9	At least ½	50/10 minutes	Use discretion in planning heavy exercise.
Yellow	85-87.9	At least 1	45/15 minutes	Suspend strenuous exercise during the first 3 weeks of training. Training activities may be continued on a reduced scale after the second week of training. Avoid activity in the direct sun.
Red	88-89.9	At least 1½	30/30 minutes	Curtail strenuous exercise for all personnel with less than 12 weeks of hot-weather training.
White	78-81.9	At least ½	Continuous	
Green	82-84.9	At least ½	50/10 minutes	Use discretion in planning heavy exercise.
Black	90 and up	More than 2	20/40 minutes	Suspend physical training and strenuous exercise. Essential operational commitments (not for training), where the risk of heat casualties may be warranted, are excluded from suspension. Enforce water intake to minimize expected heat injuries.

*MOPP gear or body armor adds 10°F to the wet-bulb gradient-temperature (WBGT) index.
**An acclimatized soldier is one who has worked in the given heat condition for 10 to 14 days.

NOTE: "Rest" means minimal physical activity. Rest should be accomplished in the shade, if possible. Any activity requiring only minimal physical activity can be performed during rest periods (for examples, training by lecture or demonstration, minor maintenance procedures on vehicles or weapons, and personal-hygiene activities such as skin and foot care).

Step 1 - Remove the LCE

8-49. Remove the LCE by unfastening/unbuttoning all connectors or tie straps and then place the equipment in a plastic bag. Place the plastic bag in the designated storage area for later decon.

Step 2 - Decon the Patient's Mask and Hood

8-50. **Begin the clothing removal process.** After the patient has been triaged and treated (if necessary) by the senior medic in the patient-decon station, the clothing removal process begins.

8-51. **Decon the hood.** Use either an IEDK or a 5 percent chlorine solution (or household bleach) to wipe down the front, sides, and top of the hood.

8-52. **Remove the hood.** Remove the hood by cutting it with scissors or by loosening it from the mask attachment points for the quick-doff hood or other similar hoods (see Figure 8-3, page 8-6). Before cutting the hood, dip the scissors in a 5 percent chlorine solution. Cut the neck cord and the small

To remove the BDU, follow the procedures for removing the protective overgarment as described in paragraphs 8-124 and 8-125. Do not remove the patient's identification tags.

Step 5 - Decon the Patient's Skin

8-130. **Spot decon.** Wash exposed skin surfaces with soap and warm water, to include areas around the neck, wrists, and lower parts of the face. Decon the patient's identification tags and chain, if necessary.

8-131. **Combat medic care.** The combat medic gently cuts away the bandage. He decontaminates the area around the wound and irrigates it with a 0.5 percent chlorine solution. If bleeding begins, he replaces the bandage with a clean one. He replaces the old tourniquet by placing a new one ½ to 1 inch above the old one. He then removes the old tourniquet and decontaminates the patient's skin with a 0.5 percent chlorine solution. He does not remove a splint. He decontaminates the splint by thoroughly rinsing it, to include the padding and cravats, with a 0.5 percent chlorine solution.

WARNING

Do not use an SDK around any wounds. Do not remove splints. The splint will not be removed until the patient has been evacuated to a forward MTF or hospital.

8-132. **Completeness of decon check.** Check the patient with M8 detector paper or the CAM for completeness of decon.

NOTE: Other monitoring devices may be used, if available.

8-133. **Contaminated-waste disposal.** Dispose of contaminated bandages and coverings by placing them in a contaminated-waste bag. Seal the bag and place it in the contaminated-waste dump.

Step 6 - Transfer the Patient to the Hot Line

8-134. Move the patient to the hot line. Two medics from the clean side of the hot line move the patient into the clean treatment area.

NOTE: Before decontaminating another patient, each decon team member drinks about half a quart of water. The exact amount of water consumed is increased or decreased according to the work level and temperature (see Table 8-2, page 8-11).

AMBULATORY PATIENT

8-135. The decon procedures for an ambulatory patient that is contaminated with radiation are discussed in the following paragraphs. Some procedures can be done with one soldier, while others require more than one.

8-136. All ambulatory patients requiring EMT or AMT in the clean area of the BAS will be decontaminated. Stable patients not requiring treatment at the BAS, but requiring evacuation to the medical company's clearing station or a corps hospital for treatment (for example, a patient with a broken arm), should be evacuated in their protective overgarments and masks by any

articles in a plastic bag, label the bag with the patient's identification, and seal the bag. If the articles are not contaminated, they are returned to the patient. If the articles are contaminated, place them in the contaminated holding area until they can be decontaminated, then return them to the patient.

8-57. **Remove the overgarment jacket.** Have the patient stand with his feet spread apart at shoulder width. Unsnap the front flap of the jacket and unzip the jacket. If the patient can extend his arms, have him clinch his fist and extend his arms backward at about a 30° angle. Move behind the patient, grasp his jacket collar at the sides of the neck, and peel the jacket off the shoulders at a 30° angle down and away from the patient. Avoid any rapid or sharp jerks, which spread contamination. Gently pull the inside sleeves over the patient's wrists and hands.

8-58. If the patient cannot extend his arms, you must cut the jacket to aid in its removal. Before cutting the overgarment jacket, dip the scissors in a 5 percent chlorine solution to prevent contamination of the patient's BDU or underclothing. As with the litter patient, make two cuts, one up each sleeve from the wrist up to the shoulder and then across the shoulder through the collar. Cut around bandages, tourniquets, and splints, leaving them in place. Do not allow your gloves to touch the patient along the cut line. Peel the jacket back and downward to avoid spreading contamination. Ensure that the outside of the jacket does not touch the patient or his inner clothing.

8-59. **Cut and remove the overgarment trousers.** Unfasten or cut all ties, buttons, or zippers before grasping the trousers at the waist and peeling them down over the patient's combat boots. Again, the trousers are cut to aid in removal. If necessary, cut both trouser legs starting at the ankle. Keep the cuts near the inside of the legs, along the inseam, to the crotch. Cut around all bandages, tourniquets, and splints. Continue to cut up both sides of the zipper to the waist and allow the narrow strip with the zipper to drop between the legs. Peel or allow the trouser halves to drop to the ground. Have the patient step out of the trouser legs one at a time. Place the trousers in the contaminated-waste bag. Place the scissors in a 5 percent chlorine solution.

8-60. **Remove the outer gloves.** Grasp the fingers of the gloves, roll the cuffs over the fingers, and turn the gloves inside out. Do not remove the inner cotton gloves at this time. Drop the gloves into the contaminated-waste bag. Do not allow the patient to touch his clothing or other contaminated objects with his hands.

8-61. **Remove the overboots.** Cut the overboot laces and fold the lacing eyelets flat on the ground. Step on the toe and heel eyelets to hold the overboot on the ground and have the patient step out of it. Repeat this procedure for the other overboot. If the overboots are in good condition, they can be decontaminated and reissued.

8-62. **Remove the patient's cotton glove liners.** Instruct the patient to remove his cotton glove liners to reduce the possibility of spreading contamination. Have the patient grasp the heel of one glove liner with the other gloved hand, peeling it off of his hand. Hold the removed glove by the inside and grasp the heel of the other glove, peeling it off of his hand. Place both gloves in the contaminated-waste bag.

Step 5 - Check the Patient for Contamination

8-63. After the patient's overgarment has been removed, check his BDU by using M8 detector paper or the CAM. Carefully survey the patient, paying particular attention to discolored areas, damp spots, and tears on the uniform; areas around the neck, wrists, and ears; and bandages, tourniquets, and splints. Remove contaminated spots by using an SDK or a 0.5 percent chlorine solution or, if possible, by cutting away the contaminated area. Always dip the scissors in a 5 percent chlorine solution after each cut. Recheck the area with the detection equipment. If significant contamination is found on the BDU, then remove it and spot decon the skin. To remove the BDU, follow the procedures for removing the overgarment as described in paragraphs 8-57 through 8-59. Do not remove the patient's identification tags.

Step 6 - Decon the Patient's Skin

8-64. **Spot decon.** Use an SDK or a 0.5 percent chlorine solution to spot decon the skin and areas of potential contamination, to include areas around the neck, wrists, and lower parts of the face. Decon the patient's identification tags and chain, if necessary.

8-65. Have the patient hold his breath, close his eyes, and lift, or assist him with lifting, his mask at the chin. Wipe his face and exposed areas of the skin with an SDK or a 0.5 percent chlorine solution. Starting at the top of the ear and quickly wiping downward, wipe all folds in the skin, ear lobes, upper lip, chin, dimples, and nose. Continue up the other side of the face to the top of the other ear. Wipe the inside of the mask where it touches the face. Have the patient reseal and check his mask.

CAUTION

Keep the decon solution out of the patient's eyes.

8-66. **Combat medic care.** The combat medic gently cuts away the bandage. He decontaminates the area around the wound and irrigates it with a 0.5 percent chlorine solution. If bleeding begins, he replaces the bandage with a clean one. He replaces the old tourniquet by placing a new one ½ to 1 inch above the old one. He then removes the old tourniquet and decontaminates the patient's skin with an SDK or a 0.5 percent chlorine solution. He does not remove a splint. He decontaminates the splint by thoroughly rinsing it, to include the padding and cravats, with a 0.5 percent chlorine solution.

8-67. **Contaminated-waste disposal.** Dispose of contaminated bandages and coverings by placing them in a contaminated-waste bag. Seal the bag and place it in the contaminated-waste dump.

Step 7 - Proceed Through the Shuffle Pit to the Clean Treatment Area

8-68. Have the decontaminated patient proceed through the shuffle pit to the clean treatment area. To ensure that the patient's boots are well decontaminated, have him stir the contents of the shuffle pit with his boots as he crosses it. The patient's combat boots and protective mask will be removed at the entrance of the CPS or clean treatment area.

BIOLOGICAL-AGENT PATIENT DECON PROCEDURES

8-69. The decon station as established for chemical-agent patients is also used for biologically contaminated patients. The 8-man, patient-decon team is required for biologically contaminated patient-decon procedures.

LITTER PATIENT

8-70. The decon procedures for a litter patient that is contaminated with a biological agent are described in the following paragraphs. Some procedures can be done with one soldier, while others require more than one.

Step 1 - Decon the Patient's Mask and Hood

8-71. **Move the patient to the clothing removal station.** After the patient has been triaged and stabilized (if necessary) by the senior medic in the patient-decon area, move him to the litter stands at the clothing removal station.

8-72. **Decon the hood.** Use a 0.5 percent chlorine solution to wipe down the front, sides, and top of the hood.

8-73. **Remove the hood.** Remove the hood by cutting it with scissors or by loosening it from the mask attachment points for the quick-doff hood or other similar hoods (see Figure 8-3, page 8-6). Before cutting the hood, dip the scissors in a 5 percent chlorine solution. Cut the neck cord and the small string under the voicemitter. Release or cut the hood shoulder straps and unzip the hood zipper. Cut the hood upward to the top of the eye-lens outsert, staying close to the filter-inlet cover and eye-lens outsert, then across the forehead to the outer edge of the other eye-lens outsert. Proceed downward toward the patient's shoulder, staying close to the eye-lens outsert and filter-inlet cover, then across the lower part of the voicemitter to the zipper. After dipping the scissors in the 5 percent chlorine solution, cut the hood from the center of the forehead over the top of the head. Fold the left and right sides of the hood to the side of the patient's head, laying the sides of the hood on the litter.

8-74. **Decon the protective mask and exposed skin**. Use a 0.5 percent chlorine solution to wipe the external parts of the mask. Cover the mask's air inlets with gauze or your hand to keep the mask filter dry. Wash the exposed areas of the patient's face, to include the neck and behind the ears, with soap and warm water or wipe them with a 0.5 percent chlorine solution. Do not remove the protective mask.

8-75. **Remove the FMC.** Cut the patient's FMC tie wire, allowing the FMC to fall into a plastic bag. Seal the plastic bag and rinse the outside of the bag with a 0.5 percent chlorine solution. Place the plastic bag under the back of the protective mask's head straps. The FMC will remain with the patient in the contaminated area and a clean copy will be made before the patient is moved to the clean area.

Step 2 - Remove Gross Contamination From the Patient's Protective Overgarment

8-76. Remove all visible contamination spots from the overgarment by using a 0.5 percent chlorine solution.

Step 3 - Remove the Patient's Personal Effects and Protective Overgarment

8-77. **Remove the patient's personal effects.** Remove the patient's personal effects from his protective overgarment and BDU pockets. Place the articles in a plastic bag, label the bag with the patient's identification, and seal the bag. If the articles are not contaminated, they are returned to the patient. If the articles are contaminated, place them in the contaminated holding area until they can be decontaminated, then return them to the patient.

> **CAUTION**
>
> Bandages may have been applied to control severe bleeding; therefore, they are treated like tourniquets. Only medical personnel will remove bandages, tourniquets, and splints.

NOTE: The overgarment jacket and trousers will be cut simultaneously. One soldier cuts the jacket, while another soldier cuts the trousers.

8-78. **Cut and remove the overgarment jacket.** Before cutting the overgarment jacket and trouser, dip the scissors in a 5 percent chlorine solution to prevent contamination of the patient's BDU or undergarments. Make two cuts, one up each sleeve from the wrist up to the shoulder and then across the shoulder through the collar (Figure 8-4, page 8-7). Cut around bandages, tourniquets, and splints, leaving them in place. Do not allow your gloves to touch the patient along the cut line. Keep the cuts close to the inside of the arms so that most of the sleeve material can be folded outward. Unzip the jacket and roll the chest sections to the respective sides, with the inner surface outward. Continue by tucking the clothing between the arm and chest. Roll the cut sleeves away from the arms, exposing the black liner.

8-79. **Cut and remove the overgarment trousers.** Cut both trousers legs starting at the ankle as shown in Figure 8-5, page 8-8. Keep the cuts near the inside of the legs, along the inseam, to the crotch. Cut the left leg all the way to the waist, avoiding the pockets. Cut the right leg across at the crotch to the left leg cut. Cut around bandages, tourniquets, and splints, leaving them in place. Place the scissors in a 5 percent chlorine solution. Fold the cut trouser halves away from the patient and allow the halves to drop to the litter with the contaminated (green) side down. Roll the inner leg portion under and between the legs.

8-80. **Remove the outer gloves.** Before touching the patient, the patient-decon team decontaminates its gloves with a 5 percent chlorine solution. Lift the patient's arms up and out of the cutaway sleeves unless detrimental to his condition. Grasp the fingers of the gloves, roll the cuffs over the fingers, and turn the gloves inside out. Do not remove the inner cotton gloves at this time. Carefully lower the patient's arms across the chest after the outer gloves have been removed (Figure 8-6, page 8-8). Do not allow the patient's arms to come into contact with the exterior of his overgarment. Drop his gloves into the contaminated-waste bag. The team members decon their gloves with the 5 percent chlorine solution.

8-81. **Remove the overboots.** Cut the overboot laces and fold the lacing eyelets flat outward. While standing at the foot of the litter, hold the patient's heel with one hand. Pull the overboot downward, then toward you to remove it. Remove the two overboots simultaneously. This reduces the likelihood of contaminating one of the combat boots.

8-82. While holding the patient's heels off the litter, have a team member wipe the end of the litter with a 0.5 percent chlorine solution to neutralize any contamination that was transferred to the litter from the overboots. Lower the patient's heels onto the decontaminated litter. Place the overboots in the contaminated-waste bag. The team members decontaminate their gloves with the 5 percent chlorine solution.

Step 4 - Remove the Patient's BDU and Undergarments

8-83. **Cut and remove the BDU.** To cut and remove the BDU jacket and trousers, follow the procedures (described in paragraphs 8-78 and 8-79) for removing the protective overgarment.

8-84. **Remove the combat boots.** Cut the bootlaces along the tongue. Remove the boots by pulling them toward you. Place the boots in the contaminated-waste bag. Do not touch the patient's skin with your contaminated gloves when removing his boots.

8-85. **Cut and remove the undergarments.** Follow the procedures for cutting away the protective overgarment and rolling it away from the patient (see paragraphs 8-78 and 8-79). If the patient is wearing a brassiere, cut it between the cups. Cut both shoulder straps where they attach to the cups and lay them back off of the shoulders. Remove the socks and cotton gloves. Do not remove the identification tags.

Step 5 - Transfer the Patient to a Decon Litter

8-86. After the patient's clothing has been cut away, transfer him to a decon litter or a canvas litter with a plastic-sheeting cover. Three decon team members decon their gloves and aprons with a 5 percent chlorine solution. One member places his hands under the patient's legs at the thighs and Achilles tendons, a second member places his arms under the patient's back and buttocks, and a third member places his arms under the patient's shoulders and supports the head and neck. They carefully lift the patient using their knees (not their backs) to minimize back strain. While the patient is elevated, another decon team member removes the litter from the litter stands and replaces it with a decontaminated (clean) litter. The patient is carefully lowered onto the clean litter. The clothing and overgarments are placed in a contaminated-waste bag and moved to the contaminated-waste dump. The dirty litter is rinsed with the 0.5 percent chlorine solution and placed in the litter storage area.

Step 6 - Decon the Patient's Skin

8-87. **Spot decon.** With the patient in a supine position, spot decon the skin by washing it with soap and warm water or by wiping it with a 0.5 percent chlorine solution. Decon areas of potential contamination, to include areas around the neck, wrists, and lower parts of the face. Decon the patient's identification tags and chain, if necessary.

NOTES:

1. Use a 0.5 percent chlorine solution to decon patients suspected of being contaminated with mycotoxins.

2. A complete body wash is not appropriate and may be harmful to the patient. During a complete body wash, the patient would have to be rolled over to reach all areas of the skin. This is not necessary for an adequate decon.

8-88. **Combat medic care.** The combat medic gently cuts away the bandage. He decontaminates the area around the wound and irrigates it with a 0.5 percent chlorine solution. If bleeding begins, he replaces the bandage with a clean one. He replaces the old tourniquet by placing a new one ½ to 1 inch above the old one. He then removes the old tourniquet and decontaminates the patient's skin with a 0.5 percent chlorine solution. He does not remove a splint. He decontaminates the splint by thoroughly rinsing it, to include the padding and cravats, with a 0.5 percent chlorine solution.

WARNING

Do not use an SDK around any wounds. Do not remove splints. The splint will not be removed until the patient has been evacuated to a forward MTF or hospital.

8-89. **Completeness of decon check.** Check the patient with M8 detector paper or the CAM for completeness of decon.

NOTE: Other monitoring devices may be used, if available.

8-90. **Contaminated-waste disposal.** Dispose of contaminated bandages and coverings by placing them in a contaminated-waste bag. Seal the bag and place it in the contaminated-waste dump.

Step 7 - Transfer the Patient to the Hot Line

8-91. The patient's clothing has been cut away and his skin, bandages, and splints have been decontaminated. Transfer the patient to the hot line and place the litter on the litter stands. A third team member will assist with transferring the patient to a clean treatment litter at the hot line. Decon personnel rinse or wipe down their aprons and gloves with a 5 percent chlorine solution.

8-92. Three team members lift the patient off the decon litter (see step 5 for lifting procedures). While the patient is elevated, another team member removes the litter from the stands and returns it to the decon area. A medic from the clean side of the hot line replaces the litter with a clean one. The patient is lowered onto the clean litter. Two medics from the clean side of the hot line move the patient to the clean treatment area. The patient is treated in this area or awaits processing into the CPS. The litter is wiped down with a 0.5 percent chlorine solution in preparation for reuse. Once the patient is in the air lock of the CPS and the air lock has been purged, his protective mask is removed. Place the mask in a plastic bag and seal it.

NOTE: Before decontaminating another patient, each decon team member drinks about half a quart of water. The exact amount of water consumed is increased or decreased according to the work level and temperature (see Table 8-2, page 8-11).

AMBULATORY PATIENT

8-93. The decon procedures for an ambulatory patient that is contaminated with a biological agent are described in the following paragraphs. Some procedures can be done with one soldier, while others require more than one.

8-94. All ambulatory patients requiring EMT or AMT in the clean treatment area of the BAS will be decontaminated. Stable patients not requiring treatment at the BAS, but requiring evacuation to the medical company's clearing station or a corps hospital for treatment (for example, a patient with a broken arm), should be evacuated in their protective overgarments and masks by any available transportation. However, before evacuation, spot remove all thickened agents from their protective overgarment.

NOTE: Place the bandage scissors that are used in this procedure in a container of 5 percent chlorine solution when not in use. Most ambulatory patients will be treated in the contaminated treatment area and returned to duty. Upon removal of an ambulatory patient's clothing, he becomes a litter patient. The BAS and clearing station do not have clothing to replace those that are cut off during the decon process. The patient must be placed in a PPW for protection during evacuation (Figure 8-7, page 8-12).

Step 1 - Remove the LCE

8-95. Remove the LCE by unfastening/unbuttoning all connectors or tie straps and then place the equipment in a plastic bag. Place the plastic bag in the designated storage area for later decon.

Step 2 - Decon the Patient's Mask and Hood

8-96. **Begin the clothing removal process.** After the patient has been triaged and treated (if necessary) by the senior medic in the patient-decon station, the clothing removal process begins.

8-97. **Decon the hood.** Use a 0.5 percent chlorine solution to wipe down the front, sides, and top of the hood.

8-98. **Remove the hood.** Remove the hood by cutting it with scissors or by loosening it from the mask attachment points for the quick-doff hood or other similar hoods (Figure 8-3, page 8-6). Before cutting the hood, dip the scissors in a 5 percent chlorine solution. Cut the neck cord and the small string under the voicemitter. Release or cut the hood shoulder straps and unzip the hood zipper. Cut the hood upward to the top of the eye-lens outsert, staying close to the filter-inlet cover and eye-lens outsert, then across the forehead to the outer edge of the other eye-lens outsert. Proceed downward toward the patient's shoulder, staying close to the eye-lens outsert and filter-inlet cover, then across the lower part of the voicemitter to the zipper. After dipping the scissors in the 5 percent chlorine solution, cut the hood from the center of the forehead over the top of the head. Fold the left and right sides of the hood away from the patient's head and remove the hood.

8-99. **Decon the protective mask and exposed skin.** Use a 0.5 percent chlorine solution to decon the external parts of the mask. Cover the mask's air inlets with gauze or your hands to keep the mask filters dry. Wash the exposed areas of the patient's face, to include the neck and behind the ears, with soap and warm water or wipe them with a 0.5 percent chlorine solution. Do not remove the protective mask.

8-100. **Remove the FMC.** Cut the patient's FMC tie wire, allowing the FMC to fall into a plastic bag. Seal the plastic bag and rinse the outside of the bag with a 5 percent chlorine solution. Place the plastic bag under the back of the protective mask's head straps. The FMC will remain with the patient in the contaminated area and a clean copy will be made before the patient is moved to the clean area.

Step 3 - Remove Gross Contamination From the Patient's Protective Overgarment

8-101. Remove all visible contamination spots from the overgarment by using a 0.5 percent chlorine solution.

Step 4 - Remove the Patient's Personal Effects and Protective Overgarment

8-102. **Remove the patient's personal effects.** Remove the patient's personal effects from his protective overgarment and BDU pockets. Place the articles in a plastic bag, label the bag with the patient's identification, and seal the bag. If the articles are not contaminated, they are returned to the patient. If the articles are contaminated, place them in the contaminated holding area until they can be decontaminated, then return them to the patient.

8-103. **Remove the overgarment jacket.** Have the patient stand with his feet spread apart at shoulder width. Unsnap the front flap of the jacket and unzip the jacket. If the patient can extend his arms, have him clinch his fist and extend his arms backward at about a 30° angle. Move behind the patient, grasp his jacket collar at the sides of the neck, and peel the jacket off the shoulders at a 30° angle down and away from the patient. Avoid any rapid or sharp jerks, which spread contamination. Gently pull the inside sleeves over the patient's wrists and hands.

8-104. If the patient cannot extend his arms, you must cut the jacket to aid in its removal. Before cutting the overgarment jacket, dip the scissors in a 5 percent chlorine solution to prevent contamination of the patient's BDU or undergarment. As with the litter patient, make two cuts, one up each sleeve from the wrist up to the shoulder and then across the shoulder through the collar. Cut around bandages, tourniquets, and splints, leaving them in place. Do not allow your gloves to touch the patient along the cut line. Peel the jacket back and downward to avoid spreading contamination. Ensure that the outside of the jacket does not touch the patient or his inner clothing.

8-105. **Cut and remove the overgarment trousers.** Unfasten or cut all ties, buttons, or zippers before grasping the trousers at the waist and peeling them down over the patient's combat boots. Again, the trousers are cut to aid in removal. If necessary, cut both trouser legs starting at the ankle. Keep the cuts near the inside of the legs, along the inseam, to the crotch. Cut around all bandages, tourniquets, and splints. Continue to cut up both sides of the zipper to the waist and allow the narrow strip with the zipper to drop between the legs.

Peel or allow the trouser halves to drop to the ground. Have the patient step out of the trouser legs one at a time. Place the trousers in the contaminated-waste bag. Place the scissors in a 5 percent chlorine solution.

8-106. **Remove the outer gloves.** Grasp the fingers of the gloves, roll the cuffs over the fingers, and turn the gloves inside out. Do not remove the inner cotton gloves at this time. Drop the gloves into the contaminated-waste bag. Do not allow the patient to touch his clothing or other contaminated objects with his hands.

8-107. **Remove the overboots.** Cut the overboot laces and fold the lacing eyelets flat on the ground. Step on the toe and heel eyelets to hold the overboot on the ground and have the patient step out of it. Repeat this procedure for the other overboot. If the overboots are in good condition, they can be decontaminated and reissued.

8-108. **Remove the patient's cotton glove liners.** Instruct the patient to remove his cotton glove liners to reduce the possibility of spreading contamination. Have the patient grasp the heel of one glove liner with the other gloved hand, peeling it off of his hand. Hold the removed glove by the inside and grasp the heel of the other glove, peeling it off of his hand. Place both gloves in the contaminated-waste bag.

Step 5 - Check the Patient for Contamination

8-109. After the patient's overgarment has been removed, check his BDU by using M8 detector paper or the CAM. Carefully survey the patient, paying particular attention to discolored areas, damp spots, and tears on the uniform; areas around the neck, wrists, and ears; and bandages, tourniquets, and splints. Remove contaminated spots by washing with soap and water, by wiping with a 0.5 percent chlorine solution or, if possible, by cutting away the contaminated area. Always dip the scissors in a 5 percent chlorine solution after each cut. Recheck the area with the detection equipment. If significant contamination is found on the BDU, then remove it and spot decon the skin. To remove the BDU, follow the procedures for removing the protective overgarment as described in paragraphs 8-103 through 8-105. Do not remove the patient's identification tags.

Step 6 - Decon the Patient's Skin

8-110. **Spot decon.** Use soap and water or a 0.5 percent chlorine solution to spot decon the skin and areas of potential contamination, to include areas around the neck, wrists, and lower parts of the face. Decon the patient's identification tags and chain, if necessary.

8-111. Have the patient hold his breath, close his eyes, and lift, or assist him with lifting, his mask at the chin. Wash his face and exposed areas of the skin with soap and water or wipe them with a 0.5 percent chlorine solution. Starting at the top of the ear and quickly wiping downward, wipe all folds in the skin, ear lobes, upper lip, chin, dimples, and nose. Continue up the other side of the face to the top of the other ear. Wipe the inside of the mask where it touches the face. Have the patient reseal and check his mask.

CAUTION

Keep the decon solution out of the patient's eyes.

8-112. **Combat medic care.** The combat medic gently cuts away the bandage. He decontaminates the area around the wound and irrigates it with a 0.5 percent chlorine solution. If bleeding begins, he replaces the bandage with a clean one. He replaces the old tourniquet by placing a new one ½ to 1 inch above the old one. He then removes the old tourniquet and decontaminates the patient's skin with an SDK or a 0.5 percent chlorine solution. He does not remove a splint. He decontaminates the splint by thoroughly rinsing it, to include the padding and cravats, with a 0.5 percent chlorine solution.

NOTE: Use a 0.5 percent chlorine solution to decon ambulatory patients suspected of being contaminated with mycotoxins.

8-113. **Contaminated-waste disposal.** Dispose of contaminated bandages and coverings by placing them in a contaminated-waste bag. Seal the bag and place it in the contaminated-waste dump.

Step 7 - Cross the Hot Line to the Clean Treatment Area

8-114. Have the decontaminated patient cross the hot line to the clean treatment area. The patient's boots must be decontaminated at the hot line before enters the clean area. The patient's combat boots and protective mask will be removed at the entrance of the CPS or clean treatment area.

RADIOLOGICAL-AGENT PATIENT DECON PROCEDURES

8-115. The decon of patients that are contaminated with radiation is easily accomplished without interfering with the required medical care. The 8-man, patient-decon team is required to perform the decon procedures on patients contaminated with radiation.

NOTE: Patients must be monitored by using a radiac meter before, during, and after each step of the decon procedure.

LITTER PATIENT

8-116. The decon procedures for a litter patient that is contaminated with radiation are discussed in the following paragraphs. Some procedures can be done with one soldier, while others require more than one soldier.

Step 1 - Decon the Patient's Mask and Hood

8-117. **Move the patient to the clothing removal station.** After the patient has been triaged and stabilized (if necessary) by the senior medic in the patient-decon area, move him to the litter stands at the clothing removal station.

8-118. **Decon the hood.** Brush contamination off or use soap and water to wipe down the front, sides, and top of the hood.

8-119. **Remove the hood.** Remove the hood by cutting it with scissors or by loosening it from the mask attachment points for the quick-doff hood or other similar hoods (see Figure 8-3, page 8-6). Before cutting the hood, dip the scissors in a 5 percent chlorine solution. Cut the neck cord and the small string under the voicemitter. Release or cut the hood shoulder straps and unzip the hood zipper. Cut the hood upward to the top of the eye-lens outsert, staying close to the filter-inlet cover and eye-lens outsert, then across the forehead to the outer edge of the other eye-lens outsert. Proceed downward toward the patient's shoulder, staying close to the eye-lens outsert and filter-inlet cover, then across the lower part of the voicemitter to the zipper. After dipping the scissors in the 5 percent chlorine solution, cut the hood from the center of the forehead over the top of the head. Fold the left and right sides of the hood to the side of the patient's head, laying the sides of the hood on the litter.

8-120. **Decon the protective mask and exposed skin.** Use soap and water to wipe the external parts of the mask. Cover the mask's air inlets with gauze or your hand to keep the mask filter dry. Wash the exposed areas of the patient's face, to include the neck and behind the ears, with soap and warm water. Do not remove the protective mask.

8-121. **Remove the FMC.** Cut the patient's FMC tie wire, allowing the FMC to fall into a plastic bag. Seal the plastic bag and rinse the outside of the bag with soap and water. Place the plastic bag under the back of the protective mask's head straps. The FMC will remain with the patient in the contaminated area and a clean copy will be made before the patient is moved to the clean area.

Step 2 - Remove Gross Contamination From the Patient's Protective Overgarment

8-122. Remove all visible contamination spots from the overgarment by brushing off or by washing with soap and water.

Step 3 - Remove the Patient's Personal Effects and Protective Overgarment

8-123. **Remove the patient's personal effects.** Remove the patient's personal effects from his protective overgarment and BDU pockets. Place the articles in a plastic bag, label the bag with the patient's identification, and seal the bag. If the articles are not contaminated, they are returned to the patient. If the articles are contaminated, place them in the contaminated holding area until they can be decontaminated, then return them to the patient.

CAUTION

Bandages may be applied to control severe bleeding; therefore, they are treated like tourniquets. Only medical personnel will remove bandages.

NOTE: The overgarment jacket and trousers will be cut simultaneously. One soldier cuts the jacket, while another soldier cuts the trousers.

8-124. **Cut and remove the overgarment jacket.** Before cutting the overgarment jacket and trousers, dip the scissors in a 5 percent chlorine solution to prevent contamination of the patient's BDU or undergarment. Make two cuts, one up each sleeve from the wrist up to the shoulder and then across the shoulder through the collar (Figure 8-4, page 8-7). Cut around bandages, tourniquets, and splints, leaving them in place. Do not allow your gloves to touch the patient along the cut line. Keep the cuts close to the inside of the arms so that most of the sleeve material can be folded outward. Unzip the jacket and roll the chest sections to the respective sides, with the inner surface outward. Continue by tucking the clothing between the arm and chest. Roll the cut sleeves away from the arms, exposing the black liner.

8-125. **Cut and remove the overgarment trousers.** Cut both trousers legs starting at the ankle as shown in Figure 8-5, page 8-8. Keep the cuts near the inside of the legs, along the inseam, to the crotch. Cut the left leg all the way to the waist, avoiding the pockets. Cut the right leg across at the crotch to the left leg cut. Cut around bandages, tourniquets, and splints, leaving them in place. Place the scissors in a 5 percent chlorine solution. Fold the cut trouser halves away from the patient and allow the halves to drop to the litter with the contaminated- (green-) side down. Roll the inner leg portion under and between the legs.

8-126. **Remove the outer gloves.** Before touching the patient, the patient-decon team decontaminates its gloves with a 5 percent chlorine solution. Lift the patient's arms up and out of the cutaway sleeves unless detrimental to his condition. Grasp the fingers of the gloves, roll the cuffs over the fingers, and turn the gloves inside out. Do not remove the inner cotton gloves at this time. Carefully lower the patient's arms across the chest after the outer gloves have been removed (Figure 8-6, page 8-8). Do not allow the patient's arms to come into contact with the exterior of his overgarment. Drop his gloves into the contaminated-waste bag. The team members decon their gloves with the 5 percent chlorine solution.

8-127. **Remove the overboots.** Cut the overboot laces and fold the lacing eyelets flat outward. While standing at the foot of the litter, hold the patient's heel with one hand. Pull the overboot downward, then toward you to remove it. Remove the two overboots simultaneously. This reduces the likelihood of contaminating one of the combat boots.

8-128. While holding the patient's heels off the litter, have a team member wipe the end of the litter with a 0.5 percent chlorine solution to neutralize any contamination that was transferred to the litter from the overboots. Lower the patient's heels onto the decontaminated litter. Place the overboots in the contaminated-waste bag. The team members decon their gloves with the 5 percent chlorine solution.

NOTE: Patients arriving at the MTF in MOPP gear will only have their MOPP gear removed. They will remain in their BDUs unless contamination is found on them. If contamination is found, follow the procedures for removing the protective overgarment as described in paragraphs 8-124 and 8-125.

Step 4 - Check the Patient for Contamination

8-129. After the patient's overgarment has been removed, check his BDU by using M8 detector paper or the CAM. Carefully survey the patient, paying particular attention to discolored areas, damp spots, and tears on the uniform; areas around the neck, wrists, and ears; and bandages, tourniquets, and splints. Remove contaminated spots by washing with soap and water, by wiping with a 0.5 percent chlorine solution or, if possible, by cutting away the contaminated area. Always dip the scissors in a 5 percent chlorine solution after each cut. Recheck the area with the detection equipment. If significant contamination is found on the BDU, then remove it and spot decon the skin.

To remove the BDU, follow the procedures for removing the protective overgarment as described in paragraphs 8-124 and 8-125. Do not remove the patient's identification tags.

Step 5 - Decon the Patient's Skin

8-130. **Spot decon.** Wash exposed skin surfaces with soap and warm water, to include areas around the neck, wrists, and lower parts of the face. Decon the patient's identification tags and chain, if necessary.

8-131. **Combat medic care.** The combat medic gently cuts away the bandage. He decontaminates the area around the wound and irrigates it with a 0.5 percent chlorine solution. If bleeding begins, he replaces the bandage with a clean one. He replaces the old tourniquet by placing a new one ½ to 1 inch above the old one. He then removes the old tourniquet and decontaminates the patient's skin with a 0.5 percent chlorine solution. He does not remove a splint. He decontaminates the splint by thoroughly rinsing it, to include the padding and cravats, with a 0.5 percent chlorine solution.

WARNING

Do not use an SDK around any wounds. Do not remove splints. The splint will not be removed until the patient has been evacuated to a forward MTF or hospital.

8-132. **Completeness of decon check.** Check the patient with M8 detector paper or the CAM for completeness of decon.

NOTE: Other monitoring devices may be used, if available.

8-133. **Contaminated-waste disposal.** Dispose of contaminated bandages and coverings by placing them in a contaminated-waste bag. Seal the bag and place it in the contaminated-waste dump.

Step 6 - Transfer the Patient to the Hot Line

8-134. Move the patient to the hot line. Two medics from the clean side of the hot line move the patient into the clean treatment area.

NOTE: Before decontaminating another patient, each decon team member drinks about half a quart of water. The exact amount of water consumed is increased or decreased according to the work level and temperature (see Table 8-2, page 8-11).

AMBULATORY PATIENT

8-135. The decon procedures for an ambulatory patient that is contaminated with radiation are discussed in the following paragraphs. Some procedures can be done with one soldier, while others require more than one.

8-136. All ambulatory patients requiring EMT or AMT in the clean area of the BAS will be decontaminated. Stable patients not requiring treatment at the BAS, but requiring evacuation to the medical company's clearing station or a corps hospital for treatment (for example, a patient with a broken arm), should be evacuated in their protective overgarments and masks by any

available transportation. However, before evacuation, spot remove all thickened agents from their protective overgarment.

NOTE: Place the bandage scissors that are used in this procedure in a container of 5 percent chlorine solution when not in use. Most ambulatory patients will be treated in the contaminated treatment area and returned to duty. Upon removal of an ambulatory patient's clothing, he becomes a litter patient. The BAS and clearing station do not have clothing to replace those that are cut off during the decon process. The patient must be placed in a PPW for protection during evacuation (Figure 8-7, page 8-12).

Step 1 - Remove the LCE

8-137. Remove the LCE by unfastening/unbuttoning all connectors or tie straps and then place the equipment in a plastic bag. Place the plastic bag in the designated storage area for later decon.

Step 2 - Decon the Patient's Mask and Hood

8-138. **Begin the clothing removal process.** After the patient has been triaged and treated (if necessary) by the senior medic in the patient-decon station, the clothing removal process begins.

8-139. **Decon the hood.** Brush off contamination or use soap and water to wipe down the front, sides, and top of the hood.

8-140. **Remove the hood.** Remove the hood by cutting it with scissors or by loosening it from the mask attachment points for the quick-doff hood or other similar hoods (see Figure 8-3, page 8-6). Before cutting the hood, dip the scissors in a 5 percent chlorine solution. Cut the neck cord and the small string under the voicemitter. Release or cut the hood shoulder straps and unzip the hood zipper. Cut the hood upward to the top of the eye-lens outsert, staying close to the filter-inlet cover and eye-lens outsert, then across the forehead to the outer edge of the other eye-lens outsert. Proceed downward toward the patient's shoulder, staying close to the eye-lens outsert and filter-inlet cover, then across the lower part of the voicemitter to the zipper. After dipping the scissors in the 5 percent chlorine solution, cut the hood from the center of the forehead over the top of the head. Fold the left and right sides of the hood away from the patient's head and remove the hood.

8-141. **Decon the protective mask and exposed skin.** Use soap and water to wipe the external parts of the mask. Cover the mask's air inlets with gauze or your hands to keep the mask filters dry. Wash the exposed areas of the patient's face, to include the neck and behind the ears, with soap and warm water. Do not remove the protective mask.

8-142. **Remove the FMC.** Cut the patient's FMC tie wire, allowing the FMC to fall into a plastic bag. Seal the plastic bag and rinse the outside of the bag with a 5 percent chlorine solution. Place the plastic bag under the back of the protective mask's head straps. The FMC will remain with the patient in the contaminated area and a clean copy will be made before the patient is moved to the clean area.

Step 3 - Remove Gross Contamination From the Patient's Protective Overgarment

8-143. Remove all visible contamination spots by brushing off or by washing with soap and water.

Step 4 - Remove the Patient's Personal Effects and Protective Overgarment

8-144. **Remove the patient's personal effects.** Remove the patient's personal effects from his protective overgarment and BDU pockets. Place the articles in a plastic bag, label the bag with the patient's identification, and seal the bag. If the articles are not contaminated, they are returned to the patient. If the articles are contaminated, place them in the contaminated holding area until they can be decontaminated, then return them to the patient.

8-145. **Remove the overgarment jacket.** Have the patient stand with his feet spread apart at shoulder width. Unsnap the front flap of the jacket and unzip the jacket. If the patient can extend his arms, have him clinch his fist and extend his arms backward at about a 30° angle. Move behind the patient, grasp his jacket collar at the sides of the neck, and peel the jacket off the shoulders at a 30^{0} angle down and away from the patient. Avoid any rapid or sharp jerks, which spread contamination. Gently pull the inside sleeves over the patient's wrists and hands.

8-146. If the patient cannot extend his arms, you must cut the jacket to aid in its removal. Before cutting the overgarment jacket, dip the scissors in a 5 percent chlorine solution to prevent contamination of the patient's BDU or undergarment. As with the litter patient, make two cuts, one up each sleeve from the wrist up to the shoulder and then across the shoulder through the collar. Cut around bandages, tourniquets, and splints, leaving them in place. Do not allow your gloves to touch the patient along the cut line. Peel the jacket back and downward to avoid spreading contamination. Ensure that the outside of the jacket does not touch the patient or his inner clothing.

8-147. **Cut and remove the overgarment trousers.** Unfasten or cut all ties, buttons, or zippers before grasping the trousers at the waist and peeling them down over the patient's combat boots. Again, the trousers are cut to aid in removal. If necessary, cut both trouser legs starting at the ankle. Keep the cuts near the inside of the legs, along the inseam, to the crotch. Cut around all bandages, tourniquets, and splints. Continue to cut up both sides of the zipper to the waist and allow the narrow strip with the zipper to drop between the legs. Peel or allow the trouser halves to drop to the ground. Have the patient step out of the trouser legs one at a time. Place the trousers in the contaminated-waste bag. Place the scissors in a 5 percent chlorine solution.

8-148. **Remove the outer gloves.** Grasp the fingers of the gloves, roll the cuffs over the fingers, and turn the gloves inside out. Do not remove the inner cotton gloves at this time. Drop the gloves into the contaminated-waste bag. Do not allow the patient to touch his clothing or other contaminated objects with his hands.

8-149. **Remove the overboots.** Cut the overboot laces and fold the lacing eyelets flat on the ground. Step on the toe and heel eyelets to hold the overboot on the ground and have the patient step out of it. Repeat this procedure for the other overboot. If the overboots are in good condition, they can be decontaminated and reissued.

8-150. **Remove the patient's cotton glove liners.** Instruct the patient to remove his cotton glove liners to reduce the possibility of spreading contamination. Have the patient grasp the heel of one glove liner with the other gloved hand, peeling it off of his hand. Hold the removed glove by the inside and grasp the heel of the other glove, peeling it off of his hand. Place both gloves in the contaminated-waste bag.

Step 5 - Check the Patient for Contamination

8-151. After the patient's overgarment has been removed, check his BDU by using M8 detector paper or the CAM. Carefully survey the patient, paying particular attention to discolored areas, damp spots, and tears on the uniform; areas around the neck, wrists, and ears; and bandages, tourniquets, and splints. Remove contaminated spots by washing with soap and water. Always dip the scissors in a 5 percent chlorine solution after each cut. Recheck the area with the detection equipment. If significant contamination is found on the BDU, then remove it and spot decon the skin. To remove the BDU, follow the procedures for removing the protective overgarment as described in paragraphs 8-145 through 8-147. Do not remove the patient's identification tags.

Step 6 - Decon the Patient's Skin

8-152. **Spot decon.** Wash exposed skin surfaces with soap and warm water, to include areas around the neck, wrists, and lower parts of the face. Decon the patient's identification tags and chain, if necessary.

8-153. Have the patient hold his breath, close his eyes, and lift, or assist him with lifting, his mask at the chin. Wipe his face and exposed areas of the skin with soap and water. Starting at the top of the ear and quickly wiping downward, wipe all folds in the skin, ear lobes, upper lip, chin, dimples, and nose. Continue up the other side of the face to the top of the other ear. Wipe the inside of the mask where it touches the face. Have the patient reseal and check his mask.

8-154. **Combat medic care.** The combat medic gently cuts away the bandage. He decontaminates the area around the wound and irrigates it with a 0.5 percent chlorine solution. If bleeding begins, he replaces the bandage with a clean one. He replaces the old tourniquet by placing a new one ½ to 1 inch above the old one. He then removes the old tourniquet and decontaminates the patient's skin with a 0.5 percent chlorine solution. He does not remove a splint. He decontaminates the splint by thoroughly rinsing it, to include the padding and cravats, with a 0.5 percent chlorine solution.

8-155. **Contaminated-waste disposal.** Dispose of contaminated bandages and coverings by placing them in a contaminated-waste bag. Seal the bag and place it in the contaminated-waste dump.

Step 7 - Transfer the Patient to the Hot Line

8-156. Have the decontaminated patient cross the hot line to the clean treatment area. The patient's boots must be decontaminated at the hot line before he enters the clean area. The patient's combat boots and protective mask will be removed at the entrance of the CPS or clean treatment area.

NOTE: Before decontaminating another patient, each decon team member drinks about half a quart of water. The exact amount of water consumed is increased or decreased according to the work level and temperature (see Table 8-2, page 8-11).

Chapter 9

Logistics

The materials that are needed to conduct decon operations are identified in this chapter.

SKIN DECON AND PERSONAL WIPE DOWN

9-1. Each soldier carries his own skin-decon supplies. The SDK is stored in the right cargo pocket of the overgarment trousers or BDU trousers. The basis of issue per soldier is two SDKs (12 packets). Each kit contains six individual decon packets. The unit maintains at least one SDK per assigned person for resupply. Personal wipe down is also done with these kits.

INDIVIDUAL-GEAR DECON

9-2. The standard decon kit for individual gear is the IEDK. Each soldier carries an IEDK in the right cargo pocket of his overgarment trousers or BDU trousers. The basis of issue per soldier is one IEDK. Each kit contains four individual decon packets, enough to decon two complete sets of individual gear. The unit maintains at least one IEDK per assigned soldier for resupply.

OPERATOR'S SPRAY DOWN

9-3. The M11 or M13 is the decon apparatus used for the operator's spray down. Fill the M11 with a 1 $^1/_3$-quart can of DS2 before use. The IEDK can be used on equipment that the DS2 may cause damage to by corrosive action.

9-4. Four nitrogen cylinders and two 1 $^1/_3$-quart cans of DS2 should be carried in the vehicle with each M11 (two nitrogen cylinders are authorized per can of DS2). Use the nitrogen cylinders to pressurize the M11 during use. Operating the M11 in below-freezing weather (below 32°F) requires two nitrogen cylinders to expend its contents.

9-5. Units maintain limited spare parts and accessories. They are not required to maintain additional stocks of 1 $^1/_3$-quart cans of DS2. The 5-gallon pails that are authorized to each unit can be used to refill empty M11s until a resupply of 1 $^1/_3$-quart cans of DS2 is received from the battalion.

9-6. The M13 comes with a 3.7 gallon DS2 container.

MOPP-GEAR EXCHANGE

9-7. Each soldier wears or carries one complete set of MOPP gear. The unit stocks a second complete set, which is sized and identified for each soldier. When a squad or platoon undergoes a MOPP-gear exchange, the unit sends a supply vehicle with replacement MOPP gear and any decontaminants (STB) to rendezvous with the contaminated element at the operational-decon site. (The procedure on how to prepare the STB dry mix is discussed in Appendix B.) Use about 40 pounds of the STB dry mix per squad-sized element. In wartime, 5-pound quantities of the STB dry mix can be put in 1-gallon airtight containers and stored in the company's supply section.

NOTE: Do not leave a 50-pound drum of STB dry mix open because it loses its effectiveness if left open to the air for extended periods. The unit is resupplied with MOPP gear and decontaminants from its battalion.

9-8. Units maintain a 5 percent overage of MOPP gear based on their personnel strength or authorization (whichever is greater) to ensure a complete range of sizes and replacement gear. All MOPP gear, serviceable and unserviceable, is handled as organizational clothing and equipment (Class II supplies). The higher HQ is responsible for resupply of these items.

VEHICLE WASHDOWN

9-9. Like the MOPP-gear exchange, a vehicle washdown is done in the unit's AO. The battalion's PDDE crew, stationed in the battalion trains, conducts the washdown. The crew moves to the operational-decon site, rendezvous with the contaminated element, and conducts the washdown. The crew uses its PDDE to spray about 100 to 150 gallons of hot, soapy water on each vehicle to wash off the gross contamination. For vehicles such as the M1-series armored fighting vehicles, about 200 gallons of water may be required per vehicle. One hundred gallons of water per vehicle provides a 2- to 3-minute wash. To speed up the operation, the crew should heat the water before arriving at the operational-decon site. See Chapters 2 and 3 and Table 9-1 for the planning and coordination of and the requirements for a vehicle washdown.

9-10. The M12A1 PDDA injects detergent into the water as it operates. It uses 2.5 quarts of detergent for every 1,200 gallons of water (about 1 quart of soap per 450 gallons of water). The battalion decon crew should maintain a basic load of liquid all-purpose detergent, which is sufficient to decon 30 percent of the battalion's organic vehicles.

DETAILED TROOP DECON

9-11. Generally, units conduct DTD in the brigade support area. Battalions conduct it in the division/corps support area. Materials for this technique usually are stocked in the battalion or brigade trains. Reconstitution operations should be closely associated with decon operations. The battalion's assessment and recovery team (ART) ensures that the material and equipment are available for the decon operation as part of the reconstitution effort.

DETAILED EQUIPMENT DECON

9-12. A chemical platoon must have access to large water sources—rivers, ponds, and public water systems—to conduct the DED. The chemical unit leader estimates the amounts of decontaminants that are needed. The platoon sets up the DED site, supplies the decontaminants, and conducts the DED. The platoon should carry enough decontaminants to service one company/team/battery. A decon platoon that is assigned to support a maneuver company (about 16 vehicles per tank company) should carry a minimum of 4 gallons of liquid detergent, 48 5-gallon cans of DS2, and 10 50-pound drums of STB. The chemical platoon is usually resupplied through its parent unit. Command-assignment relationships can change the resupply channels.

Table 9-1. Estimated Water Consumption for Decon

Required Equipment	Time	Water Consumption
		Operational Decon[1]
M12A1 PDDA or M17 LDS	1-3 minutes	100 to 150 gallons per regular vehicle 150 to 200 gallons per armored or larger vehicle Example: 15 (contaminated vehicles) x 150 (gallons of water) = 2,250 gallons[2]
		Detailed Equipment Decon[3]
M12A1 PDDA[4] or M17 LDS[4]	See Chapter 4 for the time allowed during a primary wash and rinse. Use the proper decon method.	Vehicles: Station 1 - primary wash, 250 gallons Station 4 - rinse, 200 gallons Armored or larger vehicle: Station 1 - primary wash, 300 gallons Station 4 - rinse, 200 gallons Example: 6 vehicles x 450 gallons of water = 2,700 gallons 4 tanks x 500 gallons of water = 2,000 gallons Total: 4,700 gallons[2]
		Detailed Troop Decon
30-gallon container		The initial setup requires 258 gallons of water. The water must be exchanged after 10 troops have been decontaminated through the DTD to avoid the transfer of contamination. Station 1 (120 gallons, 4 30-gallon containers).
3-gallon container		Station 2 (6 gallons, 2 3-gallon containers).
30-gallon container		Station 4 (180 gallons, 6 30-gallon containers).
3-gallon container		Station 7 (12 gallons, 4 3-gallon containers). Example: About 150 troops are to be decontaminated through the DTD. You will need 258 gallons of water per every 10 troops. Example: 150 (number of troops) ÷ 10 (required water exchange) = 15 (the amount of times the water will need to be exchanged) 15 x 258 (gallons of water per every 10 troops) = 3,870 gallons of water for 150 troops[2]

[1]To reduce contamination, conduct the operator's spray down before the operational decon. This process requires less water consumption during a thorough decon.

[2]Always include a 10 percent planning factor to the total estimate of the water consumption for the DED and the DTD. Example: 2,250 (gallons of water for the DTD) x 10 percent = 225 (additional gallons of water). A total of 2,250 + 225 = 2,475 (gallons of water required for the DTD).

[3]The planner should consider vehicle predecon action to estimate water consumption. He should consider vehicles that were processed through operational decon because they will normally increase the weatheriza-tion process and may reduce water usage.

[4]The M17 LDS with two wands uses 14 gallons of water per minute. The M17 LDS with only one wand, which is fitted with an injector, uses 25 gallons per minute. The M12A1 PDDA pumps 25 gallons of water per minute.

9-13. Chemical platoons attached to division engineers may be resupplied through the division engineer battalion instead of their parent company. See Table 9-2 for the decon resources that are available at each organizational level. For the equipment and supplies that are needed for decon operations, see Table 9-3, pages 9-5 through 9-7.

Table 9-2. Decon Resources Available at Each Organizational Level

Organizational Level	Decon Resources
Individual soldiers	2 SDKs 1 canteen of water 2 IEDKs
Operators and crews	2 SDKs 2 IEDKs 1 on-board decon apparatus (M11 or M13) Soap and water 2 50-pound drums STB
Companies	2 5-gallon pails of DS2 2 immersion heaters 2 to 3 30-gallon containers 6 3-gallon containers 6 long-handled brushes 6 sponges 300 plastic trash bags
Battalion PDDE crews	PDDE (M17 LDS) Basic load, liquid detergent
Chemical company decon squad	PDDE (M12A1 PDDA/M17 LDS) Basic load, liquid detergent
Chemical company decon platoon	PDDE (M12A1 PDDA/M17 LDS) Interior decon equipment Sufficient materials to set up a DTD

Table 9-3. Equipment and Supplies Needed for Decon Operations

Minimum Amounts of Equipment and Supplies Needed for Decon Techniques							Nomenclature	NSN	Class of Supply	Unit of Issue	Basis of Issue (See TOE for Actual Authorization)***
SD*	PW*	OS*	MGX**	VW**	DTD**	DED**					
		1				***	Decon apparatus, DS2, ABC-M11 or M13 DAP	4230-00-720-1618 4230-01-113-4124	II II	Each Each	1 per every major equipment
		2					Cylinder, nitrogen filled	4230-00-775-7541	II	Box	2 per can, DS2 5 per box
1	1	1	2		35	2	Decon kit, M291 SDK M295 IEDK	4230-01-0276-1905 6850-01-3577-8456	II II	Each Each	1 per mask 2 per soldier
		1					Fluid-filled container, DS2	4230-01-136-8888	II	Each	1 per M13 DAP
					1		Filter canister, C2, or filter canister, C2A1	4240-01-119-2315 4240-01-361-1319	II	Each	1 per M40-series mask
					1		Hood, M40 mask	4240-01-376-3152	II	Each	1 per mask
			1		2		Shears	5111-00-223-6371	II	Each	As needed
							Knife	5110-00-240-5943	II	Each	As needed
			***			***	Axe, single bit	5110-00-293-2336	II	Each	1 per most vehicles
			***		***	***	Shovel, hand, RD, PT, D handle	5120-00-293-3336	VII	Each	1 each for most vehicles
						***	NAAK, MK1	6505-01-140-6455	VII	Each	3 per individual
						***	CANA	6505-01-274-0951	VII	Each	1 per individual 5 per combat life saver 10 per combat medic
					***	3	Paper, chemical agent, detector, M9	6665-00-049-8982	II	Roll	1 per squad 3 per platoon

Table 9-3. Equipment and Supplies Needed for Decon Operations (Continued)

Minimum Amounts of Equipment and Supplies Needed for Decon Techniques							Nomenclature	NSN	Class of Supply	Unit of Issue	Basis of Issue (See TOE for Actual Authorization)***
SD*	PW*	OS*	MGX**	VW**	DTD**	DED**					
					***	***	Paper, chemical agent, detector, M8	6665-00-050-8529	II	Book	6 books per company
					***	***	Radiac meter, AN/PDR-2	6665-01-222-1425	VII	Each	Per MTOE
					2	***	Radiac meter, IM-93/UD	6665-00-752-7759	VII	Each	Per MTOE
					1	2	Alarm, chemical agent, M8A1 or M22	6665-00-935-6955 6665-01-438-6963	VII VII	Each Each	Per MTOE
					4	2	CAM	6665-01-199-4153	VII	Each	Per MTOE
					4		Detector kit, chemical agent, M256A1	6665-01-016-8399	II	Kit	1 per squad
					1		Mask sanitizing solution	6810-00-266-6979	III	Tube	4 per 10 masks
					**	1	Decontaminating agent, STB	6850-00-297-6653	III	Drum	2 50-pound drums per company
						***	Decontaminating agent, DS2	6850-00-753-4870	II	Can	5 gallons, 25 cans per decon
		1			2	24	Brush, scrub, long-handled	7920-00-141-5452	II	Each	As required
					7		Pail, metal, 14-quart	7240-00-160-0455	II	Each	As required
			1		9	4	Garbage can, galvanized, 30-gallon	7240-00-160-0440	II	Each	2 per company
						6	Mop	7920-00-224-8756		Each	As required
					5		Sponge, cellulose	7920-00-240-2559	II	Each	As required
			2		4	6	Brush, scrub	7920-00-240-7171	II	Each	As required
			2		2		Towels, paper	7920-00-823-6931	II	Box	As required
			1	1	1	1	Detergent, GP, liquid	7930-00-282-9699	II	Gallon	As required

Table 9-3. Equipment and Supplies Needed for Decon Operations (Continued)

Minimum Amounts of Equipment and Supplies Needed for the Decon Techniques							Nomenclature	NSN	Class of Supply	Unit of Issue	Basis of Issue (See TOE for Actual Authorization)***
SD*	PW*	OS*	MGX**	VW**	DTD**	DED**					
			1		1	1	Bag, plastic	8105-00-655-8286	II	Box	125 count
				**		10	TAP apron	8415-00-281-7813 through 8415-00-281-7816	II	Each	10 per platoon
			1		1		Suit, clothing, protective, See CTA 50-909 for clothing tariff	8415-01-137-1700 through 8415-01-137-1707	II	Each	2 per soldier
			1		1		Glove set, CP, See SB 10-523 for clothing tariff	8415-01-033-3517 through 8415-01-033-3520	II	Pair	1 per soldier
							Cover, helmet, chemical protective	8415-01-111-9028	II	Each	1 per soldier
			1		1		Overboots	8430-01-048-6305 through 8430-01-049-0887	II	Pair	1 per soldier
			2	2	2	2	NBC marking kit	9905-12-124-5955	II	Kit	1 per squad
					2		Immersion heater		II	Each	2 per company
*Techniques executed by individual soldiers							SD – Skin decon PW – Personal wipe down OS – Operator's spray down		Amount is consumption rate for 1 soldier		
**Techniques executed by units							MGX – MOPP-gear exchange VW – Vehicle washdown DTD – Detailed troop decon DED – Detailed equipment decon		Amount is consumption rate for 1 platoon (4 vehicles, 40 personnel) Amount is consumption rate for 1 company (20 vehicles, 150 personnel)		
***Amounts vary, depending on situation											

Chapter 10

Decon Procedures for Individual and Crew-Served Weapons

Decon procedures for light infantry crew-served weapons and some individual major weapons that are critical to the unit's mission are discussed in this chapter. For those units that are not authorized decon devices such as the M11 or M13 DAP, alternative measures are described. Decon procedures are specified for the following weapons: M60 machine gun; M249 squad automatic weapon (SAW); M47 Dragon; tube-launched, optically tracked, wire-guided (TOW) missile (M220 series); 81-millimeter mortar; 60-millimeter mortar; M203 grenade launcher; AT4 light antitank weapon (LAW); LAW (M72 series); and 66-millimeter rocket launcher.

OPERATIONAL-DECON PROCEDURES USING THE IEDK

10-1. Operational decon is conducted to sustain combat operations, remove gross contamination, and reduce the transfer/spread of contamination. Before decon, soldiers should visually inspect the weapon system for contamination. If liquid contamination is present, follow the procedures prescribed for each weapon system. The protection level required for operational-decon operations is MOPP4.

NOTE: The United States Marine Corp (USMC) does not possess the IEDK; therefore, it will have to use additional SDKs to decon individual and crew-served weapons.

10-2. The following procedures reduce the spread/transfer of liquid contamination. These procedures are unique to the IEDK. Using other decon equipment with these procedures could result in a malfunction of the weapons/weapon systems. Powder from the IEDK could affect the mechanical parts of the weapons/weapon systems if not used properly. See the maintenance manual for proper lubrication after decon.

> **WARNING**
> **Do not apply an IEDK to any optic lens. The abrasive effect of the charcoal inside the IEDK will damage the lens.**

M60 MACHINE GUN

NOTE: Ensure that the feed tray of the weapon is closed. This prevents the powder from the IEDK from falling inside the feed tray and jamming the weapon. Do not touch the barrel of the weapon with the chemical protective gloves when it is hot. The heat of the barrel could melt the gloves.

10-3. To decon the M60 machine gun, follow the steps below:

- Step 1. Decon your gloves using the IEDK.
- Step 2. Use the IEDK to remove all liquid contamination from the bipod, barrel assembly, forearm assembly, and carrying handle. Use a second IEDK if necessary.
- Step 3. Use the IEDK to decon the feed-tray cover, shoulder stock, and trigger assembly.
- Step 4. Repeat step 1.

NOTE: Do not decon the flash suppressor with the IEDK. Powder debris may fall into the barrel and cause the weapon to malfunction.

M249 SQUAD AUTOMATIC WEAPON

NOTE: Ensure that the feed tray of the weapon is closed. This prevents the powder from the IEDK from falling inside the feed tray and jamming the weapon. Do not touch the barrel of the weapon with chemical protective gloves when it is hot. The heat of the barrel could melt the gloves.

10-4. To decon the M249 SAW, follow the steps below:

- Step 1. Decon your gloves using the IEDK.
- Step 2. Use the IEDK to remove all liquid contamination from the bipod, barrel assembly, gas regulator, and carrying handle.
- Step 3. Use the IEDK to decon the feed-tray cover, magazine, trigger assembly, and shoulder stock.
- Step 4. Repeat step 1.

NOTE: Ensure that the weapon is upright when decontaminating the feed-tray cover with the IEDK. This prevents powder from the IEDK from falling into the ammunition feeder. Place the magazine in the well before the decon.

M203 GRENADE LAUNCHER

NOTE: The M203 is not a crew-served weapon; however, this weapon is considered critical for accomplishing the infantry squad's mission. Decon the M16A2 according to STP 21-24-SMCT. Before decontaminating the M203 grenade launcher, ensure that the dust cover is closed and the magazine is stored in the well. Ensure that the barrel assembly is closed.

10-5. To decon the M203 grenade launcher, follow the steps below:

- Step 1. Decon your gloves using the IEDK.
- Step 2. Use the IEDK to decon the barrel, hand guard, and receiver.
- Step 3. Use the IEDK to decon the lower receiver group and butt stock.
- Step 4. Repeat step 1.

M47 DRAGON

10-6. Use this procedure only when liquid contamination is present on encased missile or tracker components. If the launcher has been fired and chemical contamination is present, follow the procedures shown in steps 1, 4, and 5 of paragraph 10-7 to decon the tracker. Gather all the contaminated launching tubes in a single area before departing the contaminated area and mark them properly.

NOTE: Do not use the IEDK on any optic lens (exclude rubber eyecups), battery connector, electrical connectors, or lubricated components. Using the IEDK could cause corrosion if the powder from it is left on the equipment for an extended period of time. However, if these items are stored in a protective container/box, decon the storage container with the IEDK. Rinse or flush the lens with water only if it becomes contaminated. The rubber eyecups can be rinsed with water or decontaminated with the IEDK.

10-7. To decon the M47 Dragon, follow the steps below:

- Step 1. Decon your gloves using the IEDK.
- Step 2. Use the IEDK to decon the bipod.
- Step 3. Use the IEDK to decon the exterior of the launcher.
- Step 4. Rinse or flush any liquid contamination off the exterior of the day/night sight with water. Use a rag or any available towelettes to dry the sight but do not touch the lens.
- Step 5. Repeat step 1.

TOW MISSILE (M220 SERIES)

10-8. Gather all the contaminated launching tubes in a single area before departing the contaminated area and mark them properly.

NOTE: Do not use the IEDK on any optic lens (exclude rubber eyecups), battery connector, electrical connectors, or lubricated components. However, if these items are stored in a protective container/box, decon the storage container with the IEDK or spray it with the M11 DAP. Do not touch or decon the front lens of the night sight. Rinse or flush the lens with water if it becomes contaminated with liquid chemical agent. The rubber eyecups can be rinsed with water.

10-9. To decon the TOW missile (M220 series), follow the steps below:

- Step 1. Decon your gloves using the IEDK.
- Step 2. Use an M11 DAP filled with bleach or soapy water to spray down the launch tube, traversing unit, and tripod respectively. Protect or cover the day/night sights when spraying.
- Step 3. Use the IEDK to decon the exterior of the day-sight tracker and night-sight tracker. Avoid touching the lens. Once the sights are decontaminated, brush off any powder residue.
- Step 4. Use the IEDK to decon the collimator.
- Step 5. Use the IEDK to decon the exterior of the missile-guidance set and cable surface. Ensure that the electrical connectors are covered before the decon.
- Step 6. Use the IEDK to decon the power conditioner and its cable for the battery component.
- Step 7. Repeat step 1.

AT4 LIGHT ANTITANK WEAPON

NOTE: Ensure that the transport safety pin, the cocked-level safety, and the forward safety on the launcher are engaged.

10-10. To decon the AT4 LAW, follow the steps below:

- Step 1. Decon your gloves using the IEDK.
- Step 2. Use the IEDK to wipe the exterior of the launcher and the front and rear sights.
- Step 3. Repeat step 1.

66-MILLIMETER ROCKET LAUNCHER

NOTE: Ensure that the launcher is in the safe position and is not loaded before decontaminating. Omit step 2 if the launcher has not been fired in a contaminated environment.

10-11. To decon the 66-millimeter rocket launcher, follow the steps below:

- Step 1. Decon your gloves using the IEDK.
- Step 2. Extend the launcher into the firing position.
- Step 3. Use the IEDK to decon the exterior of the launcher and around the reflecting sight.
- Step 4. Repeat step 3 using the IEDK.
- Step 5. Repeat step 1.

81-MILLIMETER MORTAR

NOTE: Following the decon of the mortar with the IEDK, ensure that the mechanical parts of the mortar are dried. Lubricate the mortar when the mission permits. Do not touch the mortar tube with chemical protective gloves when it is hot. The heat of the tube could melt the gloves. Do not apply the IEDK to any optic lens due to the abrasive effect of the charcoal in the IEDK. Rinse or flush the lens with water, when required.

10-12. To decon the 81-millimeter mortar, follow the steps below:

- Step 1. Decon your gloves using the IEDK.
- Step 2. Use the IEDK to decon from the blast attenuator device to the lower half of the mortar tube.
- Step 3. Use the IEDK to decon the base plate.
- Step 4. Use the IEDK to decon the bipod.
- Step 5. Use the IEDK to decon the elbow telescope and telescope mount. Do not use the IEDK on the optic lens. Flush the lens with water if contamination is present.
- Step 6. Repeat step 1.

60-MILLIMETER MORTAR

NOTE: Following the decon of the mortar with the IEDK, ensure that the mechanical parts of the mortar are dried. Lubricate the mortar when the mission permits. Do not touch the mortar tube with chemical protective gloves when it is hot. The heat of the tube could

melt the gloves. Do not apply the IEDK to any optic lens due to the abrasive effect of the charcoal in the IEDK. Rinse or flush the lens with water, when required.

10-13. To decon the 60-millimeter mortar, follow the steps below:

- Step 1. Decon your gloves using the IEDK.
- Step 2. Use the IEDK to decon the mortar tube and handgrip. If the mortar is heavily contaminated, you may need to use another IEDK.
- Step 3. Use the IEDK to decon the bipod and base plate.
- Step 4. Use the IEDK to decon the exterior of the telescope. Do not apply the IEDK to the lens. Flush the lens with water if contamination is present.
- Step 5. Repeat step 1.

JAVELIN

10-14. The decon procedures for the Javelin and its components are listed in Table 10-1, page 10-6.

WARNING

The IEDK only removes a liquid hazard. Decontaminated items may still present a vapor hazard. Do not unmask until it has been determined safe to do so.

CAUTION

Close the lens covers on the Javelin before decontaminating because decon powders on the lens can obscure your vision. Do not use the IEDK on the eyecup because the decon powder entering the eyecup can obscure your vision.

THOROUGH-DECON PROCEDURES USING THE HTH SOLUTION

10-15. When conducting the DED, the equipment is decontaminated with a 10 percent HTH solution (see Table 10-2, page 10-6) or with pure bleach that is diluted. Use the M256A1 detector kit, CAM, or M8 detector paper to determine the effectiveness of the decon. The crew-served weapons should be decontaminated before the DTD by the individual weapons' crew (the preferred method). Alternatively, the commander could task his own decon team to decon the unit's light weapons. The contaminated unit is responsible for establishing an area about 109 to 164 yards upwind from the DTD and DED sites for decon of the crew-served weapons. Unit leaders supervise this procedure. The unit chemical NCO plans and coordinates the preparation of this area.

10-16. The preferred decontaminant for a thorough decon is a 10 percent HTH solution (see Table 10-2). Use a contact time for the HTH solution of 15 minutes for chemical and biological agents. Prepare the amount of solution required according to the number and size of the weapon systems that need to be decontaminated.

Table 10-1. Decon Procedures for the Javelin and Its Components

Round	Command Launch Unit (CLU)
Immediate Decon	
Step 1. Decon your gloves using the IEDK. Step 2. Decon the round by patting it with the IEDK, working from the front to the rear until the entire round is covered. **NOTE: When the shoulder strap is grossly contaminated, discard it.** Step 3. Repeat step 1. Step 4. Discard the IEDK.	Step 1. Decon your gloves using the IEDK. Step 2. Decon the CLU by patting it with the IEDK. Pay special attention to the handgrips, CLU carrying handle, battery cover, round interface-connector protective cover, and switch boots. Step 3. Use a knife to cut the protective-cover lanyard from the protective cover and remove it. Step 4. Repeat step 1. Step 5. Discard the IEDK.
Thorough Decon	
Predecon Round	**Predecon CLU**
Step 1. Check for contamination. Step 2. Remove the shoulder strap. Step 3. Use a knife to cut the connector boot. Step 4. Use a knife to cut the protective-cover lanyard from the protective cover and remove it. Step 5. Use a knife to scrape about 1/4 inch of the shoulder pad away from the launcher-tube assembly (LTA). Step 6. Ensure that the protective cover is installed on the round/CLU interface connector.	**NOTE: Ensure that the on/off switch is in the off position. If the connector was not covered before contamination, place the protective cover on the connector before decon. After the decon, remove the protective cover and allow the connector to weather. The foam around the connector may be cut away as required.** Step 1. Check for contamination. Step 2. Remove the rubber from the handles and remove the lanyards from the CLU. Step 3. Ensure that the protective covers are installed on the connectors.
Decon LTA	**Decon CLU**
Step 1. Submerge one half of the LTA in the decon solution. Step 2. Submerge the other half of the LTA in the decon solution. Step 3. Rinse one half of the LTA by submerging it in the rinse solution. Step 4. Rinse the other half of the LTA by submerging it in the rinse solution. Step 5. Use the CAM to check the LTA, ensuring that all contaminants have been removed. Step 6. Repeat steps 1 through 5 above if the LTA is still contaminated.	Step 1. Open the lens covers. Step 2. Submerge the CLU in the decon solution. Step 3. Rinse the CLU by submerging it in the rinse solution. Step 4. Use the CAM to check the CLU, ensuring that all contaminants have been removed. Step 5. Repeat steps 1 through 4 above if the CLU is still contaminated. If only the switch boots remain contaminated, remove them by using a knife.

Table 10-2. HTH Mixture

HTH Solution	Amount of HTH per 1 Gallon of Water	Amount of HTH per 5 Gallons of Water
5 percent	9.6 ounces	48 ounces
10 percent	12 ounces	60 ounces

NOTE: Refer to Appendix B for instruction on how to use sodium hypochlorite (household bleach) and HTH, and see Table 10-2 for instructions on how to prepare the 5 and 10 percent HTH solution.

10-17. Follow the procedures below during the DED and DTD. Always wear chemical gloves when performing these procedures.

- Use the HTH solution to decon up to 10 weapon systems. Use the required number of rags on a weapon system and then dispose of them properly to avoid transferring the contamination.
- Do not apply the HTH solution or bleach on electrical connectors and lens.
- Use bleach or a 10 percent HTH solution on the exterior of the optic systems that are not sensitive to bleach and then rinse them with water.

10-18. Once the HTH solution has been used on the weapon systems, thoroughly dry and lubricant them. Avoid spilling the HTH solution or bleach on the chemical overgarment.

MATERIALS

10-19. Listed in the following paragraphs are the different types of materials that will be needed to clean the weapon systems:

Cleaning Rags

10-20. The recommended number of clean rags to use with the HTH solution on each weapon system follows:

- M60 machine gun/M249 SAW (three rags)
  - 1: Bipod and barrel.
  - 1: Forearm, carrying handle, and rear sight.
  - 1: Feed-tray cover, shoulder stock, pistol grip, and trigger assembly.
- M203 grenade launcher (two rags)
  - 1: M16A1/2 barrel, hand guard, receiver group, and butt stock.
  - 1: M203 receiver group, sight, and exterior of the barrel.
- M47 Dragon (three rags)
  - 1: Exterior of the sight.
  - 1: Bipod (if weapon is not fired).
  - 1: Round (canister) (if weapon is not fired).

NOTE: Keep the electrical connector covered before the decon. The Dragon's rubber eyepiece can be removed and submerged in the decontaminant.

- TOW missile (M220 series) (seven rags)
  - 1: Exterior of the night sight.
  - 1: Exterior of the day sight.
  - 1: Launch tube.
  - 1: Traversing unit.
  - 1: Tripod assembly.
  - 1: Battery-power conditioner and cable. Ensure that the battery is in a waterproof container or case.
  - 1: Missile-guidance set and its cable.

NOTE: Lock the bridge clamp after removing the round. Do not get the HTH solution on the electrical connector under the bridge clamp or on the post-amplifier cable.

- 66-millimeter rocket launcher (one rag)
- AT4 LAW (one rag)
- 81-millimeter mortar (four rags)
  - 1: Blast denature device and tube.
  - 1: Bipod.
  - 1: Base plate.
  - 1: Exterior of the telescope.
- 60-millimeter mortar (three rags)
  - 1: Tube (barrel) and handgrip.
  - 1: Base plate and bipod.
  - 1: Exterior of the telescope.

Water

3-21. Use five gallons of water (as a minimum) for the HTH mix. Change the solution after 10 weapons are decontaminated.

Buckets

3-22. Use a minimum of two buckets for the procedures below.

Trash Bags

3-23. Dispose of the contaminated rags in plastic trash bags.

NBC Marking Kit

3-24. Mark the contaminated area once the decon procedures are completed.

Lubricants

3-25. Use the required lubricants for the weapon systems. Follow the procedures specified in the applicable TMs.

PROCEDURES

3-26. The procedures on how to decon weapons and weapon systems with the HTH solution are described below:

- Step 1. Submerge the rag or cloth in the HTH solution and ensure that the material is completely dampened.
- Step 2. Wipe down the weapons and weapon systems with the rag until the contamination is no longer suspected.
- Step 3. Submerge the rag or cloth into a container of clean water (use hot water when available) and wipe down the weapons and weapon systems once more to remove the HTH or bleach solution.
- Step 4. Completely dry the weapons and weapon systems once step 3 is completed.
- Step 5. Lubricate the weapons and weapon systems.

Appendix A

Conversions and Measurements

This appendix contains an English/Metric conversion chart (see Table A-1). It also contains the measurements and weights of decontaminant containers that are replacements for used containers (see Table A-2, page A-2).

Table A-1. English/Metric Conversion Chart

To Convert	To	Multiply By	To Convert	To	Multiply By
Feet	Centimeters Miles Kilometers Meters	30.48 0.00019 0.00003 0.3048	Gallons	Milliliters Cubic meters	3,785 0.00379
Cubic yards	Cubic meters	0.76456	Quarts (fluid)	Millimeters Liters	946.358 0.946333
Cubic feet	Cubic meters Liters	0.02831702 28.32	Ounces	Grains Drams Pounds Grams Kilograms	437.6 16 0.0625 28.34952 0.02835
Yards	Meters	0.9144	Pounds	Grams Kilograms Metric tons	453.6 0.453592 0.0004536
Inches	Centimeter Kilometers Millimeters	2.54000508 $2.54000508 \times 10^{-5}$ 25.4000508	Ounces (fluid)	Milliliters Liters	29.57353 0.02957
			Square yards	Square meters	0.836

NOTE: For temperatures, Celsius = 5/9 (F° - 32°) and Fahrenheit = 9/5 (C° + 32°)

Table A-2. Measurements and Weights of Decontaminant Containers

Containers	Measurements	Weights
Drum, 55 gallons, 16 gauge NSN 8110-00-597-2353	Volume—12 cubic feet Length—35 inches Width—27.5 inches	Empty—70 pounds Filled with water—526 pounds Filled with STB (slurry)—620 pounds Filled with DS2—507 pounds
Drum, 55 gallons, 18 gauge NSN 8110-00-292-9783	Volume—12 cubic feet Length—35 inches Width—27.5 inches	Empty—50 pounds Filled with water—506 pounds Filled with STB (slurry)—600 pounds Filled with DS2—487 pounds
Ash and garbage can, 32 gallons (GI can) NSN 7240-00-160-0440	Volume—7 cubic feet Length—26.5 inches Width—20 inches	Empty—33 pounds Filled with water—300 pounds Filled with STB (slurry)—353 pounds Filled with DS2—288 pounds
Gasoline can, 5 gallons NSN 7240-00-178-8286	Volume—1 cubic foot Length—18.50 inches Width—6.75 inches	Empty—10.5 pounds Filled with water—52 pounds Filled with STB (slurry)—66 pounds Filled with DS2—50 pounds
Pail, 5 gallons, DS2 NSN 7240-00-575-2243	Volume—1cubic foot Length—13.75 inches Width—11.12 inches	Empty—6 pounds Filled with water—48 pounds Filled with STB (slurry)—61 pounds Filled with DS2—46 pounds
Drum, 8 gallons, STB, 16 gauge NSN 6850-00-297-6693	Volume—1.4 cubic feet Length—14 inches Width—14 inches	Empty—11 pounds Filled with water—78 pounds Filled with STB (slurry)—91 pounds Filled with DS2—71 pounds

Appendix B

Decontaminants

The decontaminants listed in Tables B-1 through B-3, pages B-1 through B-14, are grouped according to their classification: standard, nonstandard, or natural. Each is identified by name and national stock number (NSN) (where applicable). Also listed are brief directions on preparing the decontaminants and any appropriate cautions and information on application, coverage, substitute solutions, and related effectiveness of the decontaminants.

Table B-1. Standard Decontaminants Available in the Supply System

Decontaminants	Use	Remarks	Cautions	Preparation
Decontaminating solution number 2; DS2 in a $1^1/_3$-quart can (NSN 6850-00-753-4827); DS2 in a 14-liter container for the M13 DAP; (NSN 4230-01-136-8888); DS2 in a 5-gallon pail (NSN 6850-00-753-4870)	Bio Chem	Is effective against all known toxic chemical and biological agents (except bacterial spores) if sufficient contact time is allowed. Must remain in contact with contaminated surface for about 30 minutes. Must be rinsed off with water. Recheck the item for contamination. Can be used at temperatures above -15°F. Is used with the M11 or M13 DAP or can be applied with brooms and swabs. Is most effective when the application is accompanied by scrubbing action.	Is extremely irritating to the eyes and skin. Must wear protective mask and rubber gloves. If DS2 contacts skin, wash area with water. Do not inhale vapors. Will cause a green or black color change upon contact with M8 detector paper and cause a false/positive reading with M9 detector paper. Will ignite spontaneously on contact with STB or HTH. Should not be spilled on the chemical protective overgarment. Is a combustible liquid with a flash point of 160°F. Do not confuse it with a fire extinguisher. Spraying DS2 on surfaces above 168°F will ignite it. Must not be used on individual mask (damages Mylar diaphragm in voicemitter assembly).	Mixing is not required. Issued in ready-to-use solutions.

Table B-1. Standard Decontaminants Available in the Supply System (Continued)

Decontaminants	Use	Remarks	Cautions	Preparation
Decontaminating solution number 2 (continued)			Will corrode aluminum, cadmium, tin, and zinc; will soften leather. May soften, remove, or discolor paint. Rinse well after use and oil metal surfaces. Is ineffective against bacterial spores.	
Supertropical bleach (NSN 6850-00-297-6653)	Bio Chem	Is effective against lewisite, V and G agents, and biological agents. Must remain in contact with contaminated surface for at least 30 minutes; then, wash off with clear water. Does not effectively decon mustard if it has solidified at low temperatures. Should be applied several times to porous surfaces.	Will ignite spontaneously on contact with liquid blister agent or DS2. Will give off toxic vapors on contact with G agent. Is not recommended for ship use. Store on top deck only. Is corrosive to most metals and damaging to most fabrics (rinse thoroughly and oil metal surfaces). Should not be inhaled or allowed to touch the skin. When preparing a slurry, wear a protective mask or respiratory protective device. Should be stored in an unheated warehouse away from combustibles and metals subject to corrosion.	Slurry paste: mix one 50-pound drum of STB with 6 gallons of water. Slurry paste consists of about equal parts (by weight) of STB and water. Dry mix: mix two shovels of STB to three shovels of earth or inert material (ashes). Slurry mix, chemical: mix will consist of 40 parts of STB to 60 parts of water (by weight). To mix in the M12A1 PDDA, use 1,300 pounds of STB, 225 gallons of water, 12½ pounds of antiset, 24 ounces of antifoam. Slurry mix, biological: mix will consist of 7 parts of STB to 93 parts of water (by weight). To mix in the M12A1 PDDA, use 150 pounds of STB, 225 gallons of water, 12½ pounds of antiset, and 24 ounces of antifoam. Camouflage: lampblack or dye mixes may be added for camouflage.

Table B-1. Standard Decontaminants Available in the Supply System (Continued)

Decontaminants	Use	Remarks	Cautions	Preparation
Mask sanitizing solution	Bio Chem	Use on a previously cleaned mask with filter elements/canisters removed. Place the mask face up; attach the canteen to the mask at the drinking tube. Drain one canteen full of sanitizing solution through the mask. Rinse the mask with two canteens of clear water. Immerse the mask and outserts in the sanitizing solution. Agitate the mask for 5 minutes. Rinse it twice in clear water, agitating 2 to 3 minutes each time. Dry all parts of the mask and reassemble. Use one gallon of solution for every 10 masks.		Fill a standard plastic canteen to the shoulder with water. Add a 0.5-gram tube of calcium hypochlorite from the water-purification kit (NSN 6810-00-266-6976). Cover the canteen and shake vigorously for 30 seconds. Mix bulk quantities as follows: add 2.0 grams of calcium hypochlorite from a 6-ounce jar (NSN 6810-00-255-0471) to 1 gallon of water. Use a ratio of about 1 pound of soap per gallon of water for smaller amounts of solution. Mix 2 pints of detergent to 450 gallons of water in the M12A1 PDDA.
Soaps and detergents: detergent, GP, liquid (NSN 7930-00-282-9699)	Rad Bio Chem	Scrub or wipe the contaminated surface with a hot, soapy water solution or immerse the item in the solution.	Are effective in physically removing contamination. However, casualty-producing levels of contamination may remain in the runoff water and must be considered contaminated.	Mix 75 pounds of powdered soap in 350 gallons of water. If powdered soap is not available, use bar laundry soap (75 pounds of soap cut into 1-inch pieces and dissolved in 350 gallons of hot water). Use a ratio of about 1 pound of soap per gallon of water for smaller amounts of soap solution. Mix 2 pints of detergent to 450 gallons of water in an M12A1 PDDA.

Table B-1. Standard Decontaminants Available in the Supply System (Continued)

Decontaminants	Use	Remarks	Cautions	Preparation
Calcium hypochlorite (high-test hypochlorite [HTH] or high-test bleach [HTB]) (NSN 6810-01-225-2682) (25 pounds), (NSN 6810-00-225-0472) (100 pounds)	Bio Chem	Is effective against mustards, lewisite, V agents, and all biological agents, including bacterial spores. Will react rapidly (within 5 minutes) with mustards and lewisite. Should be allowed a 15-minute contact time for biological agents. Will act faster than STB. Can be used as a dry mix or slurry.	Precautions are the same as for STB. Pure calcium hypochlorite will burn on contact with VX, HD, or DS2. Agent is more corrosive than STB. Will destroy clothing, has a toxic vapor, and will burn the skin. Protective mask and rubber gloves are the minimum protective equipment needed when handling.	Chemical: mix 5 pounds of decontaminant to 6 gallons of water (10 percent solution). Biological: mix 1 pound of decontaminant to 6 gallons of water (2 percent solution). PDDE: mix a slurry of 1 part decontaminant to 2 parts water (any heavier slurry will clog the decon apparatus). HTH should only be used if STB is not available.

Table B-2. Nonstandard Decontaminants Available in the Supply System

Decontaminants	Use	Remarks	Cautions	Preparation
Oxidizing agents (nitric acid, aqua regia, sodium dichromate, and potassium permanganate)	Rad	Are effective in dissolving surfaces containing absorbed radioactive contamination. Are applied to a surface or the item is dipped. Exposure must be limited due to the corrosive nature of the solution. Rinse thoroughly with water and detergent and then with clear water.	Are extremely corrosive. Use only under the supervision of a trained individual. Will require the use of a neoprene or rubber protective apron, gloves, boots, and safety glasses when handling. (Rubber offers only limited protection.)	Aqua regia is prepared by mixing 3 parts of concentrated hydrochloric acid and 1 part of concentrated nitric acid. Other oxidizing agents do not require mixing.
Iodine water purification tablets	Bio	Are used when it is impractical to boil drinking water. Two iodine tablets per canteen are effective against most biological agents.	None	None

Table B-2. Nonstandard Decontaminants Available in the Supply System (Continued)

Decontaminants	Use	Remarks	Cautions	Preparation
Complexing (chelating) agents (versene, citric acid, sequesterene, sodium citrate, tartanic acid, sodium oxalate, sodium tartrategoxalic acid, orthophosphoric acid, and similar agents)	Rad	Will aid in the physical removal of absorbed contamination on surfaces. Should be applied as a film over the surface using the PDDE, fire-fighting apparatus, or tree or garden sprayer. Should be allowed a 30-minute contact time and then flushed with water.	Do not neutralize contamination. Runoff will be contaminated.	Mix 3 to 5 percent of the agent (by weight) in water.
Disinfectant chlorine (NSN 6840-00-270-8172)	Bio	Use to decon utensils, mess gear, exteriors of sealed containers, and food products that can withstand soaking. Dispose of any food or vegetables that are damaged and any outer leaves that are bruised or torn. Do not cut or peel fruits and vegetables before disinfecting them. Leave items in the solution for 30 minutes and stir occasionally. Rinse thoroughly in potable water.	None	Dissolve one package of disinfectant in 20 gallons of warm potable water (100°F).

Table B-2. Nonstandard Decontaminants Available in the Supply System (Continued)

Decontaminants	Use	Remarks	Cautions	Preparation
Disinfectant chlorine (continued)	Bio	Use the solution only once. Prepare an emergency solution by mixing one level MRE spoonful of calcium hypochlorite (water disinfecting powder) to each 10 gallons of water. If liquid chlorine bleach is available, use about 1/3-canteen cup of 5 percent chlorine bleach to each 10 gallons of water. Make fresh solutions for rinsing and disinfecting utensils for each 100 persons.	None	
Ethylene oxide	Bio	Is effective against all microorganisms, to include bacterial spores. Should be applied in the strength of 30 pounds for every 1,000 cubic feet. Should be allowed a 6-hour contact time (contact time must be doubled for each 20°F drop in temperature below 75°F). Should be used in an airtight enclosure.	Is flammable. Not recommended for interior use.	None
Formalin (formaldehyde)	Bio	Agent is effective against all microorganisms, to include bacterial spores. Agent is used for interior decon of relatively close areas. Vapors should remain 16 hours in a closed structure; then, aerate until the odor is no longer objectionable.	Vapors are very toxic. Agent will curl and discolor paper. Leaves a white residue. A self-containing breathing apparatus is required in closed spaces. Personnel handling or spraying the agent should wear impermeable protective clothing.	No mixing is required. However, less residue remains and less aeration is required if the mixture of 5 parts formalin and 3 parts methanol is used. (Use this mixture at a rate of 4/5 quart per 1,000 cubic feet of space).

Table B-2. Nonstandard Decontaminants Available in the Supply System (Continued)

Decontaminants	Use	Remarks	Cautions	Preparation
Formalin (formaldehyde) (continued)	Bio	Optimum conditions for spraying formalin are 70° to 80°F with an 85 percent relative humidity. The minimum effective relative humidity is 70 percent. At 60°F, the exposure should be increased to 24 hours. Agent is applied as a vapor from standard insecticide sprayers or is vaporized by heat or a bubbling steam from a pan.	Personnel entering an area containing formalin vapors should— • Wear a protective mask. • Wear washable outer clothing, fastened to prevent vapors from entering at wrists, ankles, or neck. • Remove outer clothing after emerging from vapors. • Shower and put on clean clothing as soon as possible. Vapors are not flammable; open flame should not be used for vaporizing when methanol has been added to the agent. When steam is used, the source of the steam should be outside the area being decontaminated.	
Detrochlorite	Bio	Agent is a thickened bleach that is very useful on vertical surfaces. Coverage is 1 gallon per 8 square yards. Apply by means of the PDDA. Allow 30 minutes contact time, then rinse with water.	Agent is very corrosive. See *Preparation* column.	Mix by weight 19.3 percent diatomaceous earth, 0.5 percent anionic wetting agent, 2.9 percent calcium hypochlorite (70 percent available chlorine), and 77.3 percent water. Mix wetting agent and diatomaceous earth with water before adding the calcium hypochlorite. Mixing the wetting agent and calcium hypochlorite in a dry and undiluted state may cause an explosion.

Table B-2. Nonstandard Decontaminants Available in the Supply System (Continued)

Decontaminants	Use	Remarks	Cautions	Preparation
Peracetic acid (PAA)	Bio	Agent is effective against all microorganisms, to include bacterial spores. A contact time of 10 minutes must be allowed. Item should be wiped with a rag or swab. Remove excess acid and aerate for 10 to 15 minutes or until no objectionable odor remains. Small items should be immersed for 10 minutes. Remove excess acid and aerate until no objectionable odor remains.	Fumes are highly irritating. Prolonged exposure will damage most material. A 40 percent solution has a low flash point (105°F); a 3 percent solution is nonflammable. Burns and blistered on the skin will occur. Agent must be stored in original containers under refrigeration to prevent decomposition. Protective mask and clothing are required. A violent explosion may result if heavy metal ions come in contact with the agent. Prolonged exposure will corrode iron and deteriorate rubber, plastic, and leather.	Available as a 40 percent solution. Mix 1 quart of PAA to 3½ gallons of water (add PAA to the water).
Hyamine (benzethonium chloride)	Bio	Agent is effective against all bacteria. A contact time of 5 to 30 minutes should be allowed.	Agent is very toxic. Estimated fatal dose to man is 1 to 3 grams. Care should be taken when mixing to avoid inhalation of powder. Agent is not to be used on aircraft or ships.	Use a 0.1 to 1 percent solution (1 pound of hyamine for every 12 gallons of water yields a 1 percent solution).
Carbon dioxide and ethylene oxide mixture (with more than 87 percent ethylene oxide)	Bio	Should be applied in the strength of 30 pounds for every 1,000 cubic feet. Should be allowed a 12-hour contact time (doubled for each 20°F drop in temperature below 75°F). Is nonflammable and is recommended for interior use. Should be used in an airtight enclosure.	Will blister the skin. Aerate items next to the skin 18 to 24 hours.	None

Table B-2. Nonstandard Decontaminants Available in the Supply System (Continued)

Decontaminants	Use	Remarks	Cautions	Preparation
Sodium hypochlorite solution (household bleach)	Bio Chem	Is effective against blister and V agents and all biological agents. Will react rapidly (within 5 minutes) with blister and V agents. Should be allowed a 10- to 15-minute contact time for biological agents. Should be applied undiluted with brooms, brushes, or swabs. Is the preferred decon for ship use. A 5:1 concentration is recommended. Has a limited storage problem.	Is harmful to the skin and clothing if undiluted. Remove from the skin and clothing by flushing with water. Is corrosive to metals unless rinsed, dried, and lubricated after decon. Should be stored in a cool place.	For chemical decon, no mixing is required. For biological decon, dilute by adding 2 parts bleach to 10 parts water. For decon of cotton clothing and utensils, dilute 2 cups of bleach to 1 gallon of water. For application, dilute half and half with water and spray from the PDDE.
2-Propanone (acetone)	Chem	Freezing point is -203°F; boiling point is 133°F (evaporates rapidly). Good decontaminant for use in arctic regions. Commonly obtained as fingernail polish remover or paint thinners. Scrubbing increases its effectiveness.	Is extremely flammable. Does not neutralize agents. Is effective for dissolving and flushing agent by physically removing it.	None
Dietyl ether	Chem	Freezing point is -241°F; boiling point is 93°F. Good decontaminant for use in arctic regions. Available through medical supply facilities. Scrubbing increases its effectiveness.	Is the same as for 2-propanone.	None

Table B-2. Nonstandard Decontaminants Available in the Supply System (Continued)

Decontaminants	Use	Remarks	Cautions	Preparation
Ethylene glycol	Chem	Scrub on contaminated surfaces and rinse thoroughly.	Removes contamination, but does not neutralize it. Therefore, runoff residue must be considered contaminated.	Mix equal amounts of solution and water.
Solvents (gasoline, JP-4, diesel fuel, kerosene, and similar solvents)	Chem	Scrub on contaminated surfaces and rinse thoroughly.	Is the same as for ethylene glycol. May damage materials such as rubber and plastic.	None
Sodium hydroxide (caustic soda or lye) (NSN 6810-00-174-6581) (100 pounds)	Bio Chem	Is effective against G agents, lewisite, and all biological agents, including bacterial spores. Will neutralize G agents on contact. Should be allowed to remain in contact with the chemically contaminated surface for about 15 minutes.	Agent will damage the skin, eyes, and clothes and can cause upper respiratory or lung damage if inhaled. Full rubber protective clothing, gloves, boots, and mask are required when using. Affected area must be washed immediately with large amounts of water and flushed with diluted acetic acid or vinegar. Remove affected clothing. If eyes are involved, flush them at once with large amounts of warm water and seek medical attention. Runoff from decon operations is highly corrosive and toxic. Drain runoff into a sump and bury. Equipment must be flushed with large amounts of clear water.	Small amount—10 pounds of lye to 12 gallons of water (10 percent solution). Mix in an iron or steel container (never aluminum, zinc, or tin). Add lye to the water to prevent boiling and splattering due to heat being emitted. Do not handle mixing container with bare hands. Large amount (PDDE use)—Prepare a solution of 227 grams (½ pound) of lye for each gallon of water. Pump 350 gallons of water into the tank unit. Connect the tank unit, pump unit, and heater together. Heat the water to 122°F. Disconnect the heater unit and add 175 pounds of lye to the heated water. Circulate the solution with the pump unit until all the lye is dissolved. The temperature will increase noticeably. Use while hot.

Table B-2. Nonstandard Decontaminants Available in the Supply System (Continued)

Decontaminants	Use	Remarks	Cautions	Preparation
Sodium hydroxide (caustic soda or lye) (NSN 6810-00-174-6581) (100 pounds) (continued)			Sodium hydroxide— • Is not recommended for ship use. Store on top deck only. • Is corrosive to most metals. • Is not recommended if less toxic caustic decontaminants are available. Will cause a red color change upon contact with M8 detector paper. Agent's effectiveness is directly proportional to the strength of the solution.	Simultaneous mixing and applying—Sprinkle dry lye on the contaminated area and then dissolve it with a spray of steam or hot water. Do not wash the lye off the surface while applying the steam or hot water. Paint removal—1 pound of lye per 2½ gallons of water is capable of removing an average coat of paint from about 11 square yards of surface. This solution is effective in removing paint on which chemical contamination has absorbed. (Can substitute calcium hydroxide, potassium hydroxide, or trisodium phosphate for sodium hydroxide.)
Sodium carbonate (washing soda, soda ash, sal soda, or laundry soda)	Chem	Is effective against G agents and CN. Will react rapidly with G agents, normally within 5 minutes. Is the preferred decontaminant for ship use. The recommended concentration is 5 percent by weight. There is no storage problem. Should be used with a hot solution to decon CN effectively.	Should not be used for VX. It cannot detoxify VX and creates extremely toxic by-products. Does not dissolve mustard agents or detoxify them.	Mix 10 pounds of washing soda to 12 gallons of water (10 percent solution).
Potassium hydroxide (caustic potash)	Bio Chem	Is the same as for sodium hydroxide.	Is the same as for sodium hydroxide.	Is the same as for sodium hydroxide.
Hexachloramelamine	Chem	Is effective against mustard agents.	May require the use of a protective mask and rubber gloves when used. Is corrosive to metal.	Is not soluble in water but is soluble in organic solvents such as gasoline, kerosene, and paint thinner.

Table B-2. Nonstandard Decontaminants Available in the Supply System (Continued)

Decontaminants	Use	Remarks	Cautions	Preparation
Ammonia or ammonium hydroxide (household ammonia)	Chem	Is effective against G agents. Is slower acting than sodium hydroxide or potassium hydroxide.	May require the use of a self-contained breathing apparatus or special purpose mask.	Ammonium hydroxide is a water solution of ammonia. No further mixing is required.
Dichloramine-B and Dichloramine-T	Chem	Is effective against mustard agents.	May require the use of a protective mask and rubber gloves when used. Is corrosive to metal.	Is not soluble in water, but is soluble in certain organic solvents. Is normally mixed as a 10 percent solution in dichloroethane.
Perchloroethylene (tetrachloroethylene)	Chem	Freezing point is -8°F; boiling point is 250°F. Agent— • Is good for use in arctic climates. • Is a nonflammable, synthetic solvent widely used in dry cleaning plants. • Dissolves H and V agents but not G. • Has a low toxicity. Scrubbing increases its effectiveness.	Physically dissolves and removes contamination, but does not neutralize it.	Requires no mixing (practically insoluble in water).
Acids (sulfuric acid, hydrochloric acid, oxalic acid, and similar acids)	Rad	Are effective solvents for rust and mineral deposits holding radioactive material on metal surfaces. Should be allowed a 1-hour contact time. Must be flushed with water, scrubbed with a water-detergent solution, and flushed again with water.	Are difficult to handle and are harmful to the body, especially the eyes. Can produce boiling and splattering of the solution when mixed. May require the use of rubber boots, gloves, aprons, and goggles when used. Will require the use of respiratory protection when used in closed areas. Can cause serious injury when they come in contact with the body. Flush the area immediately with water. Use a 5 percent solution of water and baking soda (sodium bicarbonate).	None

Table B-3. Natural Decontaminants

Decontaminants	Use	Remarks	Cautions	Preparation
Water	Rad Bio Chem	Can be used to flush contamination from surfaces.	Is effective in physically removing contamination, but does not neutralize it. Do not use water on lewisite.	Hot water with soap makes water more effective in removing agents. Boiling for 15 minutes (30 minutes at high altitude) destroys biological agents.
Steam	Rad Bio Chem	Is more effective when accompanied by scrubbing.	Is effective in physically removing contamination, but does not neutralize it.	None
Absorbents (earth, sawdust, ashes, rags, and similar materials)	Chem	Are used to remove gross contamination, physically, from surfaces.	The contamination is transferred from the surface to the absorbent. The absorbent becomes contaminated and must be disposed of accordingly. Sufficient contamination to produce casualties may well remain on surfaces.	
Sealants (concrete, asphalt, earth, paint, and similar materials)	Rad Bio Chem	Sealants are used to physically seal in or shield contamination. They are effective as follows: • 12 inches of earth provides a good protection from fallout (3 inches will reduce the dose rate about one half). • 1 inch of asphalt or concrete completely absorbs alpha and beta radiation.	A break in the surface of the sealant will expose the contamination. Contaminated areas covered with sealants must be marked with appropriate NBC warning signs.	None

Figure B-3. Natural Decontaminants (Continued)

Decontaminants	Use	Remarks	Cautions	Preparation
Sealants (concrete, asphalt, earth, paint, and similar materials) (continued)	Rad Bio Chem	• 1/4 inch of grout shields alpha and bata radiation. • 4 inches of earth provides good protection from chemical contamination. Burying items contaminated with biological agents is an effective means of sealing off contamination.	A break in the surface of the sealant will expose the contamination. Contaminated areas covered with sealants must be marked with appropriate NBC warning signs.	None
Weather/time	Rad Bio Chem	UV light kills most bio-organism agents, organisms, and radiation decay overtime.	Should be used when time and the mission permits.	None
Burning	Bio Chem	Must be approved by the first general officer in the chain of command.	Creates downwind hazards. Requires that sentries be posted to keep people out of the danger area.	None

Appendix C

Decon of Specific Surfaces and Materials

Table C-1, pages C-1 through C-7, lists specific surfaces or materials and explains briefly how to best decon each for chemical, biological, or radiological contamination.

Table C-1. Decon Procedures for Specific Surfaces and Materials

Surface or Material	Types of Contamination and How to Decon		
	Chemical	**Biological**	**Radiological**
Asphalt roads (applicable to small vital areas only)	Flush with water. Spray with an STB slurry from the PDDE. Cover with STB (pure form). When liquid contamination is visible and personnel are nearby, use the STB dry mix. Weather. Cover small areas or paths across roads with 4 inches of earth.	Weather (remain masked). Wet with water (will help prevent secondary aerosols but does not decon). Apply 2 percent household-bleach solution. Spray with an STB slurry from the PDDE. Pour, spray, or spread oil on the surface (will help prevent secondary aerosols, but does not decon).	Brush or sweep. Flush with water (this may drive some of the contamination into the surface; waste must be controlled). Clean with a vacuum.
Roofs	Follow the same directions as for asphalt roads.	Follow the same directions as for asphalt roads. Apply detrochlorite; leave on at least 30 minutes, then flush with water.	Follow the same directions as for asphalt roads.
Brick and stone roads (applicable to small vital areas only)	Weather. Spray with an STB slurry from the PDDE or apply with brushes and brooms. Leave on for 24 hours, then flush with water. Wash with soapy water, preferably hot. Cover small areas or paths across roads with 4 inches of earth.	Follow the same directions as for asphalt roads.	Follow the same directions as for asphalt roads. Use abrasion (sand blasting). This provides direct and complete removal of contaminated dust; however, sand and equipment being used becomes contaminated.

Table C-1. Decon Procedures for Specific Surfaces and Materials (Continued)

Surface or Material	Types of Contamination and How to Decon		
	Chemical	**Biological**	**Radiological**
Brick and stone buildings, bunkers, gun emplacements, and tank obstacles	Spray with an STB slurry from the PDDE or apply with brushes and brooms. Leave on for 24 hours, then flush with water. Use STB (pure form) or an STB dry mix around buildings where waste water runs. Wash with soapy water, preferably hot. Weather.	Follow the same directions as for asphalt roads. Apply an STB slurry to vertical surfaces by manual means or with the PDDE. The slurry may be left on exteriors.	Follow the same directions as for brick and stone roads.
Concrete roads (applicable to small vital areas only)	Spray with an STB slurry from the PDDE. Cover with an STB slurry or STB dry mix. Weather. Cover small areas or paths across roads with 4 inches of earth. Scrape the layer of contaminated earth to the side of the road.	Follow the same directions as for asphalt roads.	Follow the same directions as for brick and stone roads.
Earth (applicable to small vital areas only): roads, gun emplacements, bivouac areas, pathways, and bomb craters	Spray with an STB slurry from the PDDE. Cover with STB (pure form). When liquid contamination is visible and personnel are nearby, use an STB dry mix. Weather. Cover small areas or paths across roads with 4 inches of earth. Scrape the layer of contaminated earth to the side of the road.	Follow the same directions as for asphalt roads. Burn.	Earth moving (removal): Try to control contaminated dust, as equipment may become contaminated. Consider waste disposal. Sealing (with earth): Does not create a waste-disposal problem; however, equipment may be contaminated.
Grass and low vegetation: fields and open terrain	Burn. Spray with an STB slurry from the PDDE. Cover with STB (pure form) or an STB dry mix. Explode drums of STB. Clear paths through area using detonating cord or other detonating devices.	Burn. Follow the same directions as for asphalt roads.	Follow the same directions as for earth.

Table C-1. Decon Procedures for Specific Surfaces and Materials (Continued)

Surface or Material	Types of Contamination and How to Decon		
	Chemical	**Biological**	**Radiological**
Undergrowth, tall grass (applicable to small vital areas only): meadows, jungles, and forests	Burn (may cause a downwind vapor hazard). Spray slurry with PDDE. Explode drums of STB. Clear a path with detonating cord, bangalore torpedoes, or demolition snakes.	Burn. Follow the same directions as for sand.	Follow the same directions as for earth.
Sand (applicable to small vital areas only): beaches and deserts	Flush with water. Use DS2. Spread STB (pure form) or spray an STB slurry over the surface. Weather. Cover paths with roofing paper. Scrape off 2 to 4 inches of contaminated top layer.	Burn.	Follow the same directions as for earth.
Fabrics: canvas, covers, tarpaulins, tentage, mask carriers, web gear, and clothing	Cotton: **NOTE: Not for MOPP gear.** Immerse in boiling, soapy water for 1 hour (1 pound of soap to 10 gallons of water) and stir. Use a 5 percent sodium-carbonate solution for G agents. Immerse in boiling water for 1 hour. Launder by standard methods. Use an STB slurry. Weather except for V agents. Woolen: Do not use DS2, as it is not recommended. Immerse in warm, soapy water (100°F) for 1 hour or longer with light agitation. Dry items slowly because the fabric may shrink.	Cotton: Boil in water for 15 minutes. Autoclave for 45 minutes at 253°F. Immerse in a 2 percent household-bleach solution for 30 minutes and rinse immediately. Launder (destroys or inactivates all but highly resistant spores). Woolen: Do not use DS2, as it is not recommended. Launder (fabric may shrink).	Cotton and woolen: Using DS2 is not recommended. Brushing removes contamination dust, but it presents a dust hazard to personnel. Laundering is the most practical procedure; however, the fabric may shrink. Try to control waste.

Table C-1. Decon Procedures for Specific Surfaces and Materials (Continued)

Surface or Material	Types of Contamination and How to Decon		
	Chemical	Biological	Radiological
Leather: boots, gloves, and other items	Scrub with hot, soapy water and rinse. Immerse in warm, soapy water at 120°F for 4 hours and rinse. Use a 5 percent sodium-carbonate solution for G agents. Air.	Immerse in a 2 percent household-bleach solution and rinse. Immerse in a 2 percent PAA for 10 minutes, rinse, and air 10 to 15 minutes. Wipe with a 2 percent PAA, remove excess agent, and air 10 to 15 minutes.	Brush. Flush with water or soapy water.
Glass, windows	Use an IEDK or DS2. Wash with hot, soapy water. Rinse with clear water or an organic solvent. Blot off surface. Air. Weather.	Use an IEDK. Wash with soap and water. Wipe with a disinfectant solution or a 2 percent PAA (see similar procedures on page C-7 for mess gear).	Use an IEDK. Wash with a detergent. Flush with water. Wipe with solvents.
Glass, lens	Wash with hot, soapy water. Rinse with clear water or an organic solvent. Blot off surface. Air. Weather.	Wash with hot, soapy water. Wipe with alcohol or household bleach.	Brush or wipe (be careful so as not to scratch the lens). Use compressed air to blow contamination from surface.
Metal (unpainted), ammunition	Wipe with soapy water. Wipe with organic solvent and dry. Air.	Wipe with soapy water. Wipe with a 2 percent household-bleach solution. Air.	Brush or wipe.
Metal (unpainted), machinery	Use DS2. Follow the same directions as for ammunition.	Use DS2. Wipe with a 2 percent PAA, rinse, and air 10 to 15 minutes.	Brush or wipe. Wash with a detergent. Flush with water.
Metals (painted): vehicles, weapons, and equipment	Use DS2 (may soften paint). Wash with hot, soapy water and rinse. Spray with an STB slurry from the PDDE then remove it in 1 hour and oil the surface. Weather. Air. Use an IEDK to decon individual gear.	Wash with a detergent and a high-pressure water system. Apply detrochlorite. Leave it on for 30 minutes then remove by washing with a stream of water. Steam clean using a detergent. Use a household-bleach solution. Use a 2 percent PAA.	Brush or wipe. Wash. Use organic solvents, caustic agents (not on aluminum or magnesium surfaces), complexing agents (of small value on weathered surfaces), or abrasives.

Table C-1. Decon Procedures for Specific Surfaces and Materials (Continued)

Surface or Material	Types of Contamination and How to Decon		
	Chemical	**Biological**	**Radiological**
Wood (unpainted surfaces): buildings, vehicle bodies, boxes, crates, and similar items	Apply an STB slurry with the PDDE, brooms, or swabs. Let the slurry remain 12 to 24 hours then flush with water. Repeat the application and flush again. Scrub with hot, soapy water and rinse. Weather.	Apply detrochlorite. Leave it on for at least 30 minutes then flush with water. Apply an STB slurry to vertical surfaces. The slurry may be left on interiors. Weather (sun and rain eliminate most microorganisms within 1 day). Burn.	Wash the interior with large amounts of water (some contamination may soak into surfaces).
Wood (painted surfaces): buildings and boxes	Apply an STB slurry with the PDDE, brooms, or swabs. Let the slurry remain 12 to 24 hours then flush with water. Scrub with hot, soapy water and rinse. Use DS2 and rinse. DS2 may soften paint. Weather.	Follow the same directions as for wood (unpainted surfaces).	Wash the interior with large amounts of water. Wipe the contamination from the surface.
Plastics (opaque): insulation, telephones, and panel boards	Use DS2 (may soften or damage some plastics). Wash with hot, soapy water and rinse. Weather. Air.	Follow the same directions as for glass lens.	Wash with a detergent. Flush with water. Wipe or brush.
Plastics (transparent): eyepieces and airplane canopies	Use DS2 (may soften or damage some plastics). Wash with hot, soapy water and rinse. Weather. Air.	Follow the same directions as for glass lens.	Wash with a detergent. Flush with water. Wipe or brush.
Nonsealed electronic equipment	Use a mild, evaporative solvent such as alcohol. Use DS2. Use a mild detergent and water. Rinse with distilled water. Keep the solution out of the adjustable switches, connectors, and relays. Use hot-air blowers.	Use a mild, evaporative solvent such as alcohol. Use DS2. Use a mild detergent and water. Rinse with distilled water. Keep the solution out of the adjustable switches, connectors, and relays. Use hot-air blowers.	Use hot-air blowers. Wipe or brush.

Table C-1. Decon Procedures for Specific Surfaces and Materials (Continued)

Surface or Material	Types of Contamination and How to Decon		
	Chemical	**Biological**	**Radiological**
Rubber (impermeable): aprons, suits, and other items	Spray with DS2 and rinse after 30 minutes. Immerse in soapy water (just below the boiling point) for 1 hour. Do not agitate. Rinse with clear water and hang to dry. Use a 10 percent sodium-carbonate solution for G agents, rinse, and air. Apply hot, soapy water with brushes and rinse. Spray with an STB slurry from the PDDE. Wash off, after a few minutes, with clear water.	Follow the same directions as for leather.	Brush. Scrub or flush with water or soapy water.
Rubber (natural and synthetic): gloves and boots	Spray with a 10 percent mixture of HTH and rinse. Immerse in STB slurry for 4 hours, rinse, and dry. Use an IEDK in emergencies. Air.	Follow the same directions as for leather.	Brush. Scrub or flush with water or soapy water.
Rubber: mask facepieces and other rubber articles coming in direct contact with the skin	Use IEDK in emergencies. Wash with warm, soapy water.	Follow the same directions as for leather.	Brush. Scrub or flush with water or soapy water.
Rubber: tires, hoses, mats, and insulation	Spray with a 10 percent mixture of HTH and rinse. Apply an STB slurry. Allow the slurry to remain at least 30 minutes then flush with clear water (may be left on tires). Apply hot, soapy water. Air. Weather.	Follow the same directions as for leather.	Brush. Scrub or flush with water or soapy water.
Water	Use trained water-purification personnel to decon.	Boil small amounts 15 minutes. Chlorinate using a chlorination kit. Add iodine water-purification tablets to small amounts.	Flocculation (requires special chemicals to remove suspended matter). Ion exchange (removes radians from solution).

Table C-1. Decon Procedures for Specific Surfaces and Materials (Continued)

Surface or Material	Types of Contamination and How to Decon		
	Chemical	**Biological**	**Radiological**
Mess gear and canned rations	Immerse in boiling, soapy water for 30 minutes and rinse. Immerse in boiling water for 30 minutes. Spray with DS2. Wash in hot, soapy water; rinse; and air.	Wash with soap and water, then immerse in disinfectant solution (disinfectant, chlorine or 1/3 cup of household bleach per 10 gallons of water). Boil in water for 15 minutes (not effective on toxins and bacterial spores.) Immerse in a 5 percent sodium-carbonate solution (4 pounds of washing soda to 10 gallons of water) and rinse with potable water. Immerse in a household-bleach solution (2 gallons of bleach to 25 gallons of water) for 30 minutes, rinse, and air for 10 to 15 minutes. Immerse in an HTH solution (½ pound to 25 gallons of water) for 30 minutes and rinse. Immerse in a 2 percent PAA for 10 minutes, rinse, and air 10 to 15 minutes.	Wash with soap and water and rinse. Brush and then wipe contamination from surfaces and containers.
Food: not canned or protected by impermeable container	Do not consume food that is suspected to be contaminated with chemical agents until veterinary personnel approve it.	Boil water 15 minutes. Cook food thoroughly. Immerse in or spray with a 2 percent household-bleach solution. (Packaged food or food that is peeled or pared can be immersed or sprayed with the solution.)	Wash or trim contamination from unpackaged food.
Food: canned, bottled, or protected by impermeable container	Follow the same directions as for mess gear and canned rations.	Follow the same directions as for mess gear and canned rations.	Follow the same directions as for mess gear and canned rations.
Paper currency	Destroy by burning. Do not decon.	Destroy by burning. Do not decon.	Destroy by burning. Do not decon.
Coins	Wipe with soapy water. Wipe with an organic solvent.	Use UV rays. Wash with soapy water. Wipe with a 2 percent household-bleach solution. Air.	Brush or wipe. Wash with a detergent. Flush with water.

Appendix D

Decon Kits, Apparatuses, and Equipment

Various materials and equipment are used in decon operations. Some are simple to use and are readily available to individual soldiers. Others are very complex to use and are available only to specially trained teams. Table D-1, pages D-1 and D-2, lists the decon equipment and materials for the different use levels; Table D-2, pages D-3 through D-5, lists the detection equipment and materials for the different use levels.

Table D-1. Decon Equipment and Materials

Item and Description	Use	Limitations	Reference
	Individual		
Decontaminating kit, skin, M291 SDK, (NSN 4230-01-276-1905) (20 kits per box)	To decon your skin, completely, through physical removal, absorption, and neutralization of toxic agent with no long-term effects **NOTE: Use this kit for both actual combat and training purposes.**	Is for external use only. **WARNING** **Keep decon powder out of eyes, cuts, or wounds. It could slightly irritate the skin or eyes.**	TM 3-4230-229-10
Decon kit, individual equipment, M295, IEDK, (NSN 6850-01-357-8456)	To decon your chemical protective gloves, mask, hood, overboots, LCE, and weapon.	**WARNING** **Do not use for skin decon. Keep off the skin and out of wounds, eyes, and mouth.**	TM 3-4230-235-10
Decontaminating apparatus, portable, M13 DAP, (3.7 gal), (NSN 4230-01-133-4124). It has a 3.7-gallon disposable DS2 container. The apparatus can be mounted to a standard 5-gallon fuel can mount on vehicles and equipment.	To spray DS2 on surfaces of vehicles and equipment.	Cannot refill DS2 container. See Table A-2, page A-2, for a list of replacement containers.	TM 3-4230-214-12&P TM 43-0001-26-1

Table D-1. Decon Equipment and Materials (Continued)

Item and Description	Use	Limitations	Reference
Battalion Decon Crew and Chemical Company			
Decontaminating apparatus, power driven, skid mounted, multipurpose, integral, 500 gallon, M12A1 PDDA, (NSN 4230-00-926-9488), LIN F81880. Apparatus includes pump unit, tank unit, and M2 water heater (all mounted on skids).	To spray decon agent, STB slurries, and solutions as well as hot, soapy water rinses during field decon operations. To pump water or foam to fight fires, de-ice items, wash vehicles, and pump various fluids.	Do not use with defoliants, herbicides, or insecticides.	LO 3-4230-209-12 LO 5-2805-259-12 TM 3-4230-209-20&P TM 43-0001-26-1
Decontaminating system, lightweight, M17 LDS, (NSN 4230-01-251-8702). The M17 is a portable pump and water-heating unit for producing hot water and steam. The system incorporates a 1,580- to 3,000-gallon collapsible water tank, two wand assemblies, and connecting hoses.	To perform operational and thorough decon of vehicles and equipment. Can also be used for troop showers, as necessary.		TM 3-4230-228-10

Table D-2. Detection Equipment and Materials

Item and Description	Use	Limitations	Reference
		Individual	
Paper, chemical agent, detector, VGH, M8, (NSN 6665-00-050-8529). Paper is issued in a book of 25 sheets, perforated for easy removal. A color-comparison bar chart is printed on the inside front cover.	To detect the presence of liquid V, G, and H chemical agents	Cannot be used to detect vapors or chemical agents in water, DS2, or petroleum products. May cause false readings.	TM 3-6665-254-12 TM 3-6665-268-10 TM 3-6665-307-10 TM 43-0001-26-1
Paper, chemical agent, detector, M9, (NSN 6665-01-226-5589). Paper is issued in a 7-ounce dispenser box that contains one 30-foot roll of 2-inch-wide detector paper and plastic storage bags. The paper has an adhesive back for attaching to equipment and clothing.	To detect the presence of liquid V, G, and H chemical agents	Cannot be used to detect vapors or chemical agents in water. Will not stick to dirty, oily, or greasy surfaces. Contamination indications cannot be read under red light or by color-blind soldier. The following can cause false readings: • Temperatures above 125°F. • Brake fluid. • Aircraft cleaning compound. • DS2. • Petroleum products. • Insect repellent.	TM 3-6665-311-10
		Company	
Chemical-agent alarm, automatic, M22 ACAA, (NSN 6665-01-438-6983)	To detect chemical nerve agents in the air.	**WARNING** **Radiation hazard. Contains beta emitters.**	TM 3-6665-321-12&P
Chemical-agent alarm, automatic, M8A1 ACAA, (NSN 6665-01-105-5623). It can be vehicle-mounted, back-packed, or ground-emplaced.	To detect chemical nerve agents in the air.	**WARNING** **Radiation hazard. Contains Americium (AM241).**	TM 3-6665-312-12&P

Table D-2. Detection Equipment and Materials (Continued)

Item and Description	Use	Limitations	Reference
Chemical-agent monitor system (CAM) (NSN 6665-01-199-4153)	To search out areas; to search and locate contamination on personnel, equipment, ship's structure, aircraft, land vehicles, buildings, and terrain; and to monitor for effectiveness of decon. Can also be used for monitoring collective protection. The CAM responds to nerve- and blister-agent vapors down to the lowest concentration that could affect personnel over a short period.	Cannot be used as a detector. The CAM is a monitor and not a detector. Since it is a monitor, it can become contaminated and overloaded (saturated). The CAM can only report conditions at the front of the inlet probe. It is, therefore, a point monitor only and cannot give a realistic assessment of the vapor hazard over an area from one position. **WARNING** **Beta radiation hazard.**	TM 3-6665-331-10
Detector kit, chemical agent, M256A1 (NSN 6665-01-438-6983). It contains M8 detector paper for liquid agents and samplers/detectors for vapors.	To detect liquid G, V, and H chemical agents using M8 detector paper and to detect and determine the type of vapor (G, V, H, and AC) using samplers/detectors.		TM 3-6665-307-10

Table D-2. Detection Equipment and Materials (Continued)

Item and Description	Use	Limitations	Reference
Radiac set, AN/VDR-2 (NSN 6665-01-222-1425)	To measure gamma radiation dose rates from 0.01 uGy/hr to 100 Gy/hr; to detect and display the level of beta particle dose rates from 0.1 uGy/hr to 5 cGy/hr; and to measure, store, and display accumulated dose rates from 0.01 uGy to 9.99 Gy. The instrument consists of a radiac meter with an internal sensor for obtaining dose rates during both mounted and dismounted operations. It has a second sensor housed in a probe and attached to the radiac meter with a cable and input connector. It is used for monitoring personnel, supplies, and equipment. The radiac set uses a presettable, an audible, and a visual warning device integral to the radiac meter. The system is air-transportable and organic to all units.		TM 11-6665-251-10
Computer indicator, radiac, CP696/PDR-75; detector, radiac/ DT236/PDR-75; radiac set, AN/PDR-75, (NSN 6665-01-211-4217)	To measure the accumulated neutron and gamma radiation dose recorded by the DT236. A person who may be exposed to radiation from tactical nuclear weapons wears the DT236 on his wrist.		

Appendix E

DTD Station Charts for Chemical Unit Leaders and NBC NCOs

This appendix provides station charts for the DTD. They can be laminated and posted at each station. They provide the station number and on the reverse side, instructions for the station supervisor/attendant. They are not meant to be all-inclusive and do not include setup procedures or equipment requirements. They are an easy guide for soldiers going through a decon and for station supervisors/attendants.

STATION 1

INDIVIDUAL-GEAR DECON

TASK: Remove contamination to a negligible risk from individual gear (LCE, mask carrier, helmet, and weapon)

- The soldier—
  - Decontaminates his own gloves with hot, soapy water or an STB slurry.
  - Decontaminates his hose and canister (if wearing the M42 mask) with either hot, soapy water and a sponge or an STB slurry mix.
  - Decontaminates his own gear by washing and scrubbing it for 4 minutes in a decontaminant container with hot, soapy water or an STB slurry mix.
- The attendant—
  - Takes the gear to the equipment checkpoint and places it on the "dirty" side of the contamination control line (engineer tape).
  - Waits as the monitor holds the detection device 1 inch from the gear and checks it for residual contamination.
  - Recycles any gear that is above one bar.
  - Places the gear on the clean side of the contamination control line if it passes the check.
  - Carries the gear to the reissue point.

NOTE: Change the slurry mix after 20 soldiers have decontaminated their gear and the rinse water after 10 soldiers or when it appears dirty (place the waste in the sump).

STATION 2

OVERBOOT AND HOOD DECON

TASK: Neutralize gross contamination on overboots, lower trouser legs, protective mask, and hood

- **The soldier walks into the shuffle pit and spreads his legs apart (double-shoulder width), bends at the waist, and uses his hands to thoroughly rub the STB dry mix or the STB slurry on his overboots and lower trouser legs.**
- **Buddy #1 cuts the shoulder straps and drawcord on Buddy #2's hood. He pulls Buddy #2's hood inside out over the front of the mask, being careful not to touch the exposed neck or head. Buddy #1 gathers Buddy #2's hood in one hand and using a cutting tool, cuts away the hood as close as possible to the eye-lens outsert, voicemitter, and inlet-valve cover. Make sure nothing is left dangling below the bottom of the mask. If a replacement hood is not available, decon Buddy #2's hood in the same manner as for a MOPP-gear exchange.**
- **The attendant—**
  - **Instructs Buddy #1 on how to roll Buddy #2's hood to ensure that it is accomplished to standard.**
  - **Will monitor the shuffle pit and add more STB after 10 soldiers have processed through it.**

STATION 3

OVERGARMENT REMOVAL

TASK: Remove contaminated overgarment before the agent penetrates material and touches the undergarments or the skin

The attendant—

- Assists the soldier in removing his overgarment.
- Cuts or removes the M9 detector paper from around the soldier's wrist.
- Unfastens the Velcro closure over the jacket zipper.
- Unties the waist cord and unfastens the Velcro straps on the soldier's jacket.
- Unfastens the back snaps.
- Cuts or removes the M9 detector paper from the soldier's trousers, if binding.
- Unfastens the Velcro straps and zipper on the cuffs of the soldier's trousers.
- Unfastens the front waist snaps and unzips the front zipper. He has the soldier lift one leg and point that foot down and bend slightly at the knees for stability.
- Grasps the cuff of the elevated boot with a hand on each side and pulls the cuff in an alternating, jerking motion until the soldier can step out of the trouser leg. The process is repeated on the other leg.

STATION 4

OVERBOOT AND GLOVE REMOVAL

TASK: Remove contaminated overboots and gloves to limit the spread of contamination (the overboots and gloves may also be decontaminated for reissue, if servicable)

- Engineer tape is used to mark the liquid-contamination control line.
- The attendant unfastens or cuts the elastic closures on the soldier's overboots.
- The soldier stands next to and faces the liquid-contamination control line.
- The soldier steps back about 12 inches from the liquid-contamination control line.
- The attendant steps on the back of the soldier's overboot and instructs him to lift his heel and work his foot out of the overboot and step across the liquid-contamination control line. Repeat the process on the other foot.
- The attendant discards the overboots.
- The soldier holds the fingertips of his gloves and partially slides his hands out.
- The attendant removes the soldier's gloves.
- The attendants—
    - Put the gloves and overboots into the containers on the "dirty" side of the control line.
    - Submerge the gloves and overboots in the container of hot, soapy water, ensuring that no water remains inside when removed from the container.
    - Submerge the gloves and overboots in the STB/HTH solution and thoroughly scrub them.
    - Rinse the scrubbed items and check for holes, tears, or punctures. If found, discard the items.
    - Place the usable items on a poncho or a plastic tarp to air dry.
    - Place the air-dried usable items into plastic trash bags along with an M256A1 detector kit. If the kit shows contamination remaining, the attendants can recycle the items or discard them. However, if the kit shows no contaminated, they can be reused.

STATION 5

MONITOR

TASK: Identify contamination on personnel (spot-decon capability and medical aid are provided, as required)

- The monitor checks the soldiers for contamination.
- The aidman checks the soldiers for any chemical-agent symptoms and treats as necessary.
- The monitor asks the soldiers about any damage to their MOPP gear that was identified at stations 2, 3, or 4.

NOTE: If all the liquid contamination is absorbed into the clothing, the M8 detector paper will be negative even though there is a hazard.

STATION 6

MASK REMOVAL

TASK: Remove the mask without contaminating the soldier

- The attendant pulls the hood over the front of the mask, grabs the mask by the voicemitter cover, and pulls the mask off the soldier while he holds his breath.
- The attendant holds the mask open so that the soldier can remove the inserts without touching the outside of the mask if the mask has optical inserts.
- The soldier walks upwind 16 feet, crosses the vapor-contamination control line, and then resumes breathing.
- The attendant brings the mask to station 7.

NOTE: No vapor hazard is expected beyond the vapor-hazard contamination control line.

- The chemical unit positions the M8A1 or M22 ACAA upwind of the site.
- The soldier moves straight ahead while his mask, which may still be emitting vapors, is held on the "dirty" side of the vapor-contamination control line and taken to the next station.

STATION 7

MASK DECON POINT

TASK: Remove all contamination from the protective masks

The attendants—

- Discard the winterization kits, if used.
- Remove the eye-lens outserts and the hoods if the hoods were not cut off at station 2.
- Remove or discard the filters or canisters and dispose of them in the properly marked containers.
- Wash the masks, hoods, eye-lens outserts, and hoses on the M42 and M43 masks in hot, soapy water.
- Rinse these items in clean water.
- Dip them in the sanitizing solution and agitate for 5 minutes.
- Rinse them in clean water again.
- Add one tube of mask sanitizing solution (calcium hypochlorite) to each quart of water.
- Wipe the masks with rags until they are almost dry.
- Discard each gallon of mask sanitizing solution into a sump after every 10 masks.
- Check for contamination with a CAM.
- Recycle the masks if they are still contaminated and then decon their gloves.
- Take the unassembled masks to the reissue point if they are not contaminated.

STATION #8

REISSUE POINT

TASK: Reissue the masks with their components to the soldiers

- The unit NBC NCO reissues the masks with their components to the soldiers.
- The soldiers will reassemble the masks in the postdecon AA.
- The unit NBC NCO provides the soldiers with replacement parts for all types of protective masks.
- The soldiers pick up individual gear and move to the postdecon AA.

Glossary

AA	assembly area
AC	hydrogen cyanide
ACAA	automatic chemical-agent alarm
AD	air defense
AFM	Air-Force Manual
AFR	Army-Force Regulation
AMT	ambulatory medical treatment
AO	area of operation
APC	armored personnel carrier
AR	Army regulation
ART	assessment and recovery team
attn	attention
AVIM	aviation intermediate maintenance facility
BAS	battalion aid station
BDO	battle-dress overgarment
BDU	battle-dress uniform
BFV	Bradley fighting vehicle
bio	biological
C^2	command and control
CAM	chemical-agent monitor
CANA	convulsant antidote for nerve agent
CARC	chemical-agent-resistant coating
CG	phosgene
cGy/hr	centigray per hour
chem	chemical
CHS	combat health support
CK	cyanogen chloride
Class II Supplies	clothing, individual equipment, tentage, organizational tool sets and kits, hand tools, maps, and administrative and housekeeping supplies and equipment

Classes of Supply	the grouping of supplies by type into 10 categories to facilitate supply management and planning
CLU	command launch unit
CMO	chemical officer
CN	chloroacetophenone
CP	chemical protective
CPS	collective protective shelter
CPU	chemical protective undergarment
CRDEC	Chemical Research Development and Engineering Center
CSS	combat service support
CTA	common table of allowance
CVCUS	combat vehicle crewman uniform system
CX	phosgene oxime
DA	Department of the Army
DAD	detailed aircraft decon
DAP	decontaminating apparatus, portable
DD	Department of Defense
DED	detailed equipment decon
DISCOM	Division-support command
DP	diphosgene
DS2	decontaminating solution number 2
DTD	detailed troop decon
DU	depleted uranium
EDTA	ethylene diamine tetra acetic acid
EIC	end-item code
EMT	emergency medical treatment
ENCU	environmental control unit
EOD	explosive ordnance disposal
F	Fahrenheit
FARP	forward arming and refueling point
FLOT	forward line of troops
FM	field manual
FMC	field medical card
FMFM	Fleet Marine Force Manual

FRAGO	fragmentary order
FSB	forward-support battalion
G4	Assistant Chief of Staff, G4 (Logistics)
G5	Assistant Chief of Staff, G5 (Civil Affairs)
GA	tabun
GB	sarin
GD	soman
GF	soman
GI	Government issue
GP	general purpose
GPM	gallons per minute
Gy	gray
HD	distilled mustard
HE	high explosive
HEMTT	heavy expanded-mobility tactical truck
HEPA	high-efficiency particulate air filter
HL	mustard-lewisite
HMMWV	high-mobility multipurpose wheeled vehicle
HN-1	nitrogen mustard
HN-2	nitrogen mustard
HN-3	nitrogen mustard
HP	horsepower
HQ	headquarters
HTB	high-test bleach
HTH	high-test hypochlorite
IEDK	individual equipment decon kit
JMAO	Joint Mortuary Affairs Office
JP8	aviation fuel
L	lewisite
LAW	light antitank weapon
LCE	load-carrying equipment
LDS	Lightweight Decontaminating System
LIN	line-item number
LO	lubrication order

LTA	launcher-tube assembly
MADCP	mortuary affairs decontamination collection point
METT-TC	mission, enemy, terrain, troops, time available, and civilian consideration
MGX	MOPP-gear exchange
MOPP	mission-oriented protective posture
MP	military police
MRE	meals, ready-to-eat
MTF	medical-treatment facility
MTOE	modification table of organization and equipment
NATO	North Atlantic Treaty Organization
NAVMC	Navy/Marine Corp
NBC	nuclear, biological, and chemical
NBC5 report	once the NBC reports are posted on the situation map, an NBC5 report showing the contaminated area is prepared by the division
NCO	noncommissioned officer
NCOIC	noncommissioned officer in charge
NSN	national stock number
OEG	operational exposure guidance
OPORD	operation order
OS	operator's spray down
PAA	peracetic acid
PALADIN	artillery howitzer
PDDA	power-driven decon apparatus
PDDE	power-driven decon equipment
PL	platoon leader
PLL	prescribed load list
PNVS	Pilot Night-Vision System
POL	petroleum, oils, and lubricants
POMCUS	prepositioning of material configured to unit sets
PPW	patient protective wrap
PSG	platoon sergeant
psi	pounds per square inch
PT	pointed
PW	personal wipe down

rad	radiological
RD	rounded
RPO	radiological protection officer
S4	Supply Officer (US Army)
SA	arsine
SAW	squad automatic weapon
SB	supply bulletin
SD	skin decon
SDK	skin decontaminating kit
SMCT	soldiers' manual of common tasks
SOI	signal operating instructions
SOP	standing operating procedure
STANAG	Standardized North Atlantic Treaty Organization Agreement
STB	supertropical bleach
STP	soldier training publication
T20	tritium oxide
TADS	Tactical Air-Defense System
TAP	toxicological apron, protective
TB	technical bulletin
TC	training circular
TGD	thickened soman
THD	thickened distilled mustard
TM	technical manual
TO	technical order
TOE	table of organization and equipment
TOW	tube launched, optically tracked, wire guided
TPU	tank and pump unit
TRADOC	United States Army Training and Doctrine Command
uGy	microgray
US	United States
USMC	United States Marine Corp
UV	ultraviolet
VDR-2	radiac meter
VR-55	soman

VW	vehicle washdown
VX	soman
WBGT	wet-bulb gradient temperature
WO	warning order

Bibliography

AR 11-9. *The Army Radiation Safety Program*. 28 May 1999.

AR 385-10. *The Army Safety Program*. 23 May 1988.

CRDEC Technical Report. CRDEC-CR-87036. *Project VALID Final Comprehensive Report. Herman Nelson Heater Utilization Study*. December 1986. AD-B108506.

CRDEC Technical Report CRDEC-TR-175. *Decontamination Effectiveness of Water Streams*. March 1991. AD-B154567.

CTA 50-909. *Field and Garrison Furnishings and Equipment*. 1 August 1993.

DA Form 2028. *Recommended Changes to Publications and Blank Forms*. 1 February 1974.

DA Form 2404. *Equipment Inspection and Maintenance Worksheet*. 1 April 1979.

DD Form 1380. *US Field Medical Card*. December 1991.

DO49 Technical Report DPB/JOD-91/025. *MOPP Exchange Evaluation, Operational*. September 1991. AB-B158128.

DO49 Technical Report DPG/TA-86-015. *Standard and Nonstandard Decontaminants, and Decontamination Efficiency (U)*. May 1987. AD-C041660.

DO49 Technical Report DPG/TA-88/018. *Effects on High Water Pressure on Composite Materials*. June 1988. AD-B124033.

DO49 Technical Report DPG/TA-88/023. *Commercial and Field Expedient Chemical Warfare Agent Decontamination*. March 1988. AD-B121231.

DO49 Technical Report DPG/TA-88/030. *Decontamination on Selected Military Equipment: US Army Ammunition Stocks*. September 1988. AD-B128325.

DO49 Technical Report DPG/TA-89024. *Decontaminative Effects of Soil*. November 1989. AD-B140334.

DO49 Technical Report DPG/TA-90/007. *Water Quality Requirements for Decontamination*. December 1989. AD-B140336.

DO49 Technical Report DPG/TA-90/014. *Mission Oriented Protective Posture (MOPP) Exchange Procedures Evaluation*. January 1990. (No DDIC Number Assigned).

DO49 Technical Report DPG/TA-90/025. *Human Factors Evaluation of the Mission Orientated Protective Posture (MOPP) Exchange Procedures*. July 1990. AD-B148421.

FM 3-3. *Chemical and Biological Contamination Avoidance (FMFM 11-7)*. 16 November 1992.

FM 3-3-1. *Nuclear Contamination Avoidance (FMFM 11-18)*. 9 September 1994.

FM 3-4. *NBC Protection (FMFM 11-9).* 29 May 1992.

FM 3-4-1. *Fixed Site Protection.* 16 August 1989.

FM 3-6. *Field Behavior of NBC Agents (Including Smoke and Incendiaries) (AFM 105-7; FMFM 7-11-H).* 3 November 1986.

FM 3-7. *NBC Field Handbook.* 29 September 1994.

FM 3-9. *Potential Military Chemical/Biological Agents and Compounds (AFR 355-7).* 12 December 1990.

FM 3-100. *Chemical Operations Principles and Fundamentals.* 8 May 1996.

FM 3-101. *Chemical Staffs and Units.* 19 November 1993.

FM 8-9. *NATO Handbook on the Medical Aspects of NBC Defensive Operations.* 1 February 1996.

FM 8-10-4. *Medical Platoon Leader's Handbook*—Tactics, Techniques, and Procedures. 16 November 1990.

FM 8-10-7. *Health Service Support in a Nuclear, Biological, and Chemical Environment.* 22 April 1993.

FM 8-33. *Control of Communicable Diseases Manual.* 9 April 1996.

FM 8-285. *Treatment of Chemical Agent Casualties and Conventional Military Chemical Injuries (FMFM 11-11).* 22 December 1995.

FM 21-10. *Field Hygiene and Sanitation.* 22 November 1988.

FM 21-11. *First Aid for Soldiers.* 27 October 1988.

FM 100-5. *Operations.* 14 June 1993.

FM 100-9. *Reconstitution.* 13 January 1992.

FM 100-10. *Combat Service Support.* 3 October 1995.

Joint Publication 3-11. *Joint Doctrine for Nuclear, Biological, and Chemical (NBC) Defense.* 10 July 1995.

Joint Publication 3-12. *Doctrine for Joint Nuclear Operations.* 18 December 1995.

Joint Publication 4-06. *Joint Tactics, Techniques, and Procedures for Mortuary Affairs in Joint Operations.* 28 August 1996.

LO 3-4230-209-12. *Decontaminating Apparatus, Power-Driven, Skid-Mounted, Multipurpose, Nonintegral, 500 Gallon, ABC-M12A1.* 21 June 1974.

LO 5-2805-259-12. *Engine, Gasoline, 20 HP, Military Standard Models 4A084-2, 4A084-3, 4A084-4, (LI-03524B-10-1A).* 30 January 1989.

SB 10-523. *Size Tariff for Clothing, equipage, and Footwear.* 1 August 1998.

STANAG 2426 NBC. *NBC Contamination Control Policy for NATO Forces (Edition 1).* 8 February 1995.

STP 21-1-SMCT. *Soldier's Manual of Common Tasks Skill Level 1.* 1 October 1994.

STP 21-24-SMCT. *Soldier's Manual of Common Tasks Skill Levels 2/3/4.* 1 October 1992.

TB 43-0116. *Identification of Radioactive Items in the Army*. 1 April 1998.

TC 3-4-1. *Chemical Agent Monitor Employment*. 17 December 1991.

TM 3-4230-209-20&P. *Unit Maintenance Manual Including Repair Parts and Special Tools List Decontaminating Apparatus: Power-Driven, Skid-Mounted, 500 Gallon, M12A1 (NSN 4230-00-926-9488)*. 28 June 1986.

TM 3-4230-214-12&P. *Operator's and Unit Maintenance Manual Including Repair Parts and Special Tools List for Decontaminating Apparatus: Decontaminating Apparatus: Portable, 14 Liter, M13 (Practice) (NSN 4230-01-345-5172)*. 15 August 1992.

TM 3-4230-228-10. *Operator's Manual for Decontaminating Apparatus: Power Driven, Lightweight, M17 (NSN 4230-01-251-8702)*. 15 May 1989.

TM 3-4230-229-10. *Operator's Manual for Decontaminating Kit, Skin: M291, (NSN 4320-01-251-8702) (SS010-AA-MMO-1-0; TO 11D1-1-131; TM 4230-10/1)*. 2 October 1989.

TM 3-4230-235-10. *Operator's Manual for Decontamination Kit, Individual Equipment: M295 (NSN 6850-01-357-8456) (EIC: Y67)*. 20 February 1994.

TM 3-6665-254-12. *Operator's and Organizational Maintenance Manual: Detector Kit, Chemical Agent, ABC-M18A2 (NSN 6665-00-903-4767)*. 16 February 1976.

TM 3-6665-268-10. *Operator's Manual for Sampling Kit, CBR Agent 34*. 6 September 1968.

TM 3-6665-307-10. *Operator's Manual for Chemical Agent Detector Kit, M256 (NSN 6665-01-016-8399) and M256A1 (NSN 6665-01-133-4964)*. 1 September 1985.

TM 3-6665-311-10. *Operator's Manual for Paper, Chemical Agent Detector: M9 (NSN 6665-01-226-5589 (TO 11H2-2-21)*. 31 August 1998.

TM 3-6665-312-12&P. *Operator's and Organizational Maintenance Manual Including Repair Parts and Special Tools List for M8A1 Automatic Chemical Agent Alarm (NSN 6665-01-105-5623) and Auxiliary Equipment M10 Power Supply (NSN 6665-00-859-2225); M10A1 Power Supply (NSN 6665-00-093-2739); M228 High Profile Mounting Kit (NSN 6665-00-859-2212) and M182 Low Profile Mounting Kit (NSN 6665-00-110-9492) (TO 11H2-17-1)*. 17 March 1985.

TM 3-6665-321-12&P. *Operator's and Unit Maintenance Manual (including Repair Parts and Special Tools List) for Alarm, Chemical Agent, Automatic: M22 (NSN 6665 01 438 6963) (EIC: Y14) and Auxiliary Equipment Power Supply, Chemical Agent Automatic Alarm: M28 (6130-01-438-6960) (EIC: Y40) Mounting Kit, Chemical Agent Automatic Alarm: M281 (NSN 6665-01-438-6959) (EIC: Y38) Alarm Unit, Chemical Agent Automatic Alarm: ABCA-M42 (NSN 6665-00-859-2215) (EIC: 399) (TO 11H2-23-1; MC 10434A-12&P; EE168-DB-OMP-010)*. March 98.

TM 3-6665-331-10. *Operator's Manual for Chemical Agent Monitor (CAM) (TO 11H2-20-1)*. 12 June 1992.

TM 10-277. *Chemical, Toxicological, and Missile Fuel Handlers Protective Clothing*. 1 November 1980.

TM 11-6665-251-10. *Operator's Manual for Radiac Set AN/VDR-2 (NSN 6665-01-222-1425)*. 1 March 1988.

TM 43-0001-26-1. *Army Equipment Data Sheets: Chemical Defense Equipment*. 12 May 1982.

NAVMC 10245. *Equipment Maintenance Request*. February 1994.

Index

*FM 3-19.15 (FM 19-15)

Field Manual
No. 3-19.15

Headquarters
Department of the Army
Washington, DC, 18 April 2005

Contents

Distribution Restriction: Approved for public release; distribution is unlimited.

*This publication supersedes FM 19-15, 25 November 1985.

Page

Preface

Field Manual (FM) 3-19.15 addresses continental United States (CONUS) and outside continental United States (OCONUS) civil disturbance operations. Today, United States (US) forces are deployed on peacekeeping, peace enforcement, and humanitarian assistance operations worldwide. During these operations, US forces are often faced with unruly and violent crowds intent on disrupting peace and the ability of US forces to maintain peace. Worldwide instability coupled with increasing US military participation in peacekeeping and related operations requires that US forces have access to the most current doctrine and tactics, techniques, and procedures (TTP) necessary to quell riots and restore public order.

In addition to covering civil unrest doctrine for OCONUS operations, FM 3-19.15 addresses domestic unrest and the military role in providing assistance to civil authorities requesting it for civil disturbance operations. It provides the commander and his staff guidance for preparing and planning for such operations. The principles of civil disturbance operations, planning and training for such operations, and the TTP employed to control civil disturbances and neutralize special threats are discussed in this manual. It also addresses special planning and preparation that are needed to quell riots in confinement facilities are also discussed.

In the past, commanders were limited to the type of force they could apply to quell a riot. Riot batons, riot control agents, or lethal force were often used. Today, there is a wide array of nonlethal weapons (NLW) available to the commander that extends his use of force along the force continuum. This manual addresses the use of nonlethal (NL) and lethal forces when quelling a riot.

Appendix A complies with current Army directives, which state that the metric system will be incorporated into all new publications.

The proponent for this publication is HQ, TRADOC. Send comments and recommendations on Department of the Army (DA) Form 2028 (*Recommended Changes to Publications and Blank Forms*) directly to Commandant, US Army Military Police School (USAMPS), ATTN: ATSJ-DD, 401 MANSCEN Loop, Fort Leonard Wood, MO 65473-8929.

Unless this publication states otherwise, masculine nouns or pronouns do not refer exclusively to men.

Chapter 1

Operational Threats of the Civil Disturbance Environment

As the Cold War ended with the fall of the Berlin Wall and the collapse of Soviet communism in 1989, the Army was no longer concerned with containing the aggressive Soviet Union. The US Army began to transform because the battlefield on which it would fight had changed. Operations Just Cause, Joint Endeavor, Desert Shield, and Desert Storm have all been recognized as containing nontraditional battlefields. Following these operations, the Army was called upon to conduct peacekeeping, peace enforcement, and humanitarian assistance operations. These operations required US forces to protect refugees and humanitarian workers and keep hostile factions separate to enforce peace.

GENERAL CAUSES FOR CIVIL UNREST

1-1. In these modern times, demonstrations, civil unrest, public disorder, and riots happen for a number of reasons. Some of these reasons are economic hardships, social injustices, ethnic differences (leading to oppression), objections to world organizations or certain governments, political grievances, and terrorist acts. An event can be triggered by a single cause or a combination of causes. For example, operations in the Balkans involving civil unrest and riots were the result of ethnic hatred, a lack of civil authority, food shortages, a revolution, and religious-based fighting factions.

1-2. Demonstrations may range from simple, nonviolent protests that address specific issues, to events that turn into full-scale riots. Gatherings in protest are recognized rights of any person or group, regardless of where US forces may be operating. This fundamental right is protected under the *Constitution of the US*. During peacekeeping or peace enforcement operations, US forces should never violate basic civil or human rights. Most protesters are law-abiding citizens who intend that their protests be nonviolent, but some protest planners insist that the event involve some kind of violence. Often, in the media, protesters can gain sympathy for their cause by prompting authorities to take physical action against them. Violence is often the result of demonstrators beginning to conduct unlawful or criminal acts and authorities (who are responsible for the safety and welfare of all) enforcing the laws of the municipality, state, or nation. The depth of violence is determined by the willingness of demonstrators to display and voice their opinions in support of their cause.

1-3. Commanders must be aware of the possibility that some individuals or groups within an organized demonstration may have the intent to cause disruption, incite violence, destroy property, and provoke the authorities. The tactical situation and actions of the crowd should dictate control and enforcement options. Agitators and criminal infiltrators within the crowd can lead to the eruption of violence. Inciting a crowd to violence or a greater intensity of violence through the use of brutish enforcement tactics should be avoided. Publicity can be detrimental to authorities and beneficial to crowds because it can further their causes.

1-4. Community unrest results in urban conflicts that arise from highly emotional social and economic issues. Economically deprived inner city residents may feel that they are treated unjustly or ignored by people in power and authority. Tensions can build quickly in a community over a variety of issues, such as hunger, poor employment opportunities, inadequate community services, poor housing, and labor issues. Tension in these areas creates the potential for violence. When tensions are high, it takes a small (seemingly minor) incident, rumor, or act of injustice to ignite groups within a crowd to riot and act violently. This is particularly true if community relations with authorities are part of the problem.

1-5. Significant ethnic differences in a community can create an atmosphere of distrust, even hatred. Unrest among ethnic groups competing for jobs, living areas, and sparse essentials can cause an eruption of civil disorder and/ or riots. As emotions run high, violence becomes likely.

1-6. Terrorist organizations may infiltrate groups within a demonstrating crowd. These terrorist groups may intend to embarrass their government or other governments. Terrorist infiltrators can be used to provoke crowds as a diversion, as part of a demonstration, or as cover for terrorist acts.

CROWD DEVELOPMENT

1-7. Crowds are a gathering of a multitude of individuals and small groups that have temporarily assembled in the same place. These small groups are usually comprised of friends, family members, or acquaintances that represent a group belief or cause. Individuals assume a sense of anonymity—they are viewed as just another face in the crowd. People in small groups are known only to companions in their group and to others in the gathering that have come from the same neighborhood or community. Commanders must consider how the individuals assembled and how they are interacting during the gathering process.

GATHERINGS

1-8. The assembly process of a gathering refers to the movement of people from different locations to a common location within a given period. This largely determines who participates. Creating a gathering from a crowd is a process with a beginning, middle, and end. As shown in *Figure 1-1* a crowd has an assembly process that leads to the gathering of the crowd, which is always followed by a dispersal process.

1-9. Gatherings are often assisted by the activities of individuals or groups with a specific agenda, such as yelling catchy slogans and cheers that everyone can easily pick up and join in on. Some groups are so well organized that they can prestage leaders to infiltrate a gathering. This creates unity, even inciting newcomers to join their cause. The first phase of the gathering is the assembly process. It can occur in one of two ways—impromptu or organized.

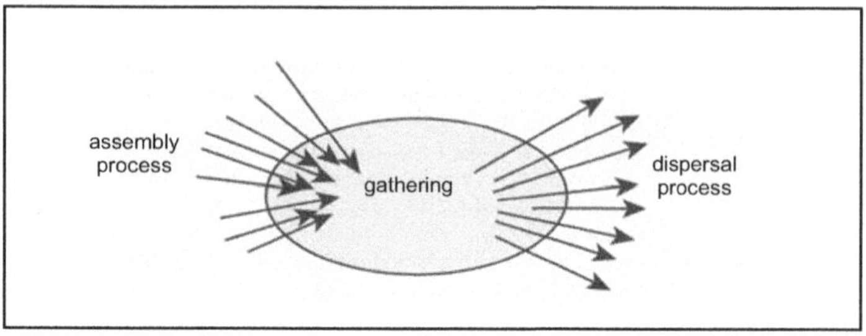

Figure 1-1. Crowd Building

IMPROMPTU GATHERINGS

1-10. Impromptu gatherings usually develop informally and are mostly done by word of mouth (one person to another or one group to another). Participants spread intelligence by telling one another when, where, and what is happening and inviting them to participate. An example of an impromptu gathering would be a gathering at a secured food distribution point after receiving information (by word of mouth) that a large truck carrying much-needed supplies and food is about to arrive. In this case, hunger would be the driving force causing the migration of people to the food distribution facility.

ORGANIZED GATHERINGS

1-11. Like an impromptu gathering, an organized gathering can also involve individuals and groups passing on information to one another. Passing intelligence of police activity and occurrences and on when and where events will take place helps organizers to prestage participants. Rarely is one represented group responsible for pulling together a gathering. Organized gatherings rely heavily on established groups that attract people to gather. Recent examples of these well-organized groups are anarchists, anti-globalization groups, and anti-free-enterprise groups. Groups representing extreme religious faiths and ethnic organizations have been common too. Some gatherings involve groups from rival, or even warring, factions.

1-12. Organized gathering processes rely more on centralized planning and organization. One or more of the groups offer organizers lists of individual names and groups, which they contact as potential participants. Modern technologies are available to these organizers through telephone banks, mailing lists, or e-mail addresses. In preparation for a long-term event, some group organizers have the means to provide logistical support of transportation, food, and water to participants.

> During the preelection period, Bosnian-Serbs (in a bid not to lose any of the territory they had won during the war) began to organize this fiercely contested city. Serbian mobs attacked US forces with bricks, rocks, metal bars, and firebombs. Organization for this altercation started days before, as many witnessed the prestaging of the items to be used in the attack. Wheelbarrows, 55-gallon metal drums, and other large containers were filled with the ammunition that would later be used to attack peacekeepers.
>
> Brcko, Bosnia
> September 1996

CROWD BUILDING

1-13. The second phase of the gathering process is the building of the crowd. Old concepts of crowds based on stereotypes are no longer true. Studies show that—

- Crowds are made of many entities, and all participants are not the same.
- Crowds are not made of isolated individuals but of a minority of individuals and a majority of small groups who may or may not be acquainted with each other.
- Groups and individuals in the crowd are not unanimous in their motivation.
- Groups and individuals in the crowd are usually not anonymous to each other, especially different groups gathered at the same place.
- Individuals within a crowd are not given to unique emotional displays; although known within a group, most individuals would not want to draw attention to themselves.
- Groups within a crowd will often act in unison in an attempt to achieve their agendas.
- Crowds are generally not unique or distinguished by violence or unlawful conduct.

DISPERSAL PROCESS OF A GATHERING

1-14. The last and final phase of any gathering of a crowd is the dispersal process. It involves the movement of people from the common location where they assembled to one or more alternate locations. The dispersal process ends the gathering of a crowd or at least begins its decline. The dispersal can occur on a routine, emergency, or coerced basis.

ROUTINE DISPERSAL

1-15. The routine dispersal may be specified in advance. It can also be included in the assembly instructions given by the organizers of an event.

EMERGENCY DISPERSAL

1-16. An emergency dispersal occurs when people evacuate an area in result of an unexpected crisis, such as a fire, explosion, bomb threat, or terrorist act. Individuals in such an emergency quickly recover from the initial shock. They keep their wits about them and improvise ways of extricating themselves and their companions from the dangerous situation. However, tragedies have occurred in some emergency dispersal situations. Therefore, it is important that forces on the scene are careful not to misinterpret what is occurring.

COERCION DISPERSAL

1-17. Coercion dispersal is caused by the use of force at some level. This is not necessarily the best way to force the dispersal of a crowd. The negotiated management of crowds is the preferred method and has proven to be highly successful in getting crowd organizers to police themselves, especially if the demonstration and/or protest leaders are available and willing to participate.

CROWD DYNAMICS

1-18. Understanding crowds and how individuals and groups form crowds through the gathering process is important because these issues apply to the dynamics of a crowd. Under most circumstances, gathered crowds are orderly and present little or no problems for authorities. Crowds consist of people who, although very motivated and passionate, are also subject to their own need for creature comforts. Rainy, cold, and nasty weather has a way of disheartening all but the few highly motivated and disciplined individuals. If problems exist, they usually fall into the following three categories:

- **Public disorder.** Public disorder is a basic breach of civic order. Individuals or small groups assembling have a tendency to disrupt the normal flow of things around them.
- **Public disturbance.** Public disturbance is designed to cause turmoil on top of the disruption. Individuals and groups assembling into a crowd begin chanting, yelling, singing, and voicing individual or collective opinions.
- **Riot.** A riot is a disturbance that turns violent. Assembled crowds become a mob that violently expresses itself by destroying property, assaulting others, and creating an extremely volatile environment.

1-19. Being part of a crowd of people has certain effects on different people. Each individual in a crowd is susceptible to behaving in a way that is contrary to their normal behavior. Some reasons for these behaviors are as follows:

- Crowds provide individuals with a sense of anonymity. With so many others, an individual realizes that he is just another face in the crowd, giving a sense of invulnerability.
- Crowd and individual behaviors are impersonal by nature. The "them-against-us" attitude affords those within the crowd the ability to freely (without hesitation or reservation) be verbally abusive, throw objects, or attack anyone who gets in the way.

- Crowds provide individuals with the idea that their moral responsibilities have shifted from themselves as a person to the crowd as a whole. Large numbers of people discourage individual behavior, and the urge to imitate others within the crowd is strong. Individuals look to others around them for cues of what to do next, disregarding their own background and training. Often, it is only the strong, well-disciplined person who can resist the prevailing behavior of a crowd.
- Crowd behavior influences the actions of both the disorderly individuals of the crowd and the authorities tasked to control them.

1-20. Individuals within a crowd are at times driven by deep-felt emotions. Emotional contagion is the most dramatic psychological factor of crowd dynamics. It provides the crowd with a temporary bond of psychological unity. Lasting long enough, this unity can push a simple organized crowd into a mob. Normal law and authority are rejected en masse under these conditions, increasing the potential for violence and panic to erupt.

1-21. Panic can erupt quickly, especially when crowds turn into mobs. Individuals within the mob can easily sense that their safety and well-being are at risk, putting them in a "fight-or-flight" mentality. Adding to the panic and confusion is the use of riot control agents (RCAs) by authorities in an attempt to gain control. Individuals in a mob, during the heat of confrontation, may attempt to leave and find that there are no escape routes and that roads are blocked. This can often lead to violent, physical attacks.

CROWD TYPES

1-22. Gaining an understanding of crowds and the dynamics that cause individuals or groups to join together in a crowd should be every leader's responsibility. Active, expressive, acquisitive, and hostile are not acceptable terms used to describe a crowd. They are the motives inferred from the actions the crowd takes. No gathering of a crowd is persistently or exclusively active, expressive, acquisitive, or hostile. For the sake of description, crowds can be identified as casual, sighting, agitated, or mob-like.

CASUAL CROWDS

1-23. Casual crowds are identified as individuals or small groups with nothing in common to bind them together. If they have an agenda, it is their own. They arrive separately and leave separately. Casual crowds are made up of individuals or small groups occupying the same common place, such as a shopping mall where these individuals or small groups meet.

SIGHTING CROWDS

1-24. Sighting crowds are similar to casual crowds with one additional element—an event. People migrate as a crowd to sporting events, are attracted to fires and accidents, and attend music concerts. Individuals or small groups gather at these events for the same purpose. It is the event and/or one's curiosity that compels a crowd to come together.

AGITATED CROWDS

1-25. Agitated crowds add responses that are based on the elements (people, space, and event). Individuals with strong emotional feelings within a crowd can quickly spread and infect the rest of the crowd. As more people within the crowd become emotionally involved, a sense of unity can develop, causing changes in the overall demeanor of the crowd. Yelling, screaming, crying, and profane name-calling are all associated with an agitated crowd.

MOB-LIKE CROWDS

1-26. Mobs have all the elements found in the first three types of crowds, with the addition of aggressive, physical, and sometimes violent actions. Under these conditions, individuals within a crowd will often say and do things they usually would not. Extreme acts of violence and property damage are often a part of mob activities. Mobs consist of (or involve) the elements of people and groups being mixed together and becoming fluid.

CROWD TACTICS

1-27. During a public disorder or disturbance, individuals and small groups within a crowd use any number of tactics to resist authority and disrupt and add turmoil in order to achieve their goals. These tactics can be unplanned or planned and violent or nonviolent. The more organized and purposeful a crowd becomes, the more likely a tactic will be used.

NONVIOLENT BEHAVIOR

1-28. As explained earlier in this chapter, most gatherings of individuals and small groups into a crowd do not involves violent behavior. A public disorder or disturbance usually involves some harmless name-calling, demonstrations to express views, corporate yelling and chanting, and even singing and dancing. In some instances, there may be property damage and the erection of makeshift barricades. All four types of crowds can be nonviolent, but mobs and agitated crowds have the greatest tendency to turn violent. Nonviolent actions of a crowd are disruptive because they are in direct conflict with what authorities want them to do, such as refusing to leave when directed, locking arms, and sitting in front of or around areas and buildings that the authorities are attempting to clear.

VIOLENT BEHAVIOR

1-29. A crowd that becomes a mob can be very violent and destructive. Although some nonviolent activities occur, violent crowds strike out physically at bystanders or others in the crowd, destroying both private and government property, setting fires, and employing bombs (in extreme cases). The only limitations for violent crowds are their own imaginations, the training of their leaders, and the materials readily available.

1-30. Violent crowds may erect barricades and physical barriers to impede the movement of authorities and prevent them from entering certain areas and/or buildings and to hide their activities. Common materials used for barricades include vehicles, felled trees, furniture, felled light poles, fencing materials, or anything else on hand and readily available.

1-31. Setting fires and preventing firefighters access to the structure by blocking it is another violent crowd tactic. Often, fires are set to create confusion in authorities or as a diversion to another activity.

RIOTS

1-32. A riot is one or more groups or individuals who are part of a larger crowd that involves threats of violence against persons or property. In some cases, a crowd will continue to gather until it evolves into a riot.

1-33. Riots vary considerably in both targets and players. A communal riot, for instance, deals with deep-seated ethnic, religious, and language differences. Commodity riots involve an attack on property by acts of vandalism, looting, and arson. Protest riots, such as the riot around the World Trade Organization (WTO) Assembly in Seattle, Washington (30 November 1999), illustrated individuals and groups aggressively and sometimes violently acting out or voicing their opposition to the assembly. The Democratic Convention in Chicago, Illinois, in 1968 reflects a riot that directly targeted police and authority in general. Celebration riots occur across the US as a result of home team victories in sporting events, among other reasons. Celebrating crowds look to make the moment more memorable through raucous acts that demonstrate their joy or happiness, for example, the riots that took place in Chicago in 1992 as a result of the Chicago Bulls winning the National Basketball Association (NBA) Championship.

1-34. Commanders and leaders must be aware that highly organized groups of protesters and/or demonstrators have developed tactics to disrupt the control force. These crowd tactics were published in handbooks for communist organizers during the Cold War, and today the Internet contains sites that are devoted to sharing these tactics with the general public and various protest groups. Although these tactics are somewhat outdated, commanders and leaders must be aware of them and their potential consequences. Many of these groups, such as anarchists, often mock the tactics employed against large crowds; for example, they may identify the police as "Darth Vader Cops" when the control force response is in full riot control gear. Commanders must be aware that well-organized crowds may attempt to engage the control force, surround it, and overpower it by sheer numbers. To avoid this, commanders must do a detailed terrain analysis (intelligence preparation of the battlefield [IPB]) of the area to include all approach and exit routes in the area. The control force flanks have to be protected to avoid being enveloped by the mob. To avoid being enveloped, the on-site commander should keep the crowd at a comfortable distance from the control force formation, which is usually the lethal zone of his NL munitions. Establishing predetermined rally points for the control force is critical in the event that this type of tactic is used by the crowd.

TACTICS USED TO DEFEAT AUTHORITIES

1-35. Organized mobs will try to defeat the control force by employing several different types of tactics. These tactics include the following:

- Constructing barricades.
- Using Molotov cocktails, smoke grenades, rocks, and slingshots.
- Feinting and flanking actions.

BARRICADES

1-36. By constructing barricades, the rioters are trying to protect themselves from assault by the control force. If the control force does advance toward the barricade, rioters hope that the barricade will split the control force and leave it vulnerable to actions by the mob.

MOLOTOV COCKTAILS, SMOKE GRENADES, ROCKS, AND SLINGSHOTS

1-37. Angry mobs will often attempt to disrupt the control force by throwing rocks and other projectiles. As the situation deteriorates, the mob may escalate the violence by using a battery of slingshots that will pellet the control force with a barrage of projectiles ranging from small rocks to marbles. They may also use smoke grenades (homemade or store-bought) to mask their movements. Finally, Molotov cocktails are used against personnel, employed vehicles, and portions of the control force. The mob will attempt to gain a position above an armored vehicle to enable them to drop a Molotov cocktail into an open hatch.

FEINTING AND FLANKING ACTIONS

1-38. Organized mobs may attempt to disrupt the movement of the control force by feinting an assault. When the control force moves to blunt the assault, the mob will assault the exposed flanks in an attempt to split the control force and envelop a portion of it.

WEAPONS

1-39. Mobs will often use various types of weapons against authorities to achieve their agendas and goals. It is extremely important that leaders train soldiers to recognize possible threats of the various weapons used by crowds or mobs, enabling the soldiers to react with the appropriate minimum force.

VERBAL ABUSE

1-40. In almost every instance of a civil disturbance or riot, verbal abuse will be an aggressive tool. Obscene language, racial remarks, taunts, ridicules, and jeers should always be expected. It is apparent that the purpose for using verbal abuse is to anger, demoralize, and provoke a physical response. Undisciplined, untrained soldiers who face such an attack could cause the situation to escalate. Just one provoked action of a soldier could be interpreted as an act of brutality by the media.

PERCEIVED INNOCENT PEOPLE OR THE WEAK

1-41. Women, children, and the elderly have often been used as pawns in civil disturbance operations. On many occasions, the innocent and the weak are moved to the front of the crowd and used as a barrier. With the innocent and the weak directly facing authorities, aggressive and violence-prone individuals or groups behind them attempt to provoke authorities to react. Often, this tactic is to exploit the situation in the eyes of the media for propaganda value. Despite the perception of sometimes being weaker, women, children, and the elderly can be just as committed to the insurgency as those behind them.

BARRICADES

1-42. Barricades are made of anything large or heavy enough to impede or prevent the movement of authorities. Trees, vehicles, fires, and furniture are all examples of items used to erect a barricade.

THROWN OBJECTS

1-43. In many cases, a crowd will throw anything in their reach that can be picked or picked up. Rotten fruit and/or vegetables, rocks, bricks, bottles, sticks, and pieces of lumber are most often used because they are readily available. More sophisticated types of thrown objects are improvised bombs, such as petrol bombs or tennis balls with nails sticking out of them. Never discount that these individuals and groups could acquire hand-thrown explosives, such as sticks of dynamite or grenades.

VEHICLES AND OTHER OBJECTS

1-44. Vehicles and other objects can be placed on higher ground (atop a hill or in a building) to disrupt civil disturbance operations. For example, a vehicle or tire can be set on fire and rolled down a hill, carts or barrels filled with hazardous or flammable materials can be pushed down a hill, or burning furniture and petrol bombs can be thrown from two-story or higher buildings or toward a formation of soldiers.

FIRE

1-45. Rioters have set buildings, vehicles, and other structures on fire to block the advances of authorities and to create confusion and diversions. If property damage is the goal of a crowd, fire is an effective tool. A sniper, for example, may set a building on fire to divert attention or provide maximum target possibilities. Other tactics used, particularly in the Balkans, were to flood the lower floors or basements of structures with flammable liquids or gas and then ignite it in an opportune moment.

FIREARMS

1-46. Sniping or massing fire from within the crowd, in buildings, or other covered positions can (at times) be effective tools for terrorists using a crowd to cover their acts. Firearms are anything that fires a projectile toward a target of opportunity.

EXPLOSIVES

1-47. A bomber's imagination is the only limiting factor in the use of explosives. Explosives can be used as simple diversions to block the advances of formations or provide an escape for rioters. The worst use of explosives is to purposely kill, injure, or demoralize authorities that are protecting and assuring the safety of everyone.

Chapter 2

Control Force Operations

Information is the key to developing civil disturbance plans. One must know who the demonstrators are; when, where, and why they are demonstrating; what their capabilities are; and what their possible course of action (COA) is. A commander's need for current, valid information cannot be overemphasized. The commander must learn as much as he can about the participants (their motivations, strategies, tactics, targets, and dedication). The more knowledge a commander has about the participants, the better equipped he is to counter their actions. He needs sound information to decide how to best use his available resources.

CIVIL DISTURBANCE PLANNING CONSIDERATIONS

2-1. The side that possesses better information and uses that information to gain understanding has a major advantage over its opponent. Forces that have this advantage can use it to bring about changes in attitudes, decisions, and actions in assembled demonstrators. In planning for crowd control or civil disturbance operations, planners must decide what data is needed to develop threat assessments. This decision is based on a multitude of information sources.

STANDARDIZED POSTURES IN COMPANY TACTICAL STANDING OPERATING PROCEDURES

2-2. Within squads, platoons, and companies, equipment may be increased or decreased, as needed. Some examples are as follows:

- Choose the M9 pistol for extraction and apprehension teams. The use of a long weapon (for example, an M16 with an M203, an M4 with an M203, or a 12-gauge shotgun) with NL munitions capability is also recommended, especially for overwatch personnel.
- Add nonstandard weapons such as shotguns for greater NL capabilities.

NOTE: The shotgun is used to protect the M203 gunner as he reloads.

- Add NL munitions to existing organic weapons systems, such as the M203.
- Arm soldiers in the front line of the formation with their standard weapon. If the weapon is a long weapon, it should be carried across the back from left to right or vice versa, with the butt up and muzzle down. Ensure that the weapon is cleared and that the magazine is in the appropriate ammunition pouch.

- Balance the mix of weapons and munitions according to the mission, enemy, terrain, troops, time available, and civilian considerations (METT-TC).
- Establish command relations and determine who has or gives the authority to fire NL munitions.
- Maximize the distance and barriers between the crowd and control formations. Use NL munitions to create a standoff distance.
- Maintain a lethal overwatch of the control force at all times.
- Maintain a reserve force at all times to reinforce the control force, as needed.
- Create NLW range cards for static positions.
- Consider environmental conditions and their effect on the performance of NL munitions. Proper storage, periodic inspections, and the rotation of NL munitions are critical to maintaining the effectiveness and viability of NL munitions.

WEAPON AND AMMUNITION CONFIGURATIONS

2-3. Individuals designated as NL shooters must have the means to transition to lethal rounds, if required. Lethal rounds are carried separately from NL rounds so that the shooter will not confuse them in the heat of confrontation. However, in the heat of confrontation where a lethal option is necessary, NL rounds can be used lethally by adjusting the point of aim to a vulnerable part of the body and closing in on the distance to the target.

2-4. Squad leaders should designate NL shooters within their squad. Ideally, the squad should not change its organization to accommodate NLW.

2-5. Commanders should not dispatch NL patrols. They should plan a combat or security patrol with NL capabilities. Soldiers are never sent in harm's way without lethal protection. NLW are only considered additional tools for the mission and not a mission in itself. There is no such thing as a NL mission.

2-6. Soldiers manning a static position should have NL capabilities depending on METT-TC. Shotguns and the M203 work well at static positions.

2-7. Recovery, apprehension, and/or extraction teams should be established before deployment. Team members should be equipped with personal protective equipment (PPE), an M9, NL munitions and weapons, and Flex-Cufs®. They should have some training in open-hand control, pain compliance, and handcuffs and/or Flex-Cufs.

2-8. Crowd control formations (see *Chapter 6*) should be well trained and well rehearsed. Rapid, coordinated movements of a well-trained and well-rehearsed control force can often be a strong enough deterrent. A lethal overwatch marksman always covers the control force.

2-9. During a NL engagement, the use of designated marksmen (DM) provides confidence and safety to those facing a riot. The DM in an overwatch position scans the crowd to identify threats and designate personnel for recovery and lethal rounds firing (if a lethal threat is presented). They are ideally suited for flank security and countersniper operations.

NONLEHTAL PLANNING

2-10. The fundamental principles of mission planning are not changed by the incorporation of NLW into a plan. NLW are intended to be operational-enhancement tools. The first concern of a commander is the successful completion of the mission. NL munitions provide a commander with a wider range of response options, but these munitions are not a replacement for lethal capabilities. Commanders must always be capable of answering a lethal attack with a lethal response.

2-11. When a commander commits his soldiers and equipment to a crowd control situation, he commits his forces with the additional tools of NL capabilities. This is an addition to the force continuum that the force commander now has available to him. Commanders that properly employ NL munitions and weapons have a tactical advantage over those who rely completely on lethal means.

2-12. When developing a tactical standing operating procedure (TSOP) (see *Chapter 6*), units should consider the following:

- Crowd control formations.
- Extraction teams.
- Apprehension teams.
- Lethal overwatch marksmen and/or observer teams.
- Reserve and/or security forces.

MILITARY WORKING DOG TEAMS

2-13. Military working dog (MWD) teams may be employed with a control force formation as a method of increasing crowd apprehension about approaching or engaging the formation. The teams should be in the rear of the formation in plain sight of the crowd, but in front of the command element and the M33A1 squad RCA disperser. The MWD teams work back and forth behind the formation as an intimidation measure. The presence of the MWD may produce a profound psychological effect on the crowd. These teams may also be used to help control individuals who have been captured by the recovery and apprehension teams.

CAUTION

Do not unleash an MWD on a crowd.

COMBAT CAMERAMEN

2-14. Video and still cameramen should make a photographic record of the individuals in the crowd who are leaders and instigators. Events must be documented to hold personnel, factions, gangs, or groups accountable for acts that violate law, destroy property, or cause physical harm. Electronically recording events aids in the prosecution of such cases and eliminates the sense of anonymity that people in large crowds often feel.

INFORMATION OPERATIONS

2-15. Information superiority helps forces to anticipate problems and requirements. It allows commanders to control situations earlier and with less force, creating the conditions necessary to achieve the optimal end state. Public affairs, psychological operations (PSYOP), and civil military operations are activities that will allow the commander to control situations earlier and with less force.

2-16. Information is available from a multitude of sources. A diversity of sources is the best approach because it prevents biased perspectives. Primary sources are as follows:

- Open sources.
  - Libraries.
  - Newspapers and news periodicals.
  - Radio and television.
  - Internet.
- Law enforcement sources.
  - Local law enforcement agencies.
  - National law enforcement agencies.
- Military sources.
  - Department of Defense (DOD) intelligence community (most restrictive source).
  - Local military intelligence (MI) field offices.

2-17. Police intelligence operations (PIO) is one of the five military police functions. This function is a process of actively and passively collecting information that is of a police, criminal, or combat nature. As military police perform the other four functions of maneuver and mobility support (MS), area security (AS), law and order (L&O), and internment and resettlement (I/R), they are gathering information that supports, enhances, and contributes to the commander's protection program, situational awareness, and battlefield visualization by portraying relevant threat information that may affect operational and tactical environments.

2-18. IPB is a continuous process for analyzing the threat and the environment of a specific geographic area. During the IPB process, the Intelligence Officer (US Army) (S2) or Assistant Chief of Staff, G2 (Intelligence) (G2) uses all available databases, intelligence sources and products, and related MI discipline to analyze the threat and the environment. The PIO function supports this process by providing the S2 with collected police, criminal, and combat information that can directly and significantly contribute to the success of the MI effort. In addition to combat information, the PIO function provides additional information on possible criminal threats and COAs. This is intended to support the S2 IPB process and can be used by the commander to upgrade force protection. (See *FM 19-10.*)

THREAT ANALYSIS

2-19. Crowd control requires its own thought process. Emphasis should be on prevention rather than confrontation. In combat, military forces are taught to fight and eliminate threats. In crowd control, military forces must deal with noncombatants that have internationally recognized rights. These rights must be respected while maintaining public order. This is an issue that law enforcement agencies (LEAs) have been struggling with for years.

2-20. Dealing with crowd control incidents is a matter of using some basic guidelines. It also includes asking the right questions in a logical manner so that key issues are not omitted. The commander must use these guidelines in context with METT-TC and the location of the incident (CONUS or OCONUS).

PREINCIDENT PLANNING

2-21. This phase of planning begins before the incident and is initiated at the operational level with guidance from the strategic level. The planning includes guidance on crowd control and addresses responsibilities, training, organization, operating procedures, use of force (CONUS), and/or rules of engagement (OCONUS). The most difficult and productive decisions are those made in the preincident planning process.

Avoid Confrontation

2-22. Crowd situations are highly unpredictable, but one thing seems certain—confrontation will likely cause crowd resistance. When pushed, people tend to resist opposition to the realization of their purposes.

Focus on Prevention

2-23. Planning should key in on the prevention of unfavorable outcomes. Experience has shown some LEAs attempting to help crowds accomplish their goals within the law that have been beneficial and even led to conceding some violations for the purpose of avoiding confrontation. However, LEAs maintain a law enforcement presence, which signifies social restraint. LEAs also decide when and where they will not compromise and the amount of force to use.

Define Goals

2-24. When defining a goal, deciding what must be accomplished is the first step. Defining the goal is fundamental. However, actually working toward and accomplishing the goal is easier said than done, as the process can drive the situation. Commanders and leaders must be aware of this. The military force must focus on what they are trying to accomplish.

BACKGROUND INFORMATION AND INTELLIGENCE

2-25. Seek to know as much as possible about social protest groups (within the limits of the law, see *Appendix B*) before an incident. This will provide insight into the organization and its functions and provide a warning as to what to expect. When you know as much about these organizations as possible, then it is possible to anticipate their next move. See *Appendix C*.

2-26. Continually assess what is to be accomplished against what others are trying to accomplish. Some groups may not have goals that conflict with those of US forces, but their actions may. Other groups may have goals and activities that do conflict. In either case, the assessment of group goals compared to the goals of US forces helps to understand and avoid potential conflicts.

CIVIL DISTURBANCE OPERATIONS PREPARATION

2-27. Counterdemonstration workgroups are a mixture of traditional and nontraditional staff proponents brought together to coordinate the resources to be employed by task force (TF) units before a potential civil disturbance. Meetings of the counterdemonstration workgroup are held on a regular basis (generally weekly) and are chaired by the Assistant Chief of Staff G3 (Operations and Plans) (G3). The following staff proponents may be involved, and others may be involved as needed:

- Public affairs office (PAO).
- Joint military commission (JMC), if organized.
- Provost marshal office (PMO).
- PSYOP.
- G2, Assistant Chief of Staff G5 (Civil Affairs) (G5), Assistant Chief of Staff G6 (Signal) (G6).
- Civil affairs (CA).
- Army airspace command and control (A2C2).
- Chaplain.
- Engineer.
- Surgeon.
- Fire support element (FSE).

RELATIONSHIPS DEVELOPMENT

2-28. Working relationships between commanders and protest group leaders are increasingly seen as the best means for preventing bad outcomes in crowd situations. This is called the negotiated management model of crowd control. LEAs in large metropolitan cities in the US and Europe practice it.

2-29. Open dialog helps develop working relationships between commanders and protest group leaders, providing an opportunity to communicate clearly. Talking allows group leaders to tell authorities and tactical commanders what they want to accomplish. It also allows authorities and commanders to tell group leaders what they are prepared to do and how they might respond to certain crowd behaviors. Such communications can do much to resolve issues and prevent violence.

2-30. Commanders may not be able to talk to all leaders before a crowd assembles. Some groups do not have recognized leaders and are ad hoc organizations. Other groups may have several leaders, but only some of those leaders will negotiate. In such cases, meetings with some leaders may tell you who the other leaders are and provide critical information. Negotiations may also encourage more moderate leaders to do things that will support the commander.

2-31. Initiate communication with group leaders to work out issues before an event. Commanders should make a concentrated effort to maintain a working relationship with the leaders of protest groups.

2-32. Communication also means persuasion. Commanders should make a concerted effort to win over demonstration leaders. Commanders may tell group leaders that they want to help them complete their mission in a safe manner. They should encourage demonstrators to protest in an acceptable way and, if possible, offer favors to get them to do just that.

NOTE: Commanders must only offer what they are willing and able to deliver.

2-33. Several rules apply in negotiations, and credibility is key. Only communicate necessary information and those actions that authorities and commanders intend to do. Negotiations are made from a position of strength, and by negotiating an agreement may be reached. However, contingencies should be developed in the event that the agreement is violated.

2-34. Working relationships between commanders and protest group leaders often result in protest groups policing themselves. This is one of the basic premises of negotiated management. Protesters are more likely to listen to their own leaders, as opposed to listening to US forces. Protest organizers are likely to buy into a cooperative effort and agree to ground rules set during preprotest negotiations. Many protest leaders seek to stay on the right side of the law.

ACCEPTABILITY

2-35. Deployed US forces will find themselves engaged in crowd control operations under difficult circumstances. The host nation (HN) may have groups that do not accept the presence of US forces, which will make imposing order and the protection of citizens difficult at best. US forces will be under intense media and political scrutiny (an environment much like what LEAs operate in on a day-to-day basis).

2-36. Winning in this environment is not like winning in combat. US forces may appear to be invincible and formidable, but they risk being portrayed as oppressors. Thus, US forces can lose by appearing to win. Groups that perceive themselves as oppressed will readily seek victimhood in an effort to gain the support of public opinion. Winning in this environment is about seizing and holding the moral high ground. US forces must maintain the authority and legitimacy of what they are doing.

2-37. Projecting a favorable image will require outreach to local leaders and citizens. It will also require developing a relationship with the media. For example, the Los Angeles Sheriffs Department often invites the media to accompany their tactical commander during crowd control situations. This shows that they have nothing to hide. It also provides an opportunity for individuals to see the commander's side of an event.

2-38. In this environment, commanders must consider how actions will play among several audiences—local, allied, US, and international. The media effect is inescapable. In some cases, it will be difficult to accommodate all of these audiences. However, the most important audience will be the US public, which is key to the continued support of US forces.

CROWD ASSESSMENT QUESTIONS

2-39. Assessing crowds requires its own set of questions. These questions should be answered before a crowd assembles or as quickly as possible if a gathering occurs without notification. In some cases, all the questions may not be answered. These questions represent a logical way to think about crowds. The questions are as follows:

- **Who are they? What is the overarching identity of the crowd? Are they strikers, ethnic factions, or social protesters? Do they identify themselves as strikers, ethnic groups, religious factions, or protesters against some perceived social injustice?** Understanding who they are will indicate what they may do. It may be possible to determine identities and goals from advance assembling instructions, leaflets distributed to bystanders, placards and banners, and chants and songs.
- **What are their goals?** What the group wants to accomplish by assembling could determine the extent to which they can be accommodated once they have assembled. They may only seek recognition for their cause (being seen and heard). If so, this goal is usually easy to accommodate. However, some groups may have more demanding goals, for example the demonstrators in Seattle who sought to stop the WTO. Goals that cannot be accommodated make confrontation very likely.
- **What is the composition of the crowd and are there any known factions?** Seattle demonstrations against the WTO were comprised of groups that were protesting environmental issues, wages, and child labor laws. Differing goals and the resulting friction were evident between such organizations as the Ruckus Society and the American Federation of Labor-Congress of Industrial Organizations (AFL-CIO). Factions within a crowd represent threats and opportunities.
- **What are they capable of doing?** Protest groups often claim that they will assemble large numbers of people to produce some disruptive action. However, there is often a difference between the claim and reality. An organization may claim that it can mass a 100,000 people, but in reality can only get 40,000. Organizers exaggerate for the following reasons: they want to boost the morale of their own people and they want the media to report that they have strength in numbers. Studying the past activities of a group may provide indications of what they are capable of doing in the future.
- **What are their traditional behaviors or cultural repertoires?** What people do during protests is not universal. It varies with the group and the culture. Social protest organizations and striking unions will carry placards and banners. Other groups will protest in a more quiet way, like the 5,000 women in Sarajevo whose standard Sunday behavior was to sit and block traffic. They were protesting the loss of male relatives in the Bosnian war. Understanding the goals of

the protesters can be helpful in deciding how to respond to their behaviors.

- **When and where will they assemble?** Every protest organizer has a time and place for assembling and perhaps a destination for the crowd to move toward. If the organizer attempts to mobilize large numbers of participants, the time and place for assembling and dispersing must be made known in the instructions. This information may be stated in the mobilization instructions or disseminated by an informal network, such as word of mouth. Such impromptu networks in densely populated areas can enable rapid assembling.
- **Where will they go?** Many crowds have destinations. Organizations may march a specified distance to ensure that their cause gets sufficient attention. Commanders need to know the route to minimize disruptions to the rest of the community. In Los Angeles, LEAs attempt to reroute traffic and prevent congestion caused by protest marches. It is also necessary to provide security along the route to prevent counterdemonstrators from confronting the marchers. This will prevent an even greater problem for the community.
- **What are the possible targets of violence?** Riots, in particular, may focus on target facilities. In the 1992 Los Angeles riots, gun stores were major targets. In developing countries, targets may be more basic. In Somalia, throngs of people stormed food supply facilities. In Haiti, riots broke out over trash dumps where US forces had discarded the remnants of meals, ready to eat (MREs).
- **What is the worst-case scenario?** The worst-case scenario must be recognized for the sake of avoidance. This may be when a peaceful crowd degenerates into a violent riot.
- **When and where will they disperse?** Crowds have a life cycle that includes how they disperse (see *Chapter 1*). Commanders must consider this. It is essential that there be one or more avenues that individuals can use to disperse. These dispersal routes should be clearly marked, visible, and open-ended. There may be a need for US forces to assist the crowd with dispersal. For example, in Los Angeles a 70,000-man march went from east to central Los Angeles. Once at their destination, they engaged in peaceful activities and then dispersed. The LEAs provided transportation back to the assembly area (AA); otherwise, the protesters would have returned to the AA in a disorganized and unsupervised manner.
- **Are there plans for subsequent gatherings?** A crowd may disperse for a short time so participants can take care of personal needs. This could be days or only a few hours.

LEGAL CONSIDERATIONS

2-40. The legal authority in which a peace operation is conducted defines the parameters of the operation. All commanders, leaders, and soldiers must know the legal authority that regulates their operations.

MANDATES

2-41. There are two types of mandates—United Nations (UN) and non-UN. Mandates are the authority under which an operation is conducted. UN mandates are created from resolutions by the UN Security Council (UNSC) or the UN General Assembly (UNGA) that authorize and define an operation. Non-UN mandates are created from treaties, accords, resolutions, or agreements evolving from other international or regional organizations.

STATUS OF FORCES AGREEMENT AND STATUS OF MISSION AGREEMENT

2-42. A status of forces agreement (SOFA) and a status of mission agreement (SOMA) are key documents that define the legal authority and responsibilities of a force and force personnel participating in an operation. A SOFA or SOMA may be a treaty or memorandum of understanding. It is an agreement negotiated between the UN and the host countries, which details the rights, privileges, immunities, and nature of services to be provided to the force and its personnel and their responsibilities and obligations.

INTERNATIONAL LAW

2-43. Legally, international law affects most stability operations and support operations (SOSO). These laws consist mainly of international agreements, treaties, and customary international law, which include the agreements and laws known as the law of war. International agreements prescribe the rights, duties, powers, and privileges of nations relative to particular undertakings. International agreements will affect US participation in SOSO in the—

- Right of US forces to enter a foreign country.
- Status of US forces while in the foreign country.
- Construction and operation of US bases of operations.
- Aircraft overflights and landing rights.
- Processing of claims for any damage to persons and property of the foreign country.

2-44. These agreements or customary international laws govern all aspects of operations carried out. In the absence of a viable HN government or proper international agreement, many aspects of the operations are controlled by domestic (US) or customary international law.

UNITED STATES LAW

2-45. Operations in peace and conflict must comply with US law, whether as a statute, executive order, regulation, or other directive from a branch or agency of the federal government. The *Uniform Code of Military Justice (UCMJ)* applies to questions of military justice. Various statutes, *Executive Order 12333; Section 3, Code of Federal Regulations (CFR);* and DOD and service regulations govern all intelligence-gathering activities. The staff judge advocate (SJA) must actively advise and participate in all activities from initial planning to redeployment.

HOST NATION LAW

2-46. If a viable HN government exists, whether at the national or the local level, all laws of the HN will apply to US forces in that country unless an international agreement provides otherwise. The types of laws that may inhibit US operations are in the fields of immigration, labor, currency exchange, the procurement of goods and services, customs and taxes, and criminal and civil liabilities. A SOFA usually covers these areas before an operation begins. Therefore, planners must understand the law to assess it and determine if it will adversely affect the operation. Assistance may be available from the local US embassy or the command judge advocate; if not, the command may have to rely on other sources for guidance. If local law hinders the operation, the commander must inform the US embassy and request that it negotiate a solution.

CIVIL AUTHORITIES

2-47. There will be times that US forces will deploy to nations where the infrastructure of civil authorities has survived the conflict and is functional. Local authorities and police are responsible for ensuring the safety of their citizens, including the UN High Commissioner for Refugees (UNHCR). US forces assist in this by providing general military security and by facilitating negotiations. In countries where the civil authority infrastructure has collapsed, the UN mandate may require the coalition of forces to help ensure the safety of citizens until their own civil authorities can be rebuilt.

RULES OF ENGAGEMENT

2-48. The rules of engagement (ROE) are the commander's rules for the use of force (RUF). Commanders are responsible for interpreting, drafting, disseminating, and training on the ROE. Ensure that the SJA assists throughout the ROE development process. This will ensure that the ROE will not improperly constrain actions, but are still consistent with domestic and international law and policies and orders of the chain of command.

2-49. Clearly stated ROE are published before Army forces are committed. The highest military authority, with input from subordinate commanders, will continually evaluate the ROE and modify them as appropriate. When the use of NLW has been authorized in the ROE, the decision to use NLW should be delegated to the lowest possible level, preferably the platoon or squad level. This requires that all personnel, not just leaders, have a clear understanding of the ROE and the commander's intent (see *FM 3-22.40*).

2-50. In peace operations, the use of force is restrained while the diplomatic solution is sought through negotiation or mediation. Accordingly, ROE are more restrictive in peace operations, making them more politically sensitive. In today's world, peace operations are usually conducted by a coalition of forces operating under the purview of the UN charter and customary international law. Therefore, the UN may mandate certain restraints. Using overwhelming force could compromise diplomatic efforts to reach a peaceful settlement. Even a single, relatively small engagement could jeopardize the legitimacy of the peacekeeping force and upset negotiations or mediations. Commanders must beware that in any confrontation ROE decisions made by soldiers can have strategic political implications on current and future operations. Therefore, it is vital that leaders and soldiers hold a common understanding of when, how, and to what degree force is to be used. This requires training.

MULTINATIONAL OPERATIONS

2-51. Multinational operations include military forces from different nations. These operations are usually interagency operations. Conducting operations with foreign military partners, like operations with civilian partners, is uncommon to many soldiers, so a clear understanding of this different environment is necessary.

2-52. The majority of US peace operations will be part of a UN peace operation. Their multinational character merits particular attention because national interests and organizational influence may compete with doctrine and efficiency. Commanders can expect contributing nations to adhere to national policies and priorities, which at times complicate the multinational effort.

2-53. When working within a multinational force and with allies, do the following:

- Train coalition partners and allies on tactics and techniques.
- Remember that teamwork and trust are essential.
- Achieve unity of effort.
- Consider equipment and communications capability and interface.
- Strive to eliminate sources of confusion and misunderstanding.
- Remember that military doctrine varies from country to country.
- Remember that national culture influences the way units operate.
- Remember that language barriers represent a significant challenge.

SCALABLE EFFECTS

2-54. Scalable effects concepts are a measured approach in response to a crowd gathering. By recognizing a use-of-force policy, soldiers must be taught and understand that they use the minimum force necessary. Without the appearance of a graduated response, the gathering crowd may consider actions as excessive, causing a possible escalation of hostilities or violence. Do the following to aid in the scalable effects process:

- Try to persuade the crowd to quietly disperse by talking with leaders.
- Use translators as necessary.
- Let the first approach be by local authorities (the mayor or police).
- Pass out handbills requesting that the crowd return home.
- Use video and still cameras to photograph individuals and events for later use at trials.
- Give warnings before moving to the next level of force.

2-55. Ideally, the force should be positioned out of sight of the crowd. Have the troops move into position with few shouted commands. For maximum effect on the crowd, have them form their formation decisively and professionally. Do the following to aid in the formation process:

- Display force in a graduated manner (for example, a UH 60, hardened vehicles, and soldiers with charged weapons).
- Exploit the psychological effects of show of force.
- Demonstrate force (do not use on unarmed civilians).
- Escalate the PSYOP message.
- Highlight the target pointer.
- Demonstrate sniper precision strike capability.
- Employ riot control measures.
- Use NL munitions.
- Use RCAs (if approved).
- Move through the crowd using riot control formations and movement techniques.
- Use the graduated response matrix (GRM)

2-56. With the increased participation of US forces in contingency operations around the world, it accentuates the need to establish procedures for applying graduated military responses to situations that threaten these missions. Numerous GRMs and similar products exist throughout the military. These products graphically portray available responses in a graduated manner. The intent of these products is to give the on-scene commanders a list of options with which to control or defuse a situation before it gets out of hand. Most threats can be eliminated without loss of life or collateral damage by effectively applying the resources available.

PLAN A GRADUATED RESPONSE MATRIX

2-57. The planning for and development of a GRM begins with the mission analysis portion of the military decision-making process (MDMP). Some missions require US forces to enforce treaties or accords to protect the lives of civilians in uncertain or hostile environments (such as noncombatant evacuation operations [NEO]) or to provide large-scale humanitarian assistance. These missions require some sort of graduated response criteria to maintain order and prevent uncertain environments from becoming hostile. The development of a GRM requires the following seven steps:

- Identify the need for a GRM.
- Establish a team to develop the GRM.
- Develop targets.
- Coordinate staff functions.
- War-game.
- Receive command approval.
- Operate deliberately.

Identify the Need for a Graduated Response Matrix

2-58. Not all missions require a GRM. The decision to use a GRM requires careful consideration. Once it is agreed that a GRM is necessary, it requires guidance from the commander regarding the response options available.

Determining the appropriate responses is based on the facts, assumptions, and limitations identified during mission analysis.

2-59. Planners (staff) must agree on the intent of the GRM. The GRM can be used as a training and rehearsal tool. It provides leaders with the most likely vignettes that can be incorporated into a COA analysis, predeployment training, and rehearsals. The GRM can also be used as a handy reference during situations that require graduated responses.

Establish a Team to Develop the Graduated Response Matrix

2-60. Establish a GRM development team with a broad range of skills, knowledge, and professionalism. The following list of individuals makes up a typical GRM team:

- Provost marshal or senior military police advisor.
- FSE.
- Brigade legal officer.
- PSYOP representative.
- Land information warfare or information operations officer.

2-61. Since the GRM is designed to give commanders graduated options for dealing with both hostile and nonhostile threats to the mission, this team composition allows for target selection, application of the ROE, and attack methods using both NL and lethal means.

Develop Targets

2-62. The FSE, in conjunction with the S2 section, develops targets for both lethal and NL attacks. In the case of stability operations, these targets are usually not the conventional specific point or piece of equipment on the ground. They are more situational than specific. The GRM identifies situations or acts that subordinate elements could face during the mission. The sample GRM in *Figure 2-1* shows three possible situations or acts that on-scene commanders could expect to encounter. From the targeting standpoint, these are groups of more specific targets.

2-63. During mission analysis, the fire support officer (FSO) identifies both lethal and NL assets available to the unit. A tactical PSYOP team (TPT) attached to the unit is an example of a NL attack asset that should not be overlooked. The following are examples of what the FSO should look for:

- RCAs.
- TPTs.
- Electronic warfare assets.
- CA teams.
- Information operations (IO) teams.
- Artillery smoke projectiles.
- Aircraft (AH-64s, OH-58Ds, and AC-130s).
- Mortars.

2-64. The lethal assets described could be used in a NL show of force or demonstration to defuse a situation before it requires lethal force. The critical element of this mission analysis is not to focus solely on lethal assets. In

<table>
<tr><th>Crowd Action</th><th colspan="4">Nonlethal Response</th></tr>
<tr><td></td><td>Command Presence/PSYOP</td><td>Show of Force</td><td>Demonstration of Force</td><td>Riot Control Means (if approved)</td></tr>
<tr><td>Unarmed Civilians</td><td rowspan="3">- Ensure that the on-site commander of the TPT directs the broadcast of the dispersal proclamation and/ or passes handbills.
- Escalate the tone of the dispersal proclamation from information to a warning of force.</td><td rowspan="3">- Display force along with escalating the dispersal proclamation.
- Display force in a graduated manner, such as a helicopter hovering over a crowd or soldiers with charged weapons.
- Exploit the psychological effect of a show of force.</td><td>- Do not demonstrate force toward unarmed civilians.</td><td rowspan="2">- Employ RCAs at the point of penetration.
- Use PSYOP to exploit the psychological effect.
- Move through the crowd using riot control formations and movement techniques.</td></tr>
<tr><td>Armed Civilians (knives, clubs)</td><td rowspan="2">- Display force along with escalating the dispersal proclamation.
- Highlight the target pointer.
- Demonstrate sniper precision strike capabilities.</td></tr>
<tr><td>Armed Crowds/ Military (firearms)</td><td>- Do not use RCAs—they may escalate the situation.</td></tr>
<tr><td colspan="5">Hostile intent/hostile act occurs by armed threat.</td></tr>
<tr><th>Crowd Action</th><th colspan="4">Lethal Response</th></tr>
<tr><td></td><td>Sniper Response</td><td>Small Arms Direct Fire</td><td>Aerial Fires</td><td>Indirect Fires</td></tr>
<tr><td>Unarmed Civilians</td><td rowspan="3">- Ensure that target leaders or troublemakers are targeted.
- Use the minimum response necessary.
- Exploit the psychological effect of an attack.</td><td rowspan="3">- Determine that sniper attack is insufficient.
- Consider demonstrating capabilities.
- Exploit the psychological effect of a lethal response.
- Escalate gradually, starting with a small caliber, single round and work up to a large caliber, automatic.</td><td>- Determine that small arms direct fire is ineffective.
- Use the minimum response necessary.
- Use a minimal precision strike initially; use subsequent fires based on the situation.
- Exploit the psychological impact of each strike.</td><td>- Determine that air assets are unavailable or ineffective.
- Use the minimum response necessary.
- Ensure that the response is directed by the on-site commander.</td></tr>
<tr><td>Armed Civilians (knives, clubs)</td><td colspan="2" rowspan="2">- CAS/indirect fires must be authorized by the MACOM commander.
- Consider requesting permission for use when—
• All lesser means have been ineffective.
• There are physical eyes on target.
• Proximity to civilians has been considered.
• Risk to friendly forces/evacuees outweighs the risk of collateral damage.</td></tr>
<tr><td>Armed Civilians/ Military (firearms)</td></tr>
</table>

Figure 2-1. Sample GRM Card

Sample

stability operations, the Army wants to prevent acts of hostility first and then be prepared, if necessary, to apply lethal force.

2-65. Graduated responses can range from command presence through the show of force, a demonstration, the use of RCAs, and other techniques (such as the application of lethal force using snipers, small arms, AC-130s, and indirect fires).

Coordinate Staff Functions

2-66. To coordinate staff functions the rest of the GRM team assembles to complete the escalation sequence for each response. PYSOP and legal representatives are critical attendees during the escalation sequencing process. In the area of psychological operations, the TPT must exploit the effects of all responses.

2-67. The legal officer evaluates each escalation of force option and graduated response to ensure that it is consistent with the ROE. The GRM is designed to recommend applications of force consistent with the ROE, yet not limit the leader or individual soldier's right of self-defense. A GRM should clearly show that if a hostile act occurs, lethal options would be first and foremost.

2-68. In the case of lethal responses, the commander's guidance must again be applied. For example, lethal responses are allowed only in self-defense.

War-Game

2-69. Once the types of escalations for each potential graduated response are determined and annotated, the GRM must then be war-gamed. The staff must walk through each act or situation from the on-scene commander's standpoint.

Receive Command Approval

2-70. Once the GRM has been war-gamed, it must be submitted to the commander for approval. This is the final check to ensure that the GRM team has applied the commander's guidance correctly and met his intent.

Operate Deliberately

2-71. Operating deliberately affords commanders ample opportunity to plan and prepare for possible civil disorder situations. Through the effective gathering of information and a working cooperation with local government and police officials, commanders can often be made aware of dates, times, locations, and what groups may assemble before the operation. The purpose is to gather as much information as possible who is involved, where they are assembling, what incident promoted the activity, and what seems to be the prevailing attitude of the assembling crowd.

2-72. Civil disturbance operations are dynamic, ever-changing environments requiring effective communications both up and down the chains of command. They require advance preparation and planning using the established troop-leading procedures (TLP) outlined in *FM 3-19.4*. The eight-step process of TLP aids commanders and their subordinate leaders in planning and preparing for an operation. Most steps can be accomplished concurrently, but rarely is there enough time to go through each step in detail. Despite this, leaders must follow these steps to ensure that nothing is left out of the planning process. As outlined in *FM 3-19.4*, the following are the eight steps of TLP:

- Receive and analyze the mission.
- Issue a warning order.
- Make a tentative plan.
- Initiate movement.
- Conduct a reconnaissance.
- Complete the plan.
- Issue the order.
- Supervise, refine, and rehearse.

CONDUCT HASTY OPERATIONS

2-73. Conducting hasty operations seems to be the norm rather than the exception. Unlike deliberate operations, hasty civil disorder operations are reactionary in nature with little or no time for planning. In these situations, the event of a crowd gathering is already underway. There is very little, if any, advanced warning of the unfolding situation, and commanders are usually put in the position of sending their soldiers into an already volatile, and perhaps hostile, environment.

2-74. Commanders should be in immediate and constant communication with local civil and police authorities. The purpose is to gather as much information as possible about who is involved, where they are assembling, what incident promoted the activity, and what seems to be the prevailing attitude of the assembling crowd.

2-75. It is essential that commanders operating in environments where civil disturbance operations are likely ensure that their soldiers receive training at all levels. Commanders and subordinate leaders need to instinctively rely on the same TLP that they would in a deliberate operation. See *FM 3-19.4*.

CONDUCT TACTICAL OPTIONS

2-76. In the peacekeeping and peace enforcement environment, the use of conventional firearms or the threat of their use may not be the solution to a situation where US forces must separate two belligerent, hostile ethnic groups or prevent a similar group from entering an area that is off limits to them.

2-77. Crowd control options are often combined. Commanders choose their options based on an evaluation of the particular crowd. Commanders select any combination of control techniques and force options they think will influence the particular situation (METT-TC). Commanders must always try to choose the response that can be expected to reduce the intensity of the situation.

Monitoring

2-78. Monitoring an assembled crowd consists of gathering necessary intelligence and watching them to determine progress and development. It is this gathered information that helps commanders select the appropriate actions. Gathered information can often be used to help defuse a situation through persuasion.

2-79. Monitoring is continuous. Without timely information, responses could cause the situation to escalate beyond what it should have or cause a response to be inadequate based on the situation.

2-80. Commanders may task teams with specific missions to monitor crowd activity and note any new developments. Monitoring done by these observation teams gives the commander up-to-the-minute information so he can gauge crowd activity and intent in relation to the overall situation.

2-81. Observation teams monitor crowd activities to gather information. They observe and report on crowd size, location, and mood and on the developing situation. An observation team may consist of a marksman, a radio operator, and an observer equipped with binoculars. They may be posted strategically on rooftops and other high terrain that overlooks the crowd. Sometimes observers use helicopter-mounted observation devices. This also affords security for the operational forces. Commanders must know where their observation teams are located so they are not mistaken for teams of snipers.

2-82. The timely flow of information may allow the commander to influence the outcome of the situation with simple negotiations. Monitoring is appropriate when more decisive action is not feasible because of crowd size or when the intensity of a situation might escalate. It is particularly useful in large, nonviolent demonstrations. Monitoring can serve as an interim measure until more control forces arrive. It includes establishing communication with crowd leaders to convey official interest and intent to the crowd. Monitoring also includes efforts to gain the cooperation of crowd leaders.

2-83. Communication with crowd leaders and participants can help a commander control a situation without more severe measures. If communications exist with crowd leaders, the authorities may be able to divert the leaders or the crowd from their stated or apparent goal. Pressure can be put on leaders to channel the crowd into an area that minimizes disruption to the community and aids crowd control operations. March routes and demonstration areas can be limited to those that will help contain the crowd and reduce their potential for disrupting the community. Pressure can be positive (offering concessions) or negative (with deterrents).

2-84. If the commander can gain the cooperation of crowd leaders, it can decrease the potential for disorder, and if crowd leaders seek cooperation from authorities, officials should try to be accommodating. Crowd leaders can be placed in liaison positions between the crowd and the control force. Crowd leaders can be made responsible for managing the crowd by policing their own activities.

2-85. Taking still pictures or videotaping the faces of individuals within a crowd may prevent or reduce unlawful and violent acts. This could lessen their sense of anonymity. If needed, photographs or videotapes can be used as evidence for prosecution. To be effective, crowd members must see their presence being recorded. The photographer or cameraman should be in uniform to let the crowd know who he or she is. Recorders must be close enough to the crowd to be seen easily, but not close enough to be endangered.

Blocking

2-86. It is not uncommon for protests and demonstrations against the US or its forces to occur outside the main entrances to military installations or US Embassies. Because of the sensitive nature of equipment and information that can be found at these locations, blocking unauthorized entry is critical.

2-87. Blocking is the physical denial of advance upon a facility or an area by a crowd. Commanders may have to task their forces to block a crowd. Blocking physically denies crowd advance. Crowd control formations, particularly line formations, along with barricades can be used to block advancing crowds. Barricades of vehicles, concertina wire, and water-filled barrels are used to block or channel the movement of the crowd.

Dispersing

2-88. Dispersing is taking deliberate actions to fragment an assembled crowd in order to prevent the destruction of property or prevent injury. It is extremely effective against smaller crowds in congested urban environments. However, this may increase and spread lawless activity rather than reduce it. Thus, the commander's forces must control dispersal routes and the areas in which the dispersal will occur. Forces must protect the facilities that could be considered likely targets of opportunity for dispersing individuals or groups. Dispersal of the crowd may require apprehension of small groups still active in the area. Proclamations, shows of force, crowd control formations, and RCAs can disperse crowds.

2-89. Avoid (if possible) the dispersion of crowds into wide-open areas because it gives the crowd the opportunity to grow rapidly in size. The direction of crowd movement is difficult to anticipate and channel.

2-90. Once the crowd has started to disperse, it may occasionally be necessary for the forces to advance, keeping the crowd on the move and in the right direction. A crowd on the move should not be hurried to avoid panic. At no time should the crowd be cornered in a position where there is the perception of no escape. This invokes the "fight-or-flight" syndrome, possibly escalating violent activity.

2-91. Issuing a proclamation (see *Figure 2-2,* page 2-20) can help disperse a crowd. Proclamations officially establish the illegal nature of crowd actions and put the populace on notice that the situation demands special measures. Proclamations prepare the people for the presence of military authority and it tends to inspire respect. Proclamations support law-abiding elements and psychologically bolster military forces trying to restore order. They also demonstrate the gravity of the situation to all concerned and are excellent ways to make a commander's intentions known to a crowd. It is also a good way to reduce crowd size before direct action is taken. Commanders can make a verbal proclamation similar to the following statement: "*In the name of the President of the United States, I command that you disperse and retire peaceably to your homes.*" Such a proclamation may even make direct action unnecessary.

I. DISPERSAL

Unlawful Gathering

"Attention! Attention! This area must be cleared at once! Further unlawful behavior will not be tolerated. Clear this area at once or the necessary force to do so will be used."*

Demonstration

"Disperse and retire peaceably! Disperse and retire peaceably! Attention all demonstrators!"

"The demonstration in which you are participating ends at ______. The permit that was agreed to by the leaders of the demonstration expires at that time. All demonstrators must depart from the ______ by _____. All persons who wish to leave voluntarily may board the buses. These buses will go to the _____. Those who wish to take buses should move to ______. Those demonstrators who do not leave voluntarily by ______ will be arrested and taken to a federal detention center. All demonstrators are urged to abide by the permit."*

Warning of Looting

"Return to your homes! Someone may be looting them at this moment! During a disturbance, criminal activity is at its peak. Your family or your property may be in danger."*

II. EMPLOYMENT OF TROOPS

Employment

"Attention! Attention! Troops are present in this area. They are preparing to advance. Order must and will be maintained. Disperse peaceably and leave this area. To avoid possible injury, leave at once."

"Disperse now and avoid possible injury! Disperse now and avoid possible injury!"*

(Repeat until troops are committed.)

When Troops Are Committed

"Troops are advancing now. They will not stop until this crowd is dispersed and order is restored. To avoid injuries, leave the area at once. Return to your homes as peaceful citizens. Troops have their orders, and they will not stop until the crowd is dispersed. Do not get hurt. Leave this area."*

III. PRESENCE OF CHILDREN

(Used in conjunction with other announcements.)

"Attention! Attention! Do not attempt to cause further disorder. Disperse now in an orderly manner and avoid possible injury to children. Return at once to your homes."*

IV. ADDITIONAL INSTRUCTIONS

*Indicate the method, streets, and direction that the crowd should use when dispersing.

Sample

Figure 2-2. Sample Proclamation

2-92. In making a proclamation, a commander may consider imposing a time limit. However, the situation may change, and not imposing a time limit leaves the commander free to choose other actions when he wants. A proclamation must be specific in its instruction.

2-93. If a time limit is stated, it must be a reasonable length of time for the crowd to comply with the instructions. When drafting a proclamation, the commander must consult closely with the SJA. He must use the simplest language possible to maximize the effects of the proclamation. If proclamations must be translated to a local language, the translation must be made with great care.

2-94. The commander at the scene may direct that a proclamation be issued over public address (PA) systems. The force of the words used in the proclamation must be gauged to the composition of the crowd. If the crowd consists of normally law-abiding citizens who are presently assembled to show disagreement with an existing situation, the proclamation requires less force. On the other hand, if the crowd consists of militant rioters, the proclamation requires more force. The text may take a number of forms, depending on the situation.

2-95. A show of force is often a useful measure for dispersing a crowd. When troops arrive, the psychological impact of their arrival can be used. Soldiers can dismount from the helicopters, buses, or trucks in plain sight of the crowd, but they must be far enough away to prevent a provoked attack of thrown objects. The first echelon to dismount from the vehicle secures the surrounding area.

2-96. When small groups are scattered throughout a large disturbance area, a show of force can be made by marching troops conducting motor marches, conducting saturation patrolling, and setting up static posts. Sometimes marching well-equipped, highly-disciplined soldiers in view of a crowd may be all that is needed to convince them to disperse and retire peaceably. On the other hand, a show of force may attract people to an event. It may also provoke a nonviolent crowd into a violent confrontation. Intelligence can help best in preparing a show-of-force response.

Containing

2-97. Containment is the process of limiting a crowd to the area they are presently occupying. It is a suitable option when the crowd must be prevented from spreading to surrounding areas and communities. Additionally, it prevents those outside the already assembled crowd from joining the gathering. Containment is useful when apprehensions become necessary for preventing those in the crowd from escaping. Crowd control formations, perimeter patrols, and barriers are effective methods to accomplish containment. In all instances, caution must be used to avoid the "fight-or-flight" syndrome common to people feeling trapped with no escape.

2-98. Armored vehicles, armored security vehicles (ASVs), and up-armored high-mobility multipurpose wheeled vehicles (HMMWVs) are adaptable to roadblock operations as they can serve as barriers. They also provide added protection for the soldiers inside and those outside manning the roadblocks because they provide an easily accessible barrier for them to crouch behind.

2-99. Military vehicles traveling at close intervals in a column formation next to a crowd are largely a psychological barrier (see *Chapter 6*). They can be used to contain a large, fast-moving crowd. The moving cordon creates a temporary obstacle between the crowd and the line beyond which they will not be allowed to cross. A well-trained driver in a mobile cordon can do a better job than dismounted soldiers. In order to execute safe, quick reversals of direction for mobile cordons on narrow roads, the soldiers executing this formation must be highly-trained vehicle operators.

2-100. By-the-number commands issued over vehicle radios are the most common method for coordinating a cordon movement. Vehicles equipped with PA systems can also prove effective to issue both commands to troops and directives to the crowd. The cordon speed is no more than 5 miles per hour with an interval of at least 20 feet. Blue and red lights, sirens, blinkers, and horns may or may not be used. Armored vehicles can also serve as mobile command posts. When used properly, vehicles provide security, communications, and mobility.

DECIDE TO CONTAIN OR DISPERSE

2-101. Current crowd control doctrine places emphasis on crowd dispersal. Forced dispersal may result in a crowd breaking up into multiple groups that scatter over a large area. This may pose even greater public order problems and may pose a continued threat to control forces. A crowd is often better controlled by means of containment (confining its activities to a given area). A crowd has a limited duration, and its numbers are likely to diminish as individual needs take precedence over those of the crowd.

2-102. During the G8 summit protest in Ottawa, Canada, in June 2002, several thousand people showed up to demonstrate against the evils of world corporate leaders. Canadian Police determined that their best response to this event was a "soft response." Throughout the two-day event, hundreds of officers in standard patrol uniforms operated in four-officer teams, prestaged throughout the city. As the demonstrators began to assemble and eventually took to marching, officers provided escorts both in the front and rear of the enormous crowd along with officers marching to the flanks of the demonstration. This "soft approach" frustrated demonstrators, especially those who had it within their agenda to cause property damage and personal injury. Even more frustrating to the demonstrators was the heavy downpour of rain they endured on both days. Their organizations quickly disbanded out of concern for their own creature comforts. In the end, 10,000 people participated in the protest, with only three arrests made in two days.

USE PATROLS

2-103. Alert, aggressive patrolling of the disturbance area deters the gathering of crowds. The use of saturation patrolling is most effective for this purpose. It allows information to be collected, and it creates the psychological impression of the control force being everywhere at once.

2-104. Standard military transport vehicles provide mobility and communications for area coverage. Soldiers must be deployed with enough vehicles to provide the flexibility to handle all situations in the disturbance area. The vehicles can be modified with sandbags, wire screens, or similar materials to protect against sniper fire and thrown objects. If sniper fire is expected or has been encountered, then up-armored HMMWVs or light armored vehicles should be used for patrolling.

2-105. Armored vehicles patrolling an area of violence provide an added psychological effect and allow troops to maneuver close to snipers to make an apprehension. They may also be used as rescue vehicles to extract hostages or people surrounded by a hostile crowd.

2-106. Varying patrol routes and times for mounted and dismounted patrols keep lawbreakers from being able to pick a safe place or time to act. Patrols are particularly useful in preventing overt arson and violence. Patrol members can also spot and promptly report fires. Whenever possible, military patrols are integrated with civil police patrols. Joint patrols conserve military forces and aid civilian military communications. They also help troops become familiar with an area quickly. In addition, the civil police are available to make apprehensions, if needed. Patrol members must practice proper standards of conduct and fair treatment of civilians at all times. They are performing an important community relation and control function.

2-107. Motorized patrols, because of their speed and mobility, provide timely reconnaissance and broad area coverage. Motorized patrols are in radio contact with TF headquarters (HQ). They can make periodic contact with foot patrols and stationary posts, while patrolling their own areas. Moreover, motorized patrols can respond quickly to calls for help from other patrols and guard posts. Motorized patrols should have at least three vehicles with three men in each. Strength in numbers is necessary for protection. When planning patrol routes, avoid areas where the patrols can be isolated or placed in jeopardy. In addition, motorized patrols are equipped with fire extinguishers to put out small fires, thus reducing the burden on the fire department.

2-108. Besides foot and motorized patrols, air patrols provide a third dimension in directing the overall control effort in the disturbance area. They can perform reconnaissance and surveillance and provide near real time formation over the disturbance area. They are an excellent means of providing timely information to the commander concerning the demonstration area and its surrounding perimeter. They can monitor the actions of rioters, the extent of damage, the status of access routes, locations and conditions of road barriers, and other important conditions.

USE DELAYING TACTICS

2-109. AOs that units are assigned to patrol are often larger than the unit has resources to manage effectively. In addition, units must manage taskings, contingency plans, and other commitments, which will quickly drain available manpower. This problem will manifest itself when factions threaten to riot or gather in large numbers to demonstrate. In order for the commander to stall for time while he shifts unit assets, he may want to delay the arrival of buses and vehicles full of demonstrators. This technique is based on the assumption that demonstrators are moving to the demonstration site by vehicle. Key and secondary routes into the demonstration site must be analyzed by the S2 when he does the IPB.

NOTE: There should be an overwatch when this technique is attempted.

2-110. As demonstrator buses and/or vehicles move toward the demonstration site, soldiers in two HMMWVs block the road just ahead of the approaching vehicles. Once the demonstrators dismount their vehicles and begin to close in on the soldiers, the soldiers leapfrog (see *Figure 2-3*) 1 kilometer farther down the road to repeat the process again. This is repeated as often as necessary. This process, if repeated enough times, should discourage the demonstrators enough that they give up and go home. It should also provide enough time to shift assets to the demonstration site.

2-111. Demonstrators are just as capable of blocking secondary and key routes leading into a demonstration or riot site to prevent relief forces from assisting with riot control as US forces are capable of preventing demonstrators access to the site. One must remember that demonstrators are often well organized and have handheld radios and other items. A great example of this was the 1999 demonstration against the WTO in Seattle, Washington. Police had erected temporary portable fences in the downtown areas to keep demonstrators from disrupting the delegates. Determined to be disruptive, the demonstrators relocated the temporary fences by not allowing police to enter the area. Police were forced to back patrol cars over their own fences to gain access.

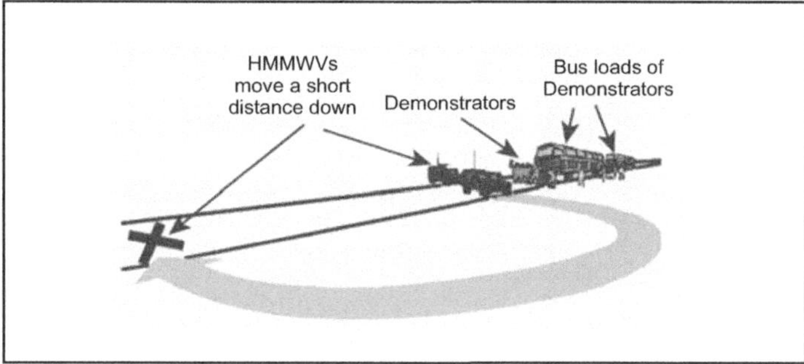

Figure 2-3. Delaying Tactics

Chapter 3

Apprehension, Search, and Detention

During the course of a civil disturbance, some members of the crowd may take part in unlawful activities, such as looting and assault. This will require the civil authorities or attached elements to apprehend, search, and detain people who are participating in the demonstration. Circumstances of the unfolding situation may require US military forces to be called upon to search, take custody of, and detain people who are participating in some form of protest and violating the law. Search, apprehension, and detention operations are conducted to halt these violations and to deter future violations.

LEGAL CONSIDERATIONS

3-1. All apprehensions are made by the civil police force, unless it is not possible for them to do so. Individuals must be known to have committed an illegal act, or there must be probable cause to believe that an individual committed such an act to be apprehended. If it becomes necessary for a military control force to apprehend or temporarily detain such violators, control force members often do so with the approval of the civil authorities. This ensures that all searches, apprehensions, and detentions (if necessary) are conducted within the parameters of the law.

3-2. Commanders operating overseas, similar to Bosnia and Kosovo, may find themselves without an established government or LEAs to work with. Commanders on the ground are then thrust into positions of authority, making them responsible for the actions, conduct, searches, apprehensions, and detentions performed by their troops.

3-3. If US military forces are called upon by the local authorities, certain policies have to be observed during search, apprehension, and detention operations. First, treat all people firmly but with reasonable courtesy and dignity. Remember that participating in a legal demonstration to express views is a right of all people, not just Americans in the US. The action, attitude, and behavior of US soldiers performing these operations are very important.

3-4. Our forces should never be seen as a military policing force on our own soil or as an occupying force in another country. Treating people with contempt, hostility, or excessive force increases the likelihood of resistance and violence. Searching people, placing them under apprehension, and detaining them without probable and just cause or without concern for their constitutional rights create problems and hinder due process. Numerous cases exist where authorities violated an individual's rights, resulting in the prosecution of police authorities or civil suits being waged against the officers and their leaders. This includes the military personnel assisting those officials.

3-5. It is very important that civil law enforcement (if available) be present and supervise all activities. Military personnel conducting a search or making an apprehension must carry out all the procedures carefully within the parameters of their training and the authority given to them. For each search and apprehension performed, the apprehension element must document the specific conduct of the person that violated the law.

3-6. Apprehending officers and supporting US servicemen must provide careful descriptive data for each subject. This description must be sufficient so that, at a later time, the suspected person can be clearly identified as the subject in question. Names and addresses of witnesses to the violation must be obtained. This information can be recorded on *Department of Defense (DD) Form 2708 (Receipt for Inmate or Detained Person)*. Apprehension teams must retain and tag each item of physical evidence, such as weapons or stolen goods, that supports the apprehension on *DA Form 4137 (Evidence/Property Custody Document)*. The team must give the owner a receipt for all property collected as evidence and retained.

TYPES OF SUBJECTS

3-7. There are two types of subjects (cooperative and uncooperative); they are as follows:

- **Cooperative.** Cooperative subjects actively and willingly comply with instructions. Cooperative subjects are usually compliant and offer no resistance to verbal directions or physical actions. However, if they provide any resistance to authority, it is usually passive. Some examples of this resistance are as follows:
  - Yelling.
  - Screaming.
  - Tightening the body.
  - Pulling away.
  - Letting the body go limp.
- **Uncooperative.** Uncooperative subjects will display preattack cues. Uncooperative subjects will actively resist being apprehended. Often, this resistance comes in the form of physical actions, such as—
  - Assaulting verbally.
  - Refusing to be handcuffed.
  - Refusing to be searched.
  - Attempting to push away, forcing a pursuit.
  - Wrestling with the apprehension team.
  - Striking out at the apprehension team with fists, hands and elbows, and feet.
  - Striking out at the apprehension team with weapons, such as knives, clubs, bats, bricks, rocks, or firearms.

APPREHENSION

3-8. US military forces will be in a supporting role to the local civilian authorities or acting as the control force in civil disturbance operations. When making apprehensions, it is best to use apprehension teams. These teams provide quick, organized responses to developing situations. Teams can be organized at the team, squad, or platoon level depending on the number of apprehensions expected. The intelligence information gathered helps to prepare for events where apprehensions may be necessary.

3-9. Each team (if possible) consists of a civilian police or apprehension officer, security element, and recorder. Team apprehension officers make the actual apprehension with the help of the security elements. If there is not a civilian police officer available to act as the apprehension officer, then it may be an officer or noncommissioned officer (NCO).

3-10. Ensuring that the apprehension process is properly conducted and documented, determine who is to be apprehended. Then inform the subject of why he is under apprehension and supervise his handling.

3-11. Security elements help the apprehension officer handle subjects. Security elements move, restrain, and search subjects under the supervision of the apprehension officer. The primary responsibility of the security elements is to watch the crowd and act as a blocking force, keeping the crowd from interfering with the apprehension. Recorders document the event, preferably by video, and complete *DD Form 2708*. If possible, recorders also photograph the apprehension officer with each subject. This aids in the identification process because it links the apprehension officer to the subject and provides the apprehension officer with documentation of the apprehension for use in court.

3-12. Electronically capturing the events as they unfold is essential in all civil disturbance operations. If personnel and equipment are available, videotape the events as they happen or, at a minimum, take still photographs to provide pictorial documentation that may be used later in court. Recording the scene before, during, and after the apprehension provides necessary evidence and can also help eliminate additional hostile and illegal activities. Having control force members take videos or pictures can help control crowd reaction. It reduces the feelings of anonymity that a crowd often enjoys and makes crowd members less prone to unruly or illegal acts.

3-13. Verbal commands given to the subject should be in a voice and manner appropriate for the purpose and the situation. Commands should be simple, concise, and in an understood language. Verbal commands should be convincing and convey an expectation that the instructions will be obeyed. With desperate, frightened people, sympathy combined with an attitude of firm assurance on the soldier's part may avoid additional violent behavior or trouble. An attitude of forceful authority may be needed with an uncooperative person. Commands given by soldiers must be done in a manner and context so that the subject has no confusion as to what the soldier is ordering them to do.

3-14. A subject will need to be transported from the area. In CONUS operations, civil authorities will usually be responsible for providing the transportation of a subject. Exceptions to this may be if the demonstration occurs on an installation. OCONUS operations may require US forces to provide vehicles, such as trucks, buses, or sedans, that will be used to transport a subject. Whenever possible, vehicles used to transport a subject should be modified with barriers to separate them from the drivers. If a barrier cannot be provided, a guard is to be placed in the vehicle. Passenger compartments are checked for items that could be used as weapons. When guards are required, the guard always sits with his weapon away from the prisoner. Prisoners must be placed where they can be best controlled, but they are not fastened to the vehicle with hand irons or Flex-Cufs. When seat belts are available they must be fastened. See *Soldier Training Publications (STPs) 19-95C1-SM and 19-95C24-SM-TG.*

3-15. Female escorts should always be used to transport a female subject. If female escorts are not available, measures must be taken to avoid false charges of molestation or sexual misconduct. Net control stations should be notified of the departure and arrival times and vehicle mileage readings (before and after). At least two male escorts must be present at all times when a female escort is not available. These procedures also apply when the male and female roles are reversed.

FLEX-CUFS

3-16. The two main positions used to place Flex-Cufs on a subject are the standing-supported and the prone.

3-17. When using the standing-supported position, the search man—

- Escorts the subject to a wall or other sturdy vertical surface (see *Figure 3-1*).

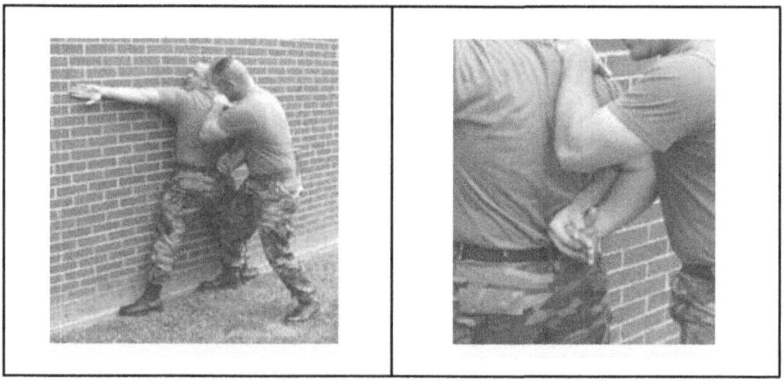

Figure 3-1. Standing-Supported Position

- Instructs the subject (verbally) to do the following:
  - Turn his head away.
  - Place his free arm (palm out, thumb down, and elbow locked) against the surface of the wall.

    - Spread his feet until instructed to stop.
    - Place his heels against the wall.
- Uses the leg closest to the controlled arm to pin the subject's leg (above the knee) to the wall.
- Reinforces verbal commands with pain compliance techniques if the subject fails to follow instructions.
- Maintains a double 90° hold on the controlled arm with his inside hand. With his free hand, the search man grasps the middle portion of the Flex-Cuf, keeping one cuff up and one cuff down.
- Places the first cuff (bottom cuff) on the subject's controlled arm, maintaining wrist manipulation (see *Figure 3-2*).
- Reaches through the free cuff with his outside hand and instructs the subject to place his free hand in the center of his back. With his outside hand, the search man takes the subject's free hand and completes the restraint by securing the cuff. He ensures that there is at least one-finger distance between the cuff and the subject's wrist (see *Figure 3-3,* page 3-6).

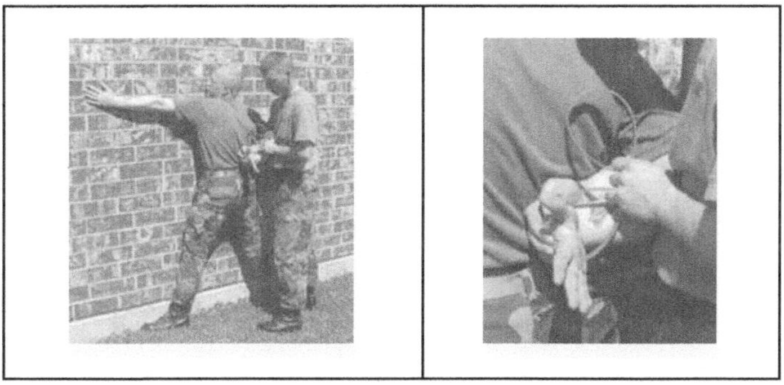

Figure 3-2. Cuffing the First Hand From the Standing Supported Position

3-18. When using the prone position, the search man—

- Grasps the Flex-Cuf with his outside hand, maintains tension on the subject's arm, and places the first cuff on the subject's controlled wrist (see *Figure 3-4,* page 3-6). Next, the search man secures the cuff, keeping at least a one-finger distance between the cuff and the wrist.
- Reaches through the free cuff with his outside hand and instructs the subject to place his free hand in the center of his back (see *Figure 3-5,* page 3-7).
- Reinforces verbal commands with pain compliance techniques if the subject fails to follow instructions. The search man bends the subject's controlled arm (using the natural break of the elbow as a hinge) toward the hand in the middle of the subject's back. With the outside hand, the search man takes the subject's free hand and completes the restraint by securing the cuff.

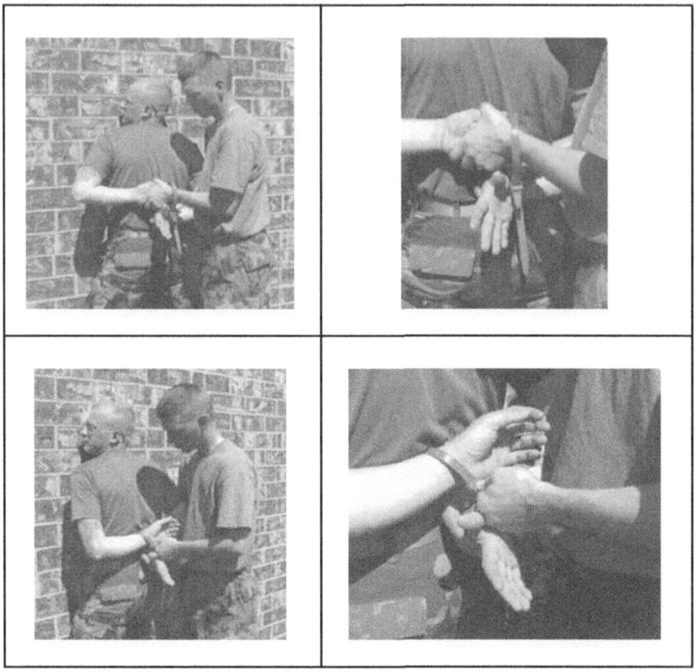

Figure 3-3. Cuffing the Second Hand From the Standing Supported Position

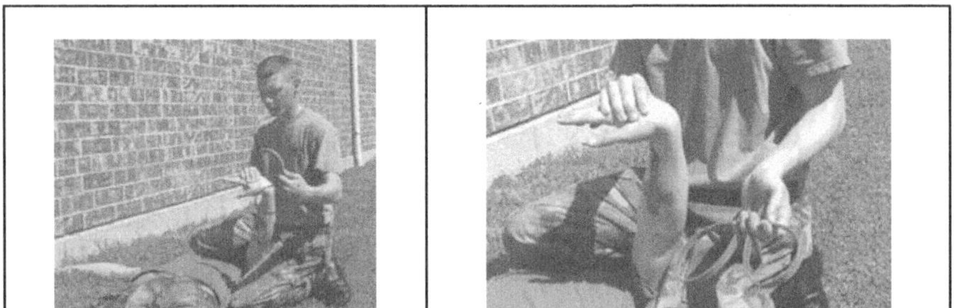

Figure 3-4. Cuffing the First Hand From the Prone Position

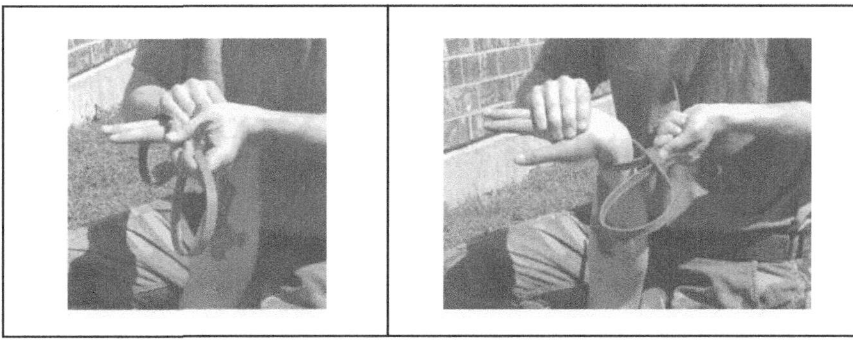

Figure 3-4. Cuffing the First Hand From the Prone Position (continued)

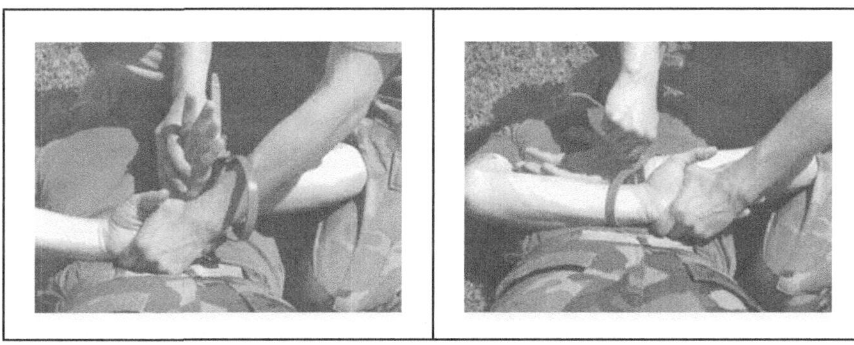

Figure 3-5. Cuffing the Second Hand From the Prone Position

SEARCHES

3-19. Initial contact with the subject is the most dangerous time during any apprehension; therefore, initial contact must be a swift and decisive team effort. Do not give the subject time to think or react. Distract his attention when possible. Apprehension teams are comprised of a search man, threat cover man, and security man (if necessary). Their responsibilities are as follows:

- The search man—
  - Issues verbal commands.
  - Conducts searches.
  - Handles radio communication.
  - Holsters his weapon before approaching a subject and informs the threat cover man that he is holstering his weapon.
  - Performs control procedures and handcuffs the subject.

  - Moves the subject to a safe area.
  - Removes weapons (if any) and sensitive evidence from the subject.
  - Disengages after the subject is restrained.
  - Controls and escorts the subject to a detention area.
- The threat cover man—
  - Provides cover for the search man.
  - Discourages escape and resistance.
  - Discourages hostile interference.
  - Alerts the search man of threats.
  - Resists distractions and maintains target acquisition or scanning.
  - Assesses threats constantly and prepares for the unexpected.
- The security man (if necessary)—
  - Provides cover for the search man.
  - Discourages hostile interference.
  - Alerts the search man of threats.
  - Resists distractions and maintains target acquisition or scanning.
  - Assesses threats constantly and prepares for the unexpected.

Search Principles

3-20. A subject should always be put at a disadvantage and searched immediately, but circumstances may dictate postponing the search until a safer place can be found. At a minimum, a quick pat down can be done en route to a safer location. The longer the apprehension takes, the greater the risk of drawing a hostile crowd and violent acts toward the control force.

3-21. Crowds often sympathize with a subject and try to help him or escalate the activities of the crowd toward violence. A subject's actions may also incite the crowd to violence. The longer a subject is the focal point of an apprehension, the easier it is to stir a crowd to hostile or violent acts. Apprehension teams should choose to move the subject out of the crowd's view. For the safety of the team, always apply Flex-Cufs or handcuffs and escort the subject to an area where a search can be done safely.

3-22. Search team members never operate alone; searches are conducted by a minimum of two team members. Initial steps must be taken to put the subject at a physical disadvantage. Before conducting a search (regardless of rank, sex, position, age, or physical condition), a team member applies the handcuffs or Flex-Cufs. Verbal commands must be short, clear, and distinct.

3-23. A search is conducted on everyone that has been apprehended to ensure that the team and those being apprehended are safe. The search man must abide by the following procedures when conducting a search:

- Search in a clear, secure area (when possible).
- Be ready to counter any resistance or an assault by the subject. Consider the subject armed until proven (by search) otherwise.
- Employ the appropriate level of force immediately, based on the actions of the subject.

- Do not step over the subject when he is at a disadvantage on the ground; go around him and remain out of sight.
- Do a 360° search of the subject's waistline for weapons. Search obvious and inconspicuous areas. Inconspicuous areas such as the groin, armpits, chest, and small of the back must be searched on both men and women; however, do not linger in these areas.

NOTE: Conduct same-gender searches when possible. If mixed gender searches are necessary for speed or security, conduct them in a respectful manner and avoid any action that could be interpreted as sexual in nature. To prevent allegations of sexual misconduct, the apprehension officer shows careful control of the soldiers who perform mixed-gender searches.

- Maintain security.
- Brief subjects quickly on what to do and who to follow.
- Remember that when conducting a search, a subject's body can be divided for a more efficient and thorough search. For example, search from head to toe or from side to side.

NOTE: The search man announces the items found and secures weapons and contraband.

- Military forces (when supporting civil authorities) may perform searches of people and property that are not usually subject to military law during a civil disturbance.
- Remember that it is always better for civilian police to search civilians and civilian property (when possible).
- Remember that if military forces are called upon to support police it is usually a last resort and there are not enough civilian officers available to accomplish the mission.
- Military personnel may search people incident to an apprehension.

3-24. The prescribed methods of searching someone are the frisk, the standing, and the prone. The method of search used largely depends on the situation and the mission. Searches are made not only of the subject and his personal effects such as purses, bags, and wallets, but also of the immediate surrounding area. This prevents the subject being searched from grabbing a weapon or destroying evidence. It may also be necessary for military forces to search private property, including vehicles, if there is—

- Reasonable belief that a person has committed or is committing a violent crime and is hiding in a building or a vehicle.
- Reasonable belief that a vehicle contains weapons or instruments of violence.
- Probable cause for searching a building or a vehicle.
- Probable cause to believe that (unless immediate action is taken) evidence of a crime will be destroyed before a warrant can be obtained.

NOTE: Search teams should always consist of two personnel—one to conduct the search, and the other to provide security and overwatch for the soldier conducting the search. Before being searched, a subject should always be restrained and put at a disadvantage by

securing his hands with handcuffs or Flex-Cufs. Ask the subject if there is anything in his pockets that could hurt the searcher, such as knives or needles.

Frisk Search

3-25. The frisk search (also known as the stop and frisk, pat down, and terry stop) is a search incident to a lawful stop. Subjects are not under apprehension at this point, but if it were an apprehension situation, handcuffs or Flex-Cufs would be applied first. Frisk searches are used when there is reasonable belief that a subject may be armed with anything that could be considered a safety threat.

3-26. Legal and regulatory circumstances for conducting a frisk search are covered in the *Manual for Courts-Martial (MCM)*. When a lawful stop is performed, the person stopped may be frisked for weapons when that person is reasonably believed to be armed and dangerous.

3-27. The primary purpose of a frisk search is safety. Frisk-search techniques are to be used only on a compliant subject. Frisk searches only cover the outer garments. Do not search the pockets of the individual unless a weapon is suspected. A frisk search begins when the search man informs the subject of the intent to search.

- During the preparation phase, the search man—
  - Tells the subject that he must perform a search for weapons for the subject's safety and his own.
  - Tells the subject to turn and face away from him and instructs the subject to keep his hands out to his sides where they are visible.

NOTE: The cover man providing security moves around the subject and positions himself out of reach and in front of the subject.

  - Instructs the subject to spread his feet until instructed to stop (more than shoulder width) and point his toes outward compromising his balance (see *Figure 3-6*).

NOTE: Do not kick at the subject's feet to spread them apart; it is unnecessary and unprofessional.

  - Tells the subject to bring both of his hands back to the small of his back with palms out, thumbs up, and the back of his hands together.
  - Instructs the subject not to move (see *Figure 3-7*).
- During the performance phase the search man—
  - Approaches the subject from behind (if possible) at a 45° angle, off center. The search man approaches with caution to ensure that his feet are to the outside of the subject's stance. See *Figure 3-8*, page 3-12.
  - Stays just inside arms reach and grips the subject's fingers with his support hand. The search man firmly grips at least one finger on each of the subject's hands (both thumbs, both index fingers, or an index and middle finger). An overhand or an underhand grip can be used effectively.

Figure 3-6. Approaching the Subject to Perform a Frisk Search

Figure 3-7. Giving Instructions to Subject Before Frisk Search

NOTE: The primary purpose of gripping the fingers is to control the subject's hands while performing the frisk search (see *Figure 3-9*, page 3-12). If the subject is going to resist or become aggressive, he will attempt to free his hands first. If the subject attempts to pull his fingers from the search man's grip, the search man must respond instantly by increasing his grip. If the grip on the fingers is lost, the search man should immediately push away from the subject and create a safe distance. This will allow the search team to respond with the necessary force to regain control of the situation.

Figure 3-8. Performing a Frisk Search

Figure 3-9. Gripping the Subject's Fingers During a Frisk Search

- Ensures that the subject always looks away from him.
- Places his strong hand on the outside of the subject's elbow (similar to the handcuffed escort position).

NOTE: The search man searches one side (the front, back, or side) at a time from head to foot.

- Divides the subject (visually) straight down his middle and does not cross the line while conducting the search.

NOTE: The search man is just feeling or searching for possible weapons, so an open hand should be used.

- Works at arm's reach. It is the sense of touch not the sense of sight that is being used. While performing the search, search men should stay behind and to the side of the subject. This provides the search man with a tactical advantage and prevention from being struck by the subject's elbows or knees if he should begin to resist.

3-28. When searching below the subject's knees, the search man takes a half step back and to the outside. He squats down, but does not drop his knees to the ground.

3-29. If no weapons are found and the subject is to be detained, the hands are secured with Flex-Cufs and the subject is escorted to the holding area. If the subject is to be released, collect information (name and address) and escort him away from the crowd.

Standing Search

3-30. Safety is the primary concern when dealing with a subject under apprehension. Therefore, the search man must protect himself by restraining subjects under apprehension before conducting the standing search. All subjects under apprehension have the propensity to turn violent or hostile. Putting a subject at a physical disadvantage provides safety for the subject and the search man. It limits the subject's ability to hide, destroy, or remove any evidence he may have on him.

3-31. Apply Flex-Cufs to the subject quickly during the apprehension. It is impossible to predict what a subject will do when put under apprehension. A subject may initially be cooperative and compliant, but as the subject becomes aware of his situation he may be overcome with emotion and begin to struggle or pull away.

3-32. There is no legal or regulatory prohibition against opposite-gender searches. Common sense prevails and whenever possible, searches should be done by soldiers of the same gender to avoid obvious questions and complaints. Once a subject is put at a physical disadvantage by cuffing and is more easily controlled, it affords the necessary time to get someone of the same gender there to conduct the search.

- During the preparation phase, the search man—

NOTE: Standing searches always begin with the subject in the handcuffed escort position (see *Figure 3-10,* page 3-14).

  - Takes his outside hand (farthest from subject) and places it on the outside of the subject's elbow while his inside hand (closest to subject) has a reverse-handshake grip of the subject's hand. At this point, both hands are on the same arm (see *Figure 3-10,* page 3-14).
  - Ensures that his position is on the same side of the subject's body being controlled (behind and to the side). When changing sides, change hand position.
  - Instructs the subject to spread his feet until instructed to stop (usually more than shoulder width) and point his toes outward, compromising the subject's balance.

Figure 3-10. Handcuffed Escort Position

NOTE: Do not kick at a subject's feet to spread them; it is unnecessary and unprofessional. If the subject becomes noncompliant or aggressive, pain compliance is usually the best tool to use, but only use enough force to gain compliance.

- Uses his outside hand to conduct the search. His inside hand is always used to maintain control during the search.

- During the performance phase, the search man—

NOTE: Only one side (the front, back, and side) of the subject is searched at a time from head to toe (see *Figure 3-11*).

- Divides the subject (visually) (straight down the middle using the spine as the centerline). He does not cross this line while conducting the search. Searches must be systematic, from head to foot. If there is an interruption, it must be started all over again.
- Searches with an open hand, using a rub down or squeezing method. Do not only pat down like in the frisk search.
- Works at arms reach. It is the sense of touch, not the sense of sight that is being used. To prevent the possibility of being struck by an elbow or knee, the search man stays behind and to the side of the subject.
- Takes a half step back and to the outside when searching below the subject's knees. The search man squats down, but does not drop his knees to the ground. Maintaining positive control is essential throughout the search, so never the reverse handshake is never released.
- Change sides when satisfied that all items have been removed from the side in which the search started. Without losing positive control, the search man conducts the search on the other side of the subject.

Figure 3-11. Performing the Standing Search

NOTE: Changing sides is accomplished by moving the outside hand (the searching hand) to the subject's uncontrolled handcuffed hand located on the outside. The outside hand and forearm will cross over the top of the reverse-handshake grip of the inside hand (see *Figure 3-12,* page 16).

- Crosses over the reverse-handshake grip with the outside hand and grips the subject's uncontrolled handcuffed hand in a reverse-handshake grip.
- Releases (quickly) the first reverse-handshake grip moving that hand to the subject's elbow as he sidesteps to the side yet to be searched. The search man repeats the standing search technique used on the first side of the subject.
- Gives strong verbal commands and uses only the force necessary to control the subject if he resists or becomes aggressive at any time during the search.
- Removes all items found during the search to be evaluated as potential evidence or a weapon. Items not considered a weapon or of evidentiary value will be returned to the subject at a later time.

Prone Search

3-33. Safety is the primary concern when dealing with a person under apprehension. Therefore, the search man must protect himself by restraining subjects under apprehension before conducting this search. Remember that all subjects under apprehension have the propensity to turn violent or hostile. That is the reason a prone search would be necessary. After performing an arm bar takedown on a noncompliant subject, it is especially important for the search man to put him at a physical disadvantage to provide for the subject's safety and his own. By doing so, the subject's ability to hide, destroy, or remove any evidence is limited.

Figure 3-12. Performing the Standing Search (Changing Sides)

3-34. The safest way to handcuff or Flex-Cuf a noncompliant subject is in the prone position, when the search man has greater control of the subject. It is usually the safest position in which to conduct a search

3-35. Handcuff the subject quickly during the apprehension. It is impossible to predict what a subject will do when put under apprehension. Often a subject may initially be cooperative and compliant; however, as the he becomes aware of his situation, he may become overcome with emotion and begin to struggle or pull away.

3-36. There is no legal or regulatory prohibition against opposite-gender searches. Common sense prevails and whenever possible, searches should be done by the same gender to avoid obvious questions and complaints. Once a subject is at a physical disadvantage by cuffing and is more easily controlled, it affords the necessary time to get someone of the same gender there to conduct the search.

NOTE: The subject is put into a physical disadvantage when he is taken facedown to the ground using an arm bar takedown and cuffed.

- During the preparation phase, the search man—
  - Ensures that the Flex-Cufs are not too tight.
  - Rotates his entire body around the subject's shoulder and center at the outside of the subject's hip with his shin over the subject's shoulder and one knee on the ground. At least one knee should remain over the subject's back (near side), with the other on the ground.
  - Pins the subject down by pressing his knee on the side of the subject's back or down on his closest elbow if resistance or aggression is encountered. He simultaneously grips the subject's closest handcuffed hand, grabs it with a reverse handshake, and applies pain for compliance.

- Uses strong, forceful verbal commands, such as "Stop resisting, stop moving around."
- Ensures that the subject's feet and ankles are crossed.

CAUTION

Stay alert. If the subject uncrosses his feet and ankles, it is usually a sign of preattack.

- Removes all items found during the search so they can be evaluated as potential evidence. Items not considered a weapon or of evidentiary value will be returned to the subject at a later time.

• During the performance phase, the search man—

NOTE: The search man must be systematic from head to foot while conducting the prone search. If interrupted during the search, it is important to start from the beginning to ensure that nothing is missed. It is better to go over the same area two or more times than to miss something that can be used as a weapon.

NOTE: The search man searches only one side (the front, back, or side) of the subject at a time until that side is complete.

- Divides the subject (visually) straight down the middle, using the spine as the centerline.

NOTE: He does not cross this line while conducting the search.

- Searches with an open hand, using a rub down or squeezing method for best results. He does not just pat down the subject like in the frisk search.
- Works at an arm's reach, searching only the area of the subject's body that is not in contact with the ground. The sense of touch, not the sense of sight, is being used.
- Stops the search as the subject's knees are reached. He taps one of the subject's legs and firmly commands him to raise his foot up and back, allowing for an inspection of the bottom of his foot. He then searches from the knee to the ankle area. This is performed on both the subject's legs and feet.

NOTE: At this point, the entire back side of has been searched. What remains to be searched is the front side of the subject (the chest and front of the hips, waist, groin, and thighs).

- Pivots back and faces the subject. Using the hand closest to the subject's head, the search man reaches across the subject's back and firmly grips the shoulder, rolling the subject up on the his side toward him to expose the areas not yet searched.
- Rolls the subject back down on his chest once all areas have been searched.

3-37. The search man must be completely satisfied that he has found everything. He must be especially sensitive to areas where weapons are often concealed like the waistband, the small of the back, ankles, pockets, shoes, the tongue of the shoes, and the groin area.

NOTE: Before removing subjects from the area, a brief summary must accompany them. The summary must include the subject's name, what he is suspected of doing, the time and date of the incident, and who saw him do the action.

DETENTION

3-38. Civil authorities must provide adequate detention facilities for all subjects. Authorities must be prepared to detain large numbers of people. They may choose to expand existing detention facilities or set up temporary facilities to accommodate the extra load. If possible, large-scale arrests are delayed until sufficient detention facilities have been set up.

3-39. If US military forces are committed to support local authorities, commanders should coordinate with them to ensure that adequate detention facilities are available and to learn their locations and capacities. If there are more detainees than civil detention facilities can handle, civil authorities may ask the military to provide support by setting up and operating temporary facilities. Army correctional facilities cannot be used to detain civilians. A temporary military detention facility can be set up if—

- Federal troops have been employed according to provisions.
- The TF commander has verified that available civilian detention facilities can no longer accommodate the number of prisoners awaiting arraignment and trial by civil courts.
- The Army Chief of Staff has granted prior approval.

3-40. Use of the temporary facility ends as soon as civil authorities can take custody of the detainees. Military forces are responsible for the custody, health, comfort, and sustenance of all detainees in their facilities until custody is transferred to civil authorities. Temporary facilities cannot be used to confine people arraigned or convicted in civil courts.

3-41. Women are detained in their own facility. Temporary facilities are supervised and controlled by officers and NCOs trained and experienced in military correctional operations. Females guard females. If two female guards are unavailable, use two male guards or one male and one female guard (see *FM 3-19.40*).

3-42. The same operational procedures that apply to the management of an installation confinement facility (see *FM 3-19.40*) apply to the management of temporary detention facilities, except for training, employment, and administrative discipline. Guards and support personnel under direct supervision and control of military officers and NCOs need not be trained or experienced in military correction operations. They must be specifically instructed and closely supervised in the proper use of force, custodial procedures, and the completion of military and civilian forms and reports that may be used. Soldiers that may be tasked to operate a temporary detention facility should be familiar with the forms and reports used for civilian apprehensions too.

3-43. Temporary facilities are set up on the nearest military installation or on suitable property under federal control. Ideally, the facility should be close enough to the disturbance area to minimize transportation and escort needs. However, it should be far enough away not to endanger those being detained. Whenever possible, existing structures are adapted for this use, but construction may be needed to provide segregation for ensuring effective control and administration.

3-44. The basic structure must include the following:

- Search areas.
- Holding areas for incoming men and women.
- A processing area.
- Holding areas for men and women who have committed misdemeanors.
- Holding areas for men and women who have committed felonies or are violent.
- A holding area for detainees' property and evidence obtained during a search.
- A holding area for administrative support and records.
- A medical station.
- Latrines.

For more information on these structures see *FM 3-19.40.*

3-45. Facility personnel must ensure that proper sanitation is maintained. When large numbers of people are detained or processed through a facility, sanitation becomes a problem. Medical personnel must conduct regular health inspections to detect unsanitary practices and conditions.

3-46. Facilities must be organized for a smooth flow of traffic. Processing stations must be set up so there is a linear or circular sequence of movement. These stations are set up out of sight of the holding areas. If possible, they are separated by a door to reduce noise. To reduce distractions, each station may be partitioned. Detainees may be more cooperative if they are out of sight of each other.

3-47. Injured people are given prompt medical treatment and transportation to medical facilities when necessary. A medical aid station for screening detainees and treating minor injuries is set up inside or next to the detention facility. Treatment areas, however, must be out of sight of the processing and holding areas. If possible, access to the medical facility bypasses the holding and processing areas. Facility personnel may also consider setting up a separate holding area for injured detainees.

3-48. Detention facility operation plans must contain emergency procedures. As a minimum, the following areas must be addressed:

- Fire evacuation.
- Disturbance control.
- Facility security.

NOTE: Depending on the situation, other emergency procedures may be needed.

3-49. Upon arrival at the detention facility, the detainee is logged in and searched, regardless of whether a complete search has already been conducted by the apprehending team in the field. Separate search areas are set up for men and women. Weapons, contraband, flame-producing devices, suspected evidence, money, and high-value items are confiscated. Medications are also confiscated. Medical personnel screen detainees that have had medications confiscated. Receipts are provided for any property or evidence that is confiscated. Confiscated items are tagged, and the items are stored in a controlled-property area.

3-50. As a detainee is brought to the facility, a file is initiated. Assigned detainee case numbers are used on all paperwork, such as logs, evidence tags, reports, and visual documentation. All paperwork (including photographs) that began outside the detention facility is marked with the case number. Facility personnel may also use hospital identification tags. Using indelible ink, the case number is written and attached to the tag on the detainee's wrist. Different colors may be used to identify different subject classifications, such as misdemeanors, felonies, or violent offenders. If opposing factions are involved, procedures are established to ensure that members of opposing factions are not processed together or detained in the same holding area.

3-51. After a detainee has been searched and classified, he is taken to a processing station where his paperwork is processed. If subjects passively resist by going limp, they may be moved by wheelchair. This reduces the number of escorts needed and the fatigue to facility personnel. Paperwork is reviewed to ensure that all information is complete, including charges, witnesses, and reasons for the apprehension.

3-52. Equipment such as height charts, scales, fingerprinting equipment, and cameras must be available for completing the police report and identifying the subject. Detainees may use aliases and not carry identification, so pay special attention to obtaining as much information as possible about the detainee's physical characteristics.

3-53. After processing is complete, the detainee is placed in the appropriate holding area. Paperwork for the detainee is forwarded to the administrative section. Files are reviewed for completeness and to determine the disposition of the detainee. Information from the file may be placed in a computer to find repeat subjects. Computers may be used to file criminal information only.

3-54. Custody transfers and release procedures must be coordinated with civil authorities and the appropriate legal counsel. Every effort must be made to arraign subjects quickly. The purpose of a detention facility is not to keep people off the streets, but to aid in processing subjects through the legal system. To speed up the release process, planners should consider issuing citations or subpoenas for minor offenses.

3-55. Military force members may have to respond to writs of habeas corpus. Writs are court orders addressed to a prisoner's custodian. Such a writ directs the custodian to bring the prisoner to court to determine the legality of the prisoner's apprehension and detention. Custodians must ensure that all documentation concerning the case is brought to court with the detainee.

3-56. Military personnel must obey writs issued by federal courts. For writs issued by a state court, the custodian or his legal advisor should respectfully reply that US authority is holding the prisoner. The Office of the SJA can answer any questions and explain the correct procedures.

Chapter 4

Riot Shield and Riot Baton Techniques

A soldier may be required to be part of a riot control formation where his primary weapon is a riot baton and his only protection a riot shield. A riot baton in the hands of a well-trained soldier is an invaluable weapon in a crowd control situation and may also be the most appropriate weapon. A riot baton is a very versatile weapon that can be used as an offensive and/or defensive weapon. The riot shield offers protection from the top of the head to just below the knees.

OVERVIEW

4-1. Use of the riot baton and riot shield is based on METT-TC and the commander's choice-of-force option. Formations committed to a riot with riot batons must have lethal overwatch, lethal weapons, and NLW employed within the formation and reserve forces with additional NL and lethal capabilities. These forces are positioned to rapidly reinforce the control force formation.

4-2. During the termination phase of the disturbance (when the violence has subsided and is expected to remain that way), soldiers are placed in a defensive posture. They may be equipped with a riot shield and a riot baton to perform their tasks. The riot shield is primarily a defensive weapon, which may be used in an offensive mode. It is primarily used as an offensive weapon when the formation is in close contact with rioters. The riot baton can also be used when formations are on the offensive and in contact with the crowd. However, crowd control elements employed with riot shields and riot batons must have lethal weapons and NLW available.

WARNING

Riot control formations are never employed with riot shields and riot batons only. There must always be lethal overwatch elements in position and lethal and NL capabilities within the riot control formation with a ready reserve force armed with lethal and NL capabilities.

RIOT SHIELD

4-3. In the hands of a properly trained soldier, the riot shield has proven to be the first line of defense for a formation. The riot shield may be used as an offensive weapon when in contact with an aggressive crowd.

4-4. The riot shield may be held with one arm. It is secured to the weak side arm with a self-sticking strap system and a handle. This system is designed to make the riot shield an extension of the soldier's arm. The riot shield is held at a slightly inward angle. This allows for debris thrown at the soldier to fall harmlessly to the ground (see *Figure 4-1*).

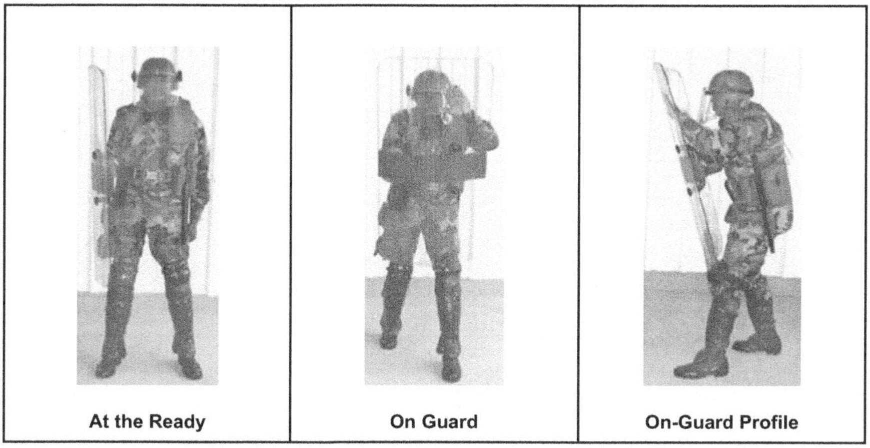

Figure 4-1. Riot Shield Positions

RIOT SHIELD CARRIES

4-5. The two basic riot shield carries are as follows:

- **At the ready.** The at-the-ready position is used while in a column formation or on the move toward hostilities. This position is designed for rapid movement. It should not be used when in contact with a crowd.
- **On guard.** The on-guard position is used when the riot control formation is in contact with the crowd or contact is eminent. It is used primarily in the following formations:
  - Line.
  - Wedge.
  - Echelon.
  - Diamond.
  - Circle.

DEFENSIVE AND RETENTION TECHNIQUES

4-6. The primary purpose of the riot shield is for defense. However, each riot shield holder must be proficient in riot shield retention techniques.

DEFENSE

4-7. The first line of defense for the formation is the riot shield. It is designed to withstand strikes from objects thrown by the crowd. In addition to protecting the riot shield bearer, it also protects other members of the formation from thrown objects.

RETENTION

4-8. When the control force is in contact with the crowd, the crowd will attempt to strip the riot shield from the bearer. When a subject grabs the top of the riot shield (see *Figure 4-2*), the riot shield holder slaps the riot shield with his strong hand near the subject's hands and gives the following commands (while continuing to slap the riot shield until the subject releases it):

- "Get back."
- "Get away."
- "Stop."

Figure 4-2. Riot Shield Retention Technique (Top Attempt)

4-9. If the riot shield is grabbed from the bottom (see *Figure 4-3,* page 4-4), then the riot shield holder forcefully drops to one knee, pinning the subject's fingers and/or hands to the ground. This must be done quickly and with force.

RIOT BATON

4-10. A riot baton in the hands of a properly trained soldier is a formidable weapon. However, it must be used in conjunction with other measures to be most effective. Soldiers must be trained with the riot baton to the point that its various techniques are automatic to them. This training must also include

Figure 4-3. Riot Shield Retention Technique (Bottom Attempt)

learning the vulnerable points of the human body so they can avoid areas that may cause permanent injury or death when struck.

> **WARNING**
>
> **The riot baton is never raised above the head to strike a subject in a club fashion. Not only is it likely to cause permanent injury; but it also gives an unfavorable image of the control force. The soldier is also vulnerable to an attack on his rib cage when his arm is raised.**

TARGET AREAS OF THE HUMAN BODY

4-11. The target areas of the human body are divided into three areas that represent the primary-, secondary-, and final-target areas. These areas are color coded as green (shown as gray in *Figure 4-4*), yellow (shown as white in *Figure 4-4*), and red (shown as patterned gray in *Figure 4-4*).

Primary-Target Areas

4-12. The primary-target areas are those areas which, when struck or restrained, cause the least amount of trauma to the body. These include the meaty or muscular areas of the body, such as the—

- Foot, shin, and instep.
- Inside and outside of the thigh.
- Lower abdominal region.
- Fore- and upper-arms.

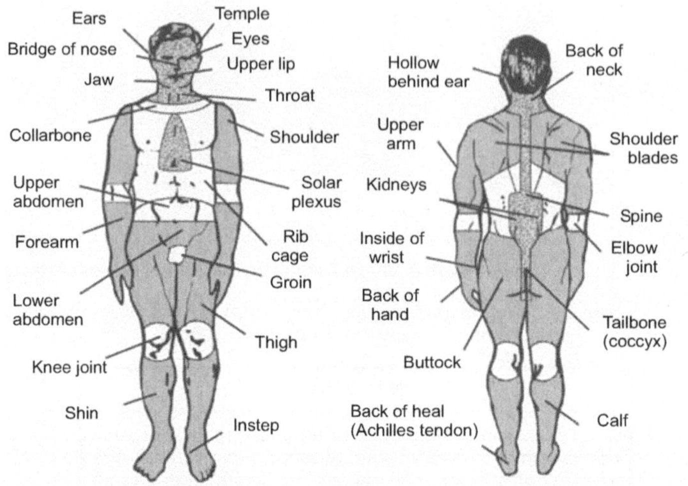

Green (lowest level of trauma) Primary-target areas:

Reasoning: Minimum level of resultant trauma. Injury tends to be temporary rather than permanent; however, exceptions can occur when applying as a restraint technique.

NOTE: When performing a block with a riot baton, the whole body is a green area, except for the head, neck, and spine.

Yellow (moderate level of trauma) Secondary-target areas:

Reasoning: Moderate level of resultant trauma. Injury tends to be more permanent, but may also be temporary when applying as a striking technique.

Red (deadly force) Final-target areas:

Reasoning: Highest level of resultant trauma. Injury tends to be permanent rather than temporary and may include unconsciousness, serious bodily injury, shock, or death.

Figure 4-4. Escalation of Trauma Chart

Secondary-Target Areas

4-13. The secondary-target areas are those areas of the body which, when struck or restrained, cause a moderate level of trauma to the body. This type of trauma tends to be longer lasting because the time to heal is longer than with other parts of the body. If these areas are struck with a riot baton, serious injury may occur. However, some of these areas are used for control or restraining holds that tend to lessen the threat of injury. Some secondary targets are the—

- Collar bone and shoulder blades.
- Elbows and knees.
- Upper abdominal region.

Final-Target Areas

4-14. The final-target areas are those areas which, when struck with a riot baton, cause serious trauma that tends to be fatal or critical. However, there are some areas that (when used for restraining only), will not cause death or serious injury. Some final-target areas are the—

- Front and back of the head (eyes, ears, nose, upper lip, and hollow behind the ear).
- Neck and throat.
- Upper chest.
- Xiphoid process.
- Spinal column.
- Groin area.

JOINTS OF THE HUMAN BODY

4-15. The joints of the human body are often described as hinges and, when manipulated in a certain fashion, allow for better control of the subject being restrained, especially if he is aggressive. When applied correctly, minimal pressure is needed to control the subject. When a subject attempts to resist, he creates additional pain for himself. By anatomically locking out these hinges, the subject will be forced to comply. When restraining a subject, emphasis must be placed on locking out two of the three joints of the upper body or two of the three joints of the lower body. The joints are as follows:

- **Shoulder.** Apply pressure to the shoulder to prevent the subject from freely moving the upper portion of his body.
- **Elbow.** Apply pressure on the elbow or just above the elbow on the ulner nerve.
- **Wrist.** Apply pressure by pancaking the hand to twist or lock out the wrist. Apply pressure at the break of the wrist and push the subject's thumb toward a stationary object (the wall or the ground) to lock the hand out.
- **Hip.** Apply pressure to the hip to knock the subject off balance or alter his center of gravity.

- **Knee.** Apply pressure to the knee by pushing outward on it to cause the subject to be placed off balance, or use the knee as a hinge for pain compliance.
- **Ankle.** Apply pressure to the ankle by clasping and turning it in a counterclockwise manner to lock the hinge out.

TYPES OF RIOT BATONS

4-16. The Army is currently fielding two types of riot batons, wooden and expandable. The most common riot baton is the 36-inch hickory riot baton with thong. There is also the 24- to 36-inch expandable riot baton, which has been added to the nonlethal capabilities set (NLCS). Regardless of which riot baton the soldiers are equipped with, all techniques for blocking and striking are identical. Similar to the riot shield, the riot baton offers soldiers both offensive and defensive capabilities.

4-17. Soldiers must be properly trained in all blocking and striking techniques. Improper use of the riot baton by an untrained soldier has the potential for creating a greater problem than what already exists.

NOMENCLATURE OF THE RIOT BATON

4-18. The riot baton is divided into four basic parts: the long end, long portion, grip portion, and grip end (see *Figure 4-5*). Although the expandable and wooden riot batons differ, their nomenclature is quite similar.

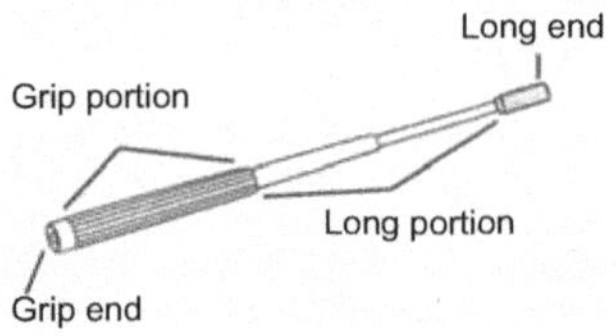

Figure 4-5. The Riot Baton

TYPES OF RIOT BATON CARRIES AND POSITIONS

4-19. As with any other weapon that the soldier will use, there are certain carries and positions from which the riot baton may be used offensively or defensively.

Two-Hand Carry

4-20. The two-hand carry is the most common carry for the riot baton. It is intended for executing all blocks and returning with strikes. When the expandable riot baton is in the collapsed position (at 24 inches), the long end of the riot baton is angled downward approximately 45°. When it is extended to 36 inches the long end is angled upward approximately 45°.

4-21. The wooden riot baton is secured with the soldier's hand by means of a leather thong. He hooks the thumb of his strong hand (palm down) in the thong, pulls the thong across the back of his hand, and grips the riot baton. This provides him with a very secure grip (see *Figure 4-6*).

4-22. The two-hand carry is intended for executing all blocks and returning with strikes. The soldier grips the riot baton with both hands. His strong hand grips the short end of the riot baton, and his weak hand grips the long portion of the riot baton, 2 to 4 inches from the end. His palm is down with his strong hand pulled in close against his strong side hip.

4-23. Wherever the long portion of the riot baton is pointed is where the soldier will strike or jab. The riot baton should be kept at a 45° angle to the ground. All blocks and two-hand jabs and strikes are executed from the two-hand carry.

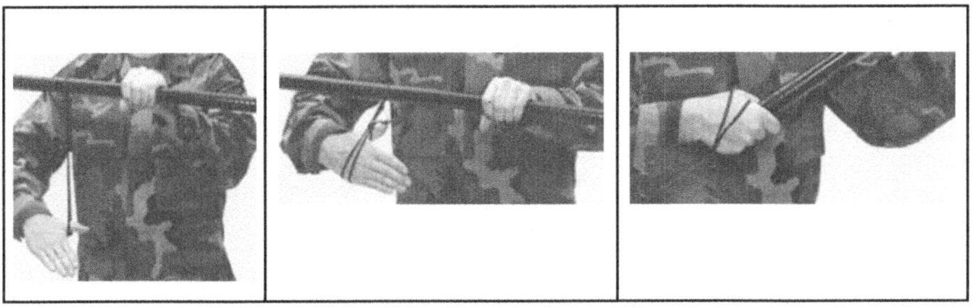

Figure 4-6. Two-Hand Carry

Outside-Arm Carry (Expandable Riot Baton Only)

4-24. The outside-arm carry (see *Figure 4-7*) is used with the expandable riot baton in the collapsed (24-inch) position only. Assuming a wide-based defensive stance (knees slightly bent and feet shoulder width apart), a soldier draws the grip portion of the riot baton with his strong hand. Without striking out or widely swinging the riot baton, he brings the long portion up against the outer part of his strong arm (the tricep area). The soldier's weak hand should be raised in a position to protect and block strikes from an aggressor. It is from this carrying position that a soldier can effectively use both the front and rear strikes. Front and rear strikes should never be performed with an expandable riot baton in the expanded position or with a 36-inch wooden riot baton.

CAUTION

The expandable and 36-inch wooden riot batons provide a greater risk of injury to the soldier's wrist.

Figure 4-7. Outside-Arm Carry

Vertical High-Profile Carry

4-25. Assuming a wide-based defensive stance, the soldier draws the riot baton with his strong hand by the grip portion. Once the riot baton has been drawn from the carrier, the soldier swings the riot baton in a downward motion, fully extending the arm and holding the riot baton down and parallel to the strong side leg. The soldier's weak hand is raised in a position to protect and block strikes from an aggressor (see *Figure 4-8*, page 4-10). Although this carry can be used with both riot baton systems, it is best suited for the expandable riot baton collapsed to 24 inches. The vertical high-profile carry clearly shows the soldier with a drawn riot baton in an effective defensive posture and ready for attack. A drawn riot baton in the hands of a trained soldier may offer just the kind of deterrence needed to stop the activity of an aggressor.

Vertical Low-Profile Carry

4-26. Assuming a wide-based defensive stance, the soldier draws the riot baton with his strong hand by the grip portion. Once the riot baton has been drawn from the carrier, the soldier swings the riot baton in a downward motion, fully extending his arm and holding the riot baton down and parallel to his strong side leg. His weak side hand is raised in a position to protect and block strikes from an aggressor (see *Figure 4-9*, page 4-10). Although this carry can be used with both types of riot batons, it is best suited for the expandable riot baton collapsed to 24 inches. This position portrays the soldier in an effective defensive position ready for an attack, but with a possible hidden weapon. The major difference between the vertical low-profile and high-profile carry is that in the low-profile carry the soldier hides the riot baton behind his strong side leg, allowing for some element of surprise to a potential aggressor. A drawn riot baton in the hands of a trained soldier may offer the deterrence needed to stop the activity of an aggressor.

Figure 4-8. Vertical High-Profile Carry

Figure 4-9. Vertical Low-Profile Carry

RIOT BATON BLOCKING TECHNIQUES

4-27. There are five basic types of riot baton blocking techniques that are executed from the two-hand carry position.

High Block

4-28. The high block is very effective in blocking a downward, vertical strike directed at the top of the head and shoulders. This type of downward, vertical strike may be from a club, pipe, or similar object. The high block is a two-step movement. To perform a high block the soldier performs the following steps:

Step 1. Raises the riot baton straight across his body until it is parallel to the ground (see *Figure 4-10*).

Step 2. Brings the riot baton straight-up in front of his body until the long portion of the riot baton is in front of his forehead (approximately 2 inches from his head). The riot baton should be in a horizontal position above his head and slightly angled down toward his weak hand, as it absorbs the shock from the strike (see *Figure 4-10*). The fingers of his weak hand should be open and behind the long portion of the riot baton when blocking the object to protect his fingers from being smashed by the strike.

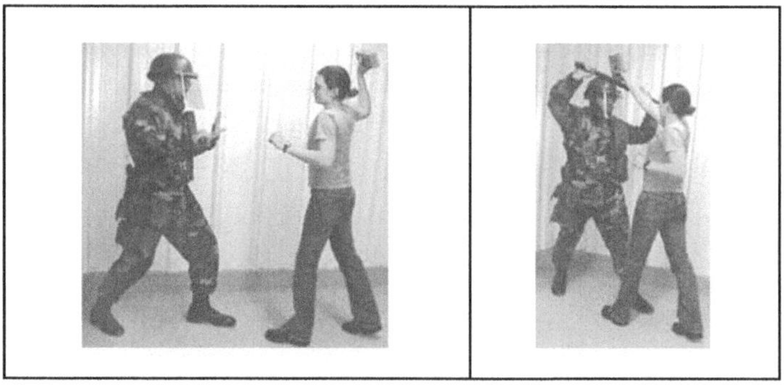

Figure 4-10. High Block (Steps 1 and 2)

Low Block

4-29. A low block is very effective in blocking an upward vertical strike directed at the groin, lower abdomen, chest, or chin. This upward vertical strike may be from an individual's foot, knee, or fist (for example, an upper cut to the chin). To perform a low block, the soldier performs the following steps:

Step 1. Raises the riot baton straight across his body until it is parallel to the ground (see *Figure 4-11*, page 4-12).

Step 2. Brings the riot baton straight down in front of his body, bending at his knees and keeping his body upright until the long portion of the riot baton is just below his knees (see *Figure 4-11*). The riot baton should be horizontal and parallel to the ground as it absorbs the shock from an upward strike. His weak hand fingers should be open and behind the long portion of the riot baton when blocking the object to protect his fingers from being smashed by the strike.

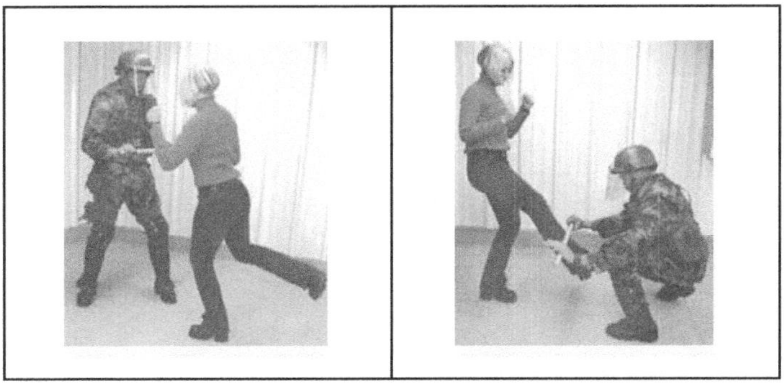

Figure 4-11. Low Block (Steps 1 and 2)

Strong Side Block

4-30. A strong side block is very effective in blocking a horizontal strike directed at the strong side of the head, neck, chest (or flank), or hip area. The horizontal strike may be from an individual's foot, fist, elbow, knee, or weapon. To perform a strong side block, the soldier performs the following steps:

Step 1. Raises the riot baton straight across his body in a vertical position where the grip portion is in his strong hand and the long portion is in his weak hand with the fingers straight up (see *Figure 4-12*).

Step 2. Turns his body laterally by rotating his hips toward his strong side and moving the riot baton across his body, keeping the riot baton vertical (see *Figure 4-12*).

4-31. The strong side block provides protection for a soldier from the front to the rear of his body. A riot baton held in the vertical position absorbs the shock of a strike coming from the strong side. The fingers of his weak hand should be open and behind the long portion of the riot baton, when blocking the object to protect his fingers from being smashed by the strike.

Weak Side Block

4-32. A weak side block is very effective in blocking a horizontal strike directed at the weak side of the head, neck, chest, or hip area. The horizontal strike may be from an individual's foot, fist, elbow, or knee. To perform a weak side block, the soldier performs the following steps:

Step 1. Moves the riot baton in a vertical position toward his weak side, ensuring that the grip portion is pointing straight down and the long portion is pointing straight up.

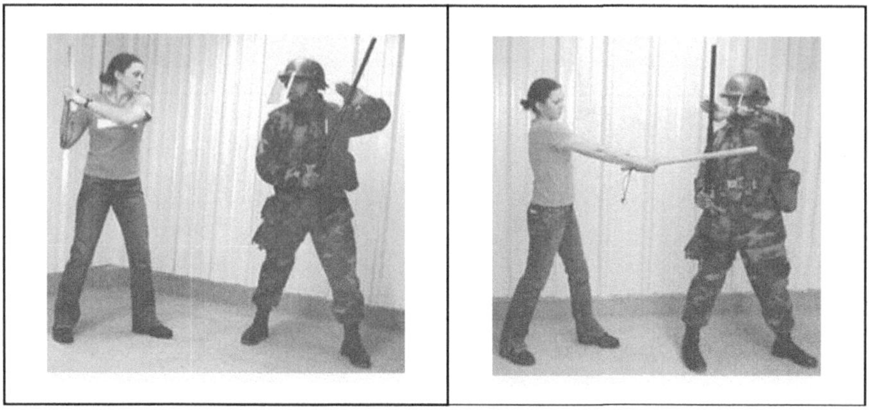

Figure 4-12. Strong Side Block (Steps 1 and 2)

Step 2. Keeps the fingers of his weak hand open and behind the long portion of the riot baton when blocking the object to protect his fingers from being smashed by the strike. At the moment of contact with the opposing strike, the blocking surface of the long portion should be at a right angle to the opposing object.

Step 3. Assess the situation following a weak side block, and take appropriate follow-up action, as needed.

Middle Block

4-33. A middle block is very effective in blocking a front horizontal strike directed at the face, throat, chest, abdomen, or hip. The horizontal strike may be from an individual trying to tackle near the waist or shove or punch at the face. To perform a middle block, the soldier performs the following steps:

Step 1. Moves the riot baton toward the middle of his body at a 45° angle, keeping the long portion pointed up and slightly forward of the grip portion (see *Figure 4-13*, page 4-14).

Step 2. Keeps the fingers of his weak hand open and behind the long portion of the riot baton when blocking the object to protect his fingers from being smashed by the strike. However, his weak hand may have to grip the long portion when pushing an individual away or forcing an individual down on the ground if the individual is trying to tackle.

RIOT BATON STRIKING TECHNIQUES

4-34. Seven strikes can be executed using the riot baton. Two of these strikes come from the outside-arm or vertical carry. The other five strikes come from the two-hand carry.

Figure 4-13. Middle Block

One-Hand Forward Strike

4-35. The one-hand forward strike is a very quick and effective offensive strike. It is usually employed as a countermeasure to an attack from the front that has been blocked effectively.

4-36. Target areas can be as high as the outside upper arm, down to the outer thigh region. An effective one-hand forward strike must be aggressively quick, with the purpose of distracting, disabling, or altering behavior. Therefore, those employing this technique must be sensitive to what part of the riot baton they use to strike the individual. Strike the target with the last 2 to 4 inches of the long end for optimal results.

4-37. A one-hand forward strike may be employed in one of the following three ways. When selecting available targets, use the principles in *Figure 4-4,* page 4-5.

- A horizontal manner.
- A downward diagonal manner to destabilize an individual.
- A vertical manner in front of the body to clear an aggressor's hands.

NOTE: When the one-hand forward strike is used in a vertical manner, keep the long portion straight up, not angled forward. By keeping the long portion up rather than angled forward, the chance of striking the individual in the face or head is significantly reduced. For the counterstrike to be effective, follow through with the technique when the baton makes contact with the aggressor and/or his object to destabilize the him. A pattern of movement such as a

forward shuffle, forward pivot, strong side step, or rear pivot may enhance this technique.

> **WARNING**
> **Do not execute the one-hand forward strike with the wooden or expandable riot baton in the extended position. It has been shown to cause injury to the wrist of the user.**

4-38. When using the vertical or outside-arm carry position (see *Figures 4-7,* page 4-9 and *4-8,* page 4-10), the soldier quickly moves the riot baton across his body using the strength and power of his hips, moving from his strong side to his weak side (see *Figure 4-14*, page 4-16). Ending this strike leaves the riot baton positioned under the weak side armpit in preparation for the one-hand reverse strike (see *Figure 4-14*). When delivering this strong side strike technique (in a horizontal or downward diagonal manner), ensure that the palm is facing up. In a vertical delivery, the palm will be toward the chest rather than up. Pause following the one-hand forward strike and assess the situation. Take appropriate follow-up action, as needed. If the situation does not call for additional strikes, avoid striking again.

One-Hand Reverse Strike

4-39. The starting position for the one-hand reverse strike is under the weak side armpit (see *Figure 4-14*). The one-hand reverse strike is used as a follow-up strike and is quick and effective.

4-40. Target areas can be as high as the outside upper arm, down to the outer thigh region. An effective one-hand reverse strike must be aggressively quick, with the purpose of distracting, disabling, or altering behavior. Therefore, those employing this technique must be sensitive to what part of the riot baton they use to strike the individual. Strike the target with the last 2 to 4 inches of the riot baton for optimal results.

4-41. A soldier may employee a reverse strike in one of the following three ways:

- A horizontal manner.
- A downward diagonal manner to destabilize an individual.
- A vertical manner.

4-42. In front of the body to clear an individual's hands. When the one-hand reverse strike is used in a vertical manner, keep the long portion straight up, not angled forward. By keeping the long portion up rather than angled forward, the chance of striking the individual in the face or head is significantly reduced. When selecting available targets, use the principles in *Figure 4-4,* page 4-5.

4-43. The one-hand reverse strike is considered a follow-up strike to the one-hand forward strike. As a follow-up strike, it is important that soldiers are taught the discipline of assessing the target. An aggressor may not require another strike, based on his condition and actions after enduring the one-hand forward strike.

Forward Strike | Target Assessment | Reverse Strike

Figure 4-14. One-Hand Forward and Reverse Strikes

4-44. When using the one-hand reverse strike, the solder quickly moves the straight baton across his body, using the strength and power of his hips (from weak side to strong side). When the strike is completed, the soldier's arm will be back in the outside-arm carry position (see *Figure 4-7,* page 4-9). The soldier ensures that his palm is facing down when delivering the counterstrike in a horizontal or diagonal manner. This will prevent injury to the user's wrist. In a vertical delivery, the palm will be toward the chest rather than up. Following the one-hand reverse strike, assess the situation and take appropriate follow-up action, as needed. If the situation does not call for additional strikes, the soldier should avoid striking again.

4-45. When the one-hand reverse strike is used in a vertical manner, keep the long portion straight up, not angled forward. By keeping the long portion up rather than angled forward, the chance of striking an individual in the face or head is significantly reduced. For the counterstrike to be effective, follow through with the technique when the baton makes contact with the aggressor and/or his object to destabilize him. A pattern of movement such as the forward shuffle, forward pivot, strong side step, or rear pivot may enhance this technique.

WARNING

This strike will not be executed with a 36-inch wooden or expandable riot baton in the extended position. It has been shown to cause injury to the wrist of the user.

Two-Hand Strong Side Horizontal Strike

4-46. Starting from the two-hand carry position, the two-hand strong side horizontal strike can be aggressively quick. The purpose of this strike is to create distance, distract, disable, or alter behavior. Strong side horizontal strikes are quick and effective offensive strikes, usually employed as a countermeasure to close frontal attacks that have been effectively blocked. When selecting available targets, use the principles in *Figure 4-4,* page 4-5.

4-47. The two-hand carry position enables soldiers to use riot batons at the 24- or 36-inch length. To effectively employ this strike, use the strength and power of the hips to thrust the riot baton in a horizontal manner. Simultaneously, pull back with the weak hand as the strong hand drives the grip end toward the target, striking a rib or the abdominal region of the aggressor (see *Figure 4-15*, page 4-18). Following the two-hand strong side horizontal strike, assess the target before following up with another strike. An aggressor may not require another strike, based on his condition and actions after enduring the one-hand forward strike.

NOTE: A pattern of movement such as a forward shuffle or forward pivot may enhance this technique.

Two-Hand Weak Side Horizontal Strike

4-48. An effective two-hand weak side horizontal strike must be aggressively quick to create distance, distract, disable, or alter behavior. The weak side horizontal strike is a quick and effective offensive strike usually employed as a follow-up strike to the strong side horizontal strike. It is a countermeasure designed to close frontal attacks that have been effectively blocked. When selecting available targets, use the principles in *Figure 4-4*. Maintaining a strong defensive stance after employing a strong side horizontal strike puts the soldier in position to execute a two-hand weak side horizontal strike.

4-49. Use of the two-hand carry position enables soldiers to use riot batons at 24 or 36 inches. To effectively employ this strike, use the strength and power of the hips to thrust the riot baton in a horizontal manner. Simultaneously, pull back with the strong hand as the weak hand drives the grip forward toward the aggressor, striking a rib or the abdominal region of the target (see *Figure 4-15*). Following the two-hand weak side horizontal strike, it is important to assess the aggressor. An aggressor may not require another strike, based on his condition and actions after enduring the one-hand forward strike.

NOTE: A pattern of movement such as a forward shuffle or forward pivot may enhance this technique.

Two-Hand Front Jab

4-50. The two-hand front jab starts from the two-hand carry position (see *Figure 4-16,* page 4-18). It is aggressively quick and creates distance, distracts, disables, and/or alters behavior. Front jabs are effective offensive strikes, usually employed as a countermeasure to charging and overpowering frontal attacks. When selecting available targets, use the principles in *Figure 4-4.*

Figure 4-15. Two-Hand, Strong Side, and Weak Side Horizontal Strikes

4-51. Use of the two-hand carry position enables soldiers to use riot batons at 24 or 36 inches. Use the strength and power of the hips to effectively employ this strike in a slightly downward or horizontal manner. Simultaneously, use both arms to shoot the long end of the riot baton straight out from the body to the target area. Immediately, pull back the riot baton to the two-hand carry position (see *Figure 4-16*).

Figure 4-16. Front Jab

4-52. Forward jabs drive the long end of the riot baton toward the aggressor, striking a rib, a hip flexor, or the abdominal region (see *Figure 4-4*, page 4-5). Following the two-hand front jab, it is important that soldiers assess the aggressor before following through with another strike, as the aggressor may not require another strike.

NOTE: A pattern of movement such as a forward shuffle will enhance the power of this technique.

Two-Hand Rear Jab

4-53. Starting from the two-hand carry position, the two-hand rear jab is aggressively quick and creates distance, distracts, disables, or alters behavior. Rear jabs are quick and effective offensive strikes usually employed as a countermeasure to charging and overpowering attacks from the rear. When selecting available targets, use the principles in *Figure 4-4,* page 4-5.

4-54. Use of the two-hand carry position enables soldiers to employ riot batons at 24 or 36 inches. To effectively employ this strike, use the strength and power of the hips to thrust the riot baton in a slightly downward or horizontal manner. Simultaneously, turn the head to the rear to see and assess the target. Use both arms to shoot the riot baton straight back from the body to the targeted area and immediately pull back the riot baton to the two-hand carry position (see *Figure 4-17*).

NOTE: A pattern of movement such as a rear shuffle will enhance the power of this technique.

Figure 4-17. Rear Jab

Two-Hand Middle Strike

4-55. The two-hand middle strike starts from the two-hand carry position. It is aggressively quick and creates distance, distracts, disables, or alters behavior. It is proven to be an effective follow-up technique to a two-hand weak or strong side block or to destabilize an individual by pushing him back and away. The two-hand middle strike is used as a two-count movement. When selecting available targets, use the principles in *Figure 4-4*.

4-56. To perform a two-hand middle strike (see *Figure 4-18*, page 4-20) the soldier does the following:

Step 1. Brings the riot baton up to chest level and slightly off parallel to the ground.

Step 2. Steps forward aggressively, simultaneously thrusting the long portion of the riot baton forward in a horizontal manner. Both arms are fully extended at the end of the movement with a quick snappy return to the step 1 position. Use the entire body to generate the power.

NOTE: It is important that soldiers assess the aggressor before applying a follow-up strike. If the soldier steps forward as he employs the middle strike, it will afford greater impact and power.

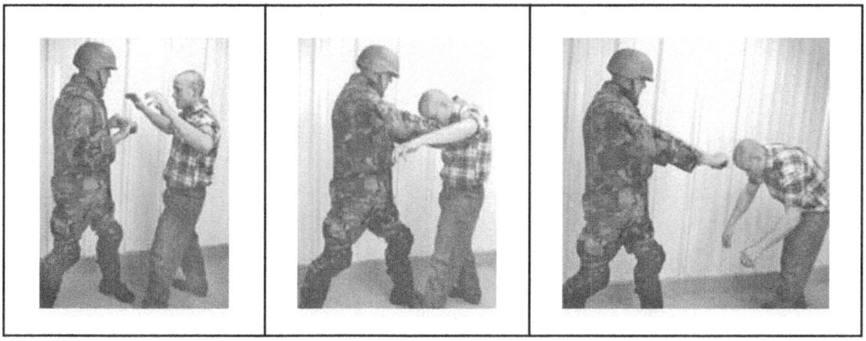

Figure 4-18. Middle Strike

Riot Baton Retention

4-57. A common technique an aggressor may use is grabbing for a soldier's riot baton or, in some cases, even taking the riot baton and using it against the soldier. This causes a disruption or distraction.

4-58. Maintaining physical control of the riot baton is essential for the safety of the soldier and the control force as a whole. Aggressors will generally attempt to grab the riot baton where they can get the strongest hold. Soldiers must resist the instinct of getting into a tug-of-war match over the riot baton, which often proves to be ineffective.

4-59. Using the momentum of the aggressor to pull the riot baton away, the soldier simply steps into or closer to the aggressor with the weak side foot. With the momentum created, the aggressor's hold on the riot baton is lessened. The soldier then rapidly uses the tracing-C technique to recover the riot baton.

4-60. A soldier using the tracing-C technique does the following:

Step 1. Pushes up on the riot baton.

Step 2. Pulls the riot baton around in an upward swing as if tracing a C in the air from bottom to top with the long end.

NOTE: This movement seems to wrap the aggressor's arms around themselves (see *Figure 4-19*).

Step 3. Drives (immediately with a quick and forceful movement) the long end of the riot baton straight down, as if slicing the C in half (see *Figure 4-19*).

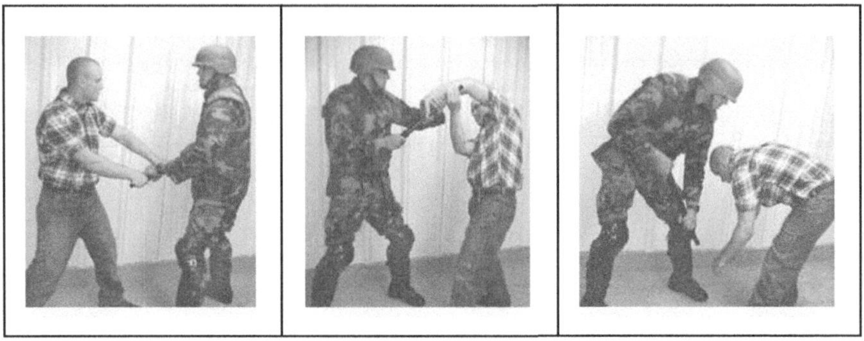

Figure 4-19. Riot Baton Retention

4-61. Once the soldier reaches the top of the C and the aggressor is tied up, it is impossible for the aggressor to maintain hold of the riot baton. Further actions such as a strike may or may not be necessary. Soldiers must be able to assess the situation and react accordingly.

WORKING AS A TEAM

4-62. Soldiers never have to rely solely on their own skills or abilities in riot control operations. Although individual skills were discussed in this chapter, it is important to remember that well-trained soldiers employing offensive, defensive, riot shield, and riot baton techniques will perform the collective task of crowd control well.

4-63. Riot control operations are dynamic and often chaotic. Soldiers armed with riot shields provide a first line of defense while those armed with riot batons and firearms provide overwatch as the formation moves toward engaging a hostile crowd. To work effectively and in concert, soldiers must be properly trained, drilled, and exercised. If an aggressor approaching a line of skirmishers violently grabs a riot shield, help is required to maintain control (see *Figure 4-20*, page 4-22). A soldier armed with a 36-inch wooden riot baton or an expandable riot baton (extended to 36 inches) can be of assistance. The soldier simply moves forward into a gap between the riot shields where he can quickly strike the top or side of the riot shield, slicing the long end across or straight down onto the hands and fingers of the aggressor (see *Figure 4-18*).

4-64. Riot batons are not the only way to effectively knock an aggressor's hands from a riot shield. Soldiers armed with 12-gauge shotguns or M16/M203s can just as effectively smash the hands or fingers of a persistent aggressor with the barrel of their weapons using the technique described above.

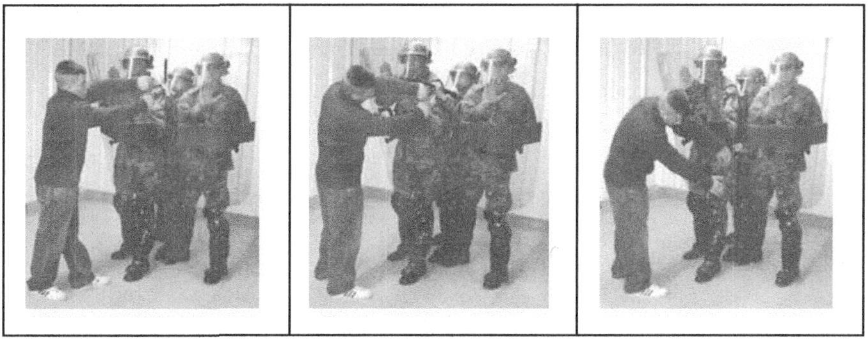

Figure 4-20. Clearing a Riot Shield

NONLETHAL MUNITIONS

4-65. With the inclusion of NL munitions, commanders now have additional tools that afford excellent standoff capabilities. This is critical because commanders now have intermediate options for dealing with a hostile crowd. No longer will a commander issue a proclamation, have it ignored by the crowd, and then be forced to use lethal munitions to backup the proclamation. Now he can use NL munitions to gain the same standoff distance but without the catastrophic results of lethal munitions.

4-66. NL munitions include (but are not limited to) a 12-gauge shotgun and 40-millimeter rounds. These blunt munitions have the capability of effectively engaging targets as close as 5 meters and as far as 50 meters. Range is dependent upon each round. These munitions are used from a distance to dissuade, discourage, and disperse the crowd. If the crowd disperses from 50 meters and soldiers are in close contact with the crowd, then the mission has been successful.

TAP-DOWN TECHNIQUE

4-67. Commanders must be familiar with the characteristics and capabilities of all rounds to use them effectively. When 12-gauge and M203 weapons are used within a line of skirmishers (usually located behind the riot shields), it is imperative that a method is developed, trained, and practiced to effectively operate as a team. Operators of these weapons may encounter problems in riot control situations when trying to identify and effectively engage legitimate targets. The tap-down technique is used to overcome target identification and engagement with NL munitions. If ordered to engage targets or the situation authorizes an engagement, soldiers armed with NL munitions may step forward, level their weapons, and fire upon legitimate targets (see *Figure 4-21*).

4-68. A soldier that is armed with a NLW and is using the tap-down technique must first tap on the shield holder's shoulder. Tapping on the shield holder's shoulder alerts him that the gunner is preparing to fire. In response to the tap, the shield holder drops to one knee while keeping the riot shield firmly affixed to their front for protection. The gunner takes a step forward and fills the gap. He then leans into the riot shield bearer's back with his knee and fires. This technique ensures that the long end of the weapon is extended beyond the riot shield.

WARNING

The long end of the weapon must be projected beyond the riot shield before firing. This is a critical safety step, which will prevent NL projectiles from hitting the riot shield and injuring soldiers.

4-69. Once soldiers firing munitions have fired, they raise their weapon back to a high port arms. The soldiers clear their weapons and tap the riot shield bearer a second time, signaling them to rise to a standing position.

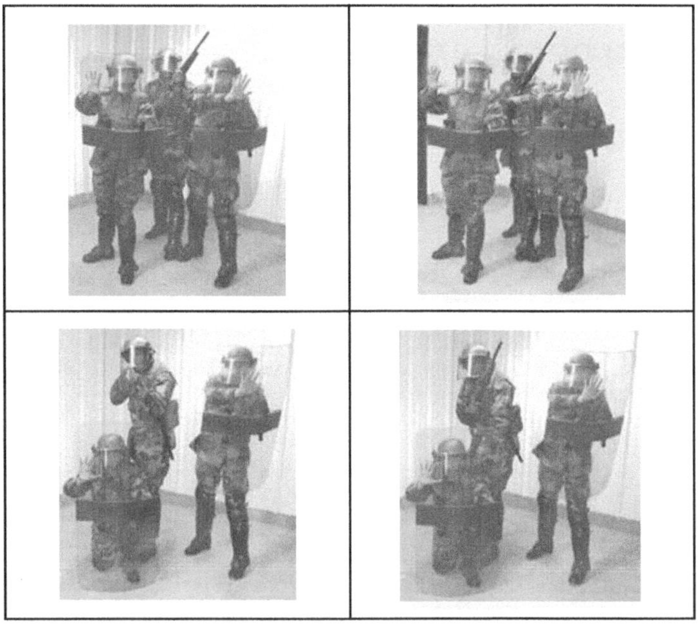

Figure 4-21. Tap-Down Technique

Chapter 5

Nonlethal Capabilities Set and Employment Considerations

During the past several decades, US forces have regularly performed peacekeeping and peace enforcement missions at a greater tempo than in the past. In the peacekeeping and peace enforcement environment, the use of conventional firearms or the threat of their use may not be the solution to a situation where US forces must separate two belligerent, hostile ethnic groups or prevent a similar group from entering an area that is off-limits to them. However, a show of force using NLW and NL munitions will cause crowds to disperse, separate, or leave the area with minimal casualities. The NLCS is a well-rounded, versatile package of both commercial off-the-shelf (COTS) and government off-the-shelf (GOTS) NL equipment and munitions. NLCS contents are divided into four distinct categories: personnel protectors, personnel effectors, mission enhancers, and training devices.

PERSONNEL PROTECTORS

5-1. Personnel protectors provide the individual soldier with added protection to the sensitive and vital areas of his body. They provide excellent protection for the individual soldier from trauma often inflicted by thrown objects (such as rocks, bricks, sticks, and bottles).

NONBALLISTIC RIOT FACE SHIELD

5-2. A nonballistic riot face shield (see *Figure 5-1,* page 5-2) has two adjustable positions and is constructed of hard durable plastic that securely attaches to the Kevlar® helmet. It may be worn with an M17 or M40 protective mask and is designed to protect the soldier's head, face, and neck from thrown objects. Nonballistic riot face shields will not protect from attacks with ballistic weapons, such as firearms. All soldiers operating in a civil disturbance environment should be outfitted with riot face shields.

NONBALLISTIC FULL-LENGTH SHIELD

5-3. A nonballistic full-length riot shield (see *Figure 5-2,* page 5-2) is a body shield constructed of hard durable plastic. Left- and right-handed personnel can easily use the nonballistic full-length riot shield. It is designed to protect the soldier's face, torso, and upper legs from thrown objects. Nonballistic full-length riot shields will not provide protection from ballistic weapons. Only the front line of skirmishers and those providing lateral protection to both flanks of the formation will be outfitted with nonballistic full-length riot shields.

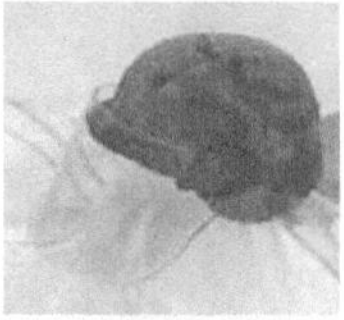

Figure 5-1. Nonballistic Riot Face Shield

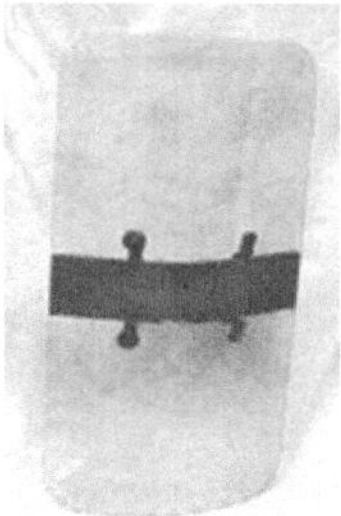

Figure 5-2. Nonballistic Full-Length Riot Shield

NONBALLISTIC SHIN GUARD

5-4. The nonballistic shin guards (see *Figure 5-3*) are constructed of hard durable plastic and are easily fastened to a soldier's legs. They are designed to protect the soldiers, feet, shins, and knees from thrown objects. They will not provide protection from ballistic weapons, such as firearms. All soldiers operating in a civil disturbance environment should be outfitted with shin guards.

BALLISTIC RIOT FACE SHIELD

5-5. The ballistic riot face shield (see *Figure 5-4*) has two adjustable positions constructed of acrylic and bullet-resistant materials that provide the wearer with Level IIIA protection. The ballistic riot face shield easily attaches to the Kevlar helmet and is capable of being worn with either an M17 or M40 protective mask. It will protect the soldier from hand-thrown objects. Because of its excessive weight (3.4 pounds), the ballistic riot face shield should not be worn for standard riot control operations. Ballistic riot face shields are best suited for use within specialized teams associated with riot control operations; for example, teams responsible for the search and apprehension of identified criminals or building- and alley-clearing operations.

Figure 5-3. Nonballistic Shin Guards

Figure 5-4. Ballistic Riot Face Shield

BALLISTIC FULL-LENGTH RIOT BODY SHIELD WITH LIGHT KIT

5-6. The ballistic full-length riot body shield (see *Figure 5-5,* page 5-4) is constructed from Spectra Shield® bullet-resistant material that provides a soldier with Level IIIA protection. It will protect the soldier from thrown objects. Because of its excessive size (18 pounds and 20 by 36 inches), the ballistic full-length riot body shield should not be used as part of a standard riot control operation. Ballistic full length riot body shields are best suited for use within specialized teams associated with riot control operations; for example, teams responsible for the search and apprehension of identified criminals or building- and alley-clearing operations.

Figure 5-5. Ballistic Full-Length Riot Body Shield

BALLISTIC SHIN GUARDS

5-7. Ballistic shin guards (see *Figure 5-6*) are constructed of Kevlar KM2 materials and provide a soldier with Level IIIA protection to the feet, shins, and upper legs. They protect the soldier from hand-thrown objects. There are three different sizes, and because of their excessive weight (7 to 10 pounds), ballistic shin guards should not be deployed as part of a standard riot control operation. Ballistic shin guards are best suited for use within specialized teams associated with riot control operations; for example, teams responsible for the search and apprehension of identified criminals or building- and alley-clearing operations.

Figure 5-6. Ballistic Shin Guards

PERSONNEL EFFECTORS

5-8. Personnel effectors are those items that provide soldiers with an NL response without the necessity of always coming into direct physical contact with aggressors. Personnel effectors are 12-gauge and 40-millimeter ammunition that inflict blunt trauma. These munitions are compatible with the shotgun and the M203 grenade launcher already in the Army inventory. There are also diversionary and rubber ball grenades, which distract and inflict blunt trauma.

WOOD RIOT BATON WITH BELT RING (36 INCHES)

5-9. The 36-inch wood riot baton (see *Figure 5-7*) is made from solid oak. When held properly, the riot baton thong provides greater retention capabilities. Intended primarily for self-defense, the 36-inch wood riot baton can be used effectively as an offensive tool if it becomes necessary to keep rioters out of arm's reach of soldiers conducting crowd control operations.

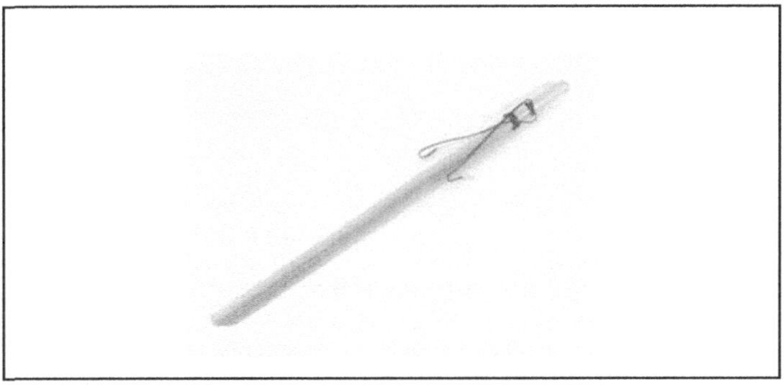

Figure 5-7. Wood Riot Baton

EXPANDABLE RIOT BATON WITH CARRIER (24 TO 36 INCHES)

5-10. The expandable riot baton (see *Figure 5-8*) provides the soldier with an NL means of crowd control and self-defense. The riot baton comes with a mounting device that attaches the riot baton to a belt. The riot baton is 24 to 36 inches in length and is primarily intended for self-defense. The expandable riot baton can be used effectively as an offensive tool if it becomes necessary to keep rioters out of arm's reach of soldiers conducting crowd control operations.

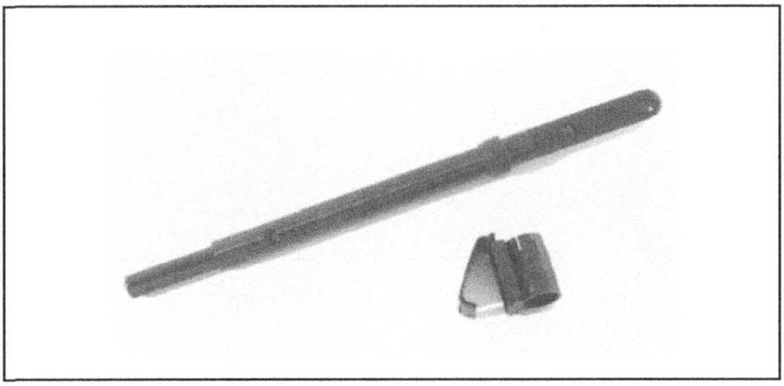

Figure 5-8. Expandable Riot Baton and Carrier

500 (3 INCH) 12-GAUGE SHOTGUN

5-11. The 12-gauge shotgun (see *Figure 5-9*) is a pump action shotgun currently in NLCS inventory. The pump action shotgun is chambered to take up to 3-inch shells. The 3-inch chamber allows for the use of M1012 and M1013 NL munitions. This shotgun also provides a visually distinct alternative to standard military weapons that may be desired based on mission considerations (force signature and acceptability).

Figure 5-9. 12-Gauge Shotgun

M7 66-MILLIMETER LAUNCHER WITH M315 INSTALLATION KIT

5-12. The M7 (see *Figure 5-10*) is a 66-millimeter vehicle-mounted, NL, grenade-launching device that is mounted on a HMMWV. It is an indirect fire support system that can deliver the M98 distraction grenade that creates a flash-bang effect, L96A1 antiriot grenade, or M99 blunt trauma grenade that creates a sting-ball effect. The M315 installation kit is used to install an M7 discharger on the turret ring of appropriate HMMWV variants. An adjustable bracket allows the launch angle to be depressed for engaging targets at ranges of 50, 75, and 100 meters. The system enforces standoff distances and deters potential threats.

Figure 5-10. M7 Launcher

MODIFIED CROWD CONTROL MUNITION-GROUND EMPLACEMENT

5-13. Similar in nature to the claymore mine, the modified crowd control munition-ground emplacement (MCCM-GE) (see *Figure 5-11*) is an NL munition that can deliver six hundred .32-caliber rubber pellets and a flash-bang effect. The MCCM-GE will stop, confuse, disorient, and/or temporarily incapacitate area targets and/or personnel at ranges out to 30 meters.

WARNING

Avoid employing the MCCM-GE within 10 meters of personnel as it may cause serious bodily harm or death.

Figure 5-11. MCCM-GE

12-GAUGE, FIN-STABILIZED POINT, NONLETHAL CARTRIDGE (M1012)

5-14. The M1012 (see *Figure 5-12*) is a single projectile round made of hard rubber that is shaped like a bomblet and designed to be fired a at a single target. With a muzzle velocity of 500 feet per second, the M1012 has the effective range of no closer than 5 meters and no further than 30 meters. Engagement inside 5 meters could cause serious injury or death. Beyond 30 meters the kinetic energy dissipates to the point where the round becomes ineffective.

Figure 5-12. M1012

12-GAUGE, CROWD DISPERSAL, NONLETHAL CARTRIDGE (M1013)

5-15. The M1013 (see *Figure 5-13*) is a multiple projectile round with .23 caliber hard rubber pellets that is designed to be fired at and employed with the purpose of affecting multiple targets. With a muzzle velocity of 900 feet per second, the M1013 has an effective range of no closer than 5 meters and no further than 30 meters. Engagement inside 5 meters could cause serious injury or death. Beyond 30 meters, the kinetic energy dissipates to the point where the rubber pellets become ineffective.

Figure 5-13. M1013

40-MILLIMETER SPONGE GRENADE (M1006)

5-16. The M1006 (see *Figure 5-14*) is a single projectile round made of pliable foam rubber with a hard plastic back. The M1006 is designed to be aimed and fired at a single target. With a muzzle velocity of 265 feet per second, the M1006 has an effective range of no closer than 10 meters and no further than 50 meters. Engagements inside 10 meters could cause serious injury or death. Beyond 50 meters, the kinetic energy dissipates to the point where the round becomes ineffective.

Figure 5-14. M1006

40-MILLIMETER CROWD DISPERSAL CARTRIDGE

5-17. The crowd dispersal cartridge (CDC) (see *Figure 5-15*) is a multiple projectile round with .48 caliber hard rubber balls that is designed to be fired and employed with the purpose of affecting multiple targets. With a muzzle velocity of 450 feet per second, the CDC has an effective range of no closer than 10 meters and no further than 30 meters. Engagement inside 10 meters could cause serious injury or death and beyond 30 meters the kinetic energy dissipates until the rubber balls become ineffective.

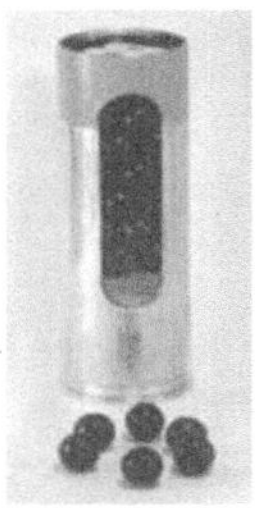

Figure 5-15. CDC

RUBBER BALL GRENADE (GG04)

5-18. The GG04 (see *Figure 5-16,* page 5-10) is a multiple projectile, flash-bang grenade with 100 .25-caliber hard rubber pellets. Each grenade has a fuse delay of 2.8 to 3 seconds with a flash measuring approximately 1 million CP and 180 decibels at 3.5 feet. At detonation, rubber pellets are dispatched at 360° with an effective range of 2 to 3 meters and a maximum engagement range of 15 to 20 meters. The rubber ball grenade is designed to be hand-thrown or muzzle-launched from a 12-gauge shotgun.

RUBBER BALL GRENADE LAUNCHING CUP AND AA30 12-GAUGE LAUNCH ROUND

5-19. The launch cup (see *Figure 5-17,* page 5-10) firmly attaches to the 12-gauge shotgun without any additional tools. It works in concert with the AA30 12-gauge launch round to propel the rubber ball grenade approximately 100 meters. The shooter simply inserts the full body and the safety lever of the grenade into the launching cup, pulls the pin, loads a 12-gauge round, and fires (approximately) at a 30° angle. The desired effect is to have the grenade explode approximately 7 to 10 feet above the target.

RELOADABLE FLASH-BANG GRENADE (M84)

5-20. The M84 (see *Figure 5-18,* page 5-10) is a reloadable flash-bang thrown grenade. Each grenade has a 1- to 2-second fuse delay, with a flash capability of 1.5 to 2.5 million CP and a bang of 168 to 175 decibels. Although it is part of the NLCS, use of the M84 is not recommended for use in crowd control situations. Specialty teams, such as security response teams, and search teams may use this device effectively.

Figure 5-16. GG04

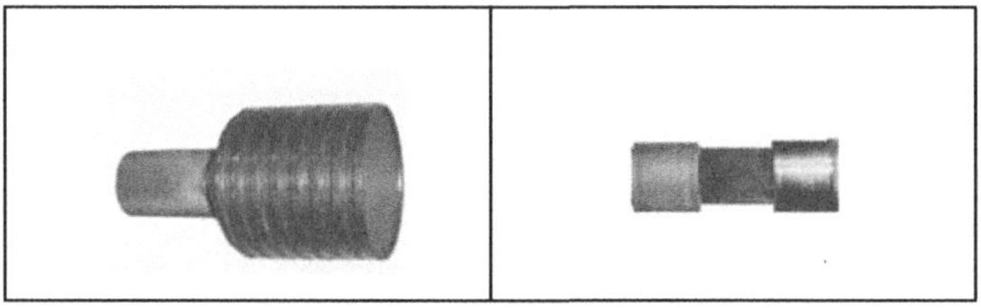

Figure 5-17. Launching Cup and AA30

Figure 5-18. M-84

FLEX-CUFS

5-21. Flex-Cufs consist of a tough pliable plastic band with a self-locking mechanism in the center for each end. When the Flex-Cuf (see *Figure 5-19*) is threaded the restraint bands circle around the wrists or ankles, impeding movement and securing the individual. They are lightweight and disposable and require a cutting tool or other instrument to remove.

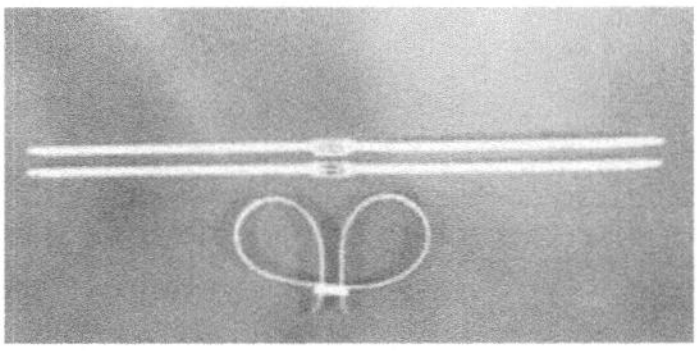

Figure 5-19. Flex-Cufs

INDIVIDUAL RIOT CONTROL AGENT DISPERSER (M36)

5-22. The M36 (see *Figure 5-20*) contains a Dibenz (F, -1 4-oxazepine [CR] solution. It can deliver 25 one-second bursts out to 12 feet. Individual disposable RCA dispersers are intended primarily for self-defense or to keep rioters out of arm's reach of soldiers conducting crowd control formations or soldiers engaged in missions where a noncombatant exists.

Figure 5-20. M36

MIDSIZE RIOT CONTROL AGENT DISPERSER (M37)

5-23. The M37 (see *Figure 5-21,* page 5-12) is the size of a standard fire extinguisher that uses compressed air to force the RCA out to a range of 30 feet. It has the capacity to employ 18 bursts for 3 seconds per charge. It is excellent for providing a wide coverage of RCAs onto a hostile crowd while maintaining excellent standoff capability. The M37 can be refilled and is rechargeable. It can be filled with CR solution (liquid agent) or CS1 (dry agent). For training purposes, CR can be substituted with water and CS1 can be substituted with talcum powder.

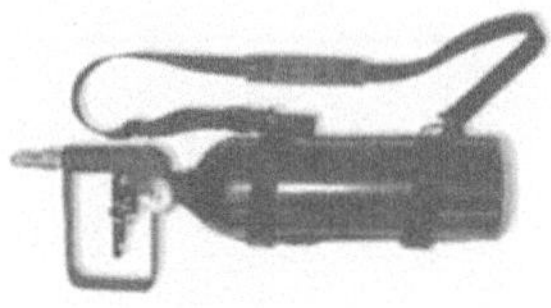

Figure 5-21. M37

SQUAD RIOT CONTROL AGENT DISPERSER (M33A1)

5-24. The M33A1 (see *Figure 5-22*) is designed to provide crowd control and protection at the squad level. It is capable of projecting a ballistic stream of RCAs beyond 25 feet in up to 25 half-second bursts. It consists of a frame and harness assembly, compressed-gas cylinder (agent container assembly), air pressure assembly, gun and hose assembly, multijet spray unit, and check valve assembly. The M33A1 can be refilled and is rechargeable. For training purposes, CR can be substituted with water and CS1 can be substituted with talcum powder.

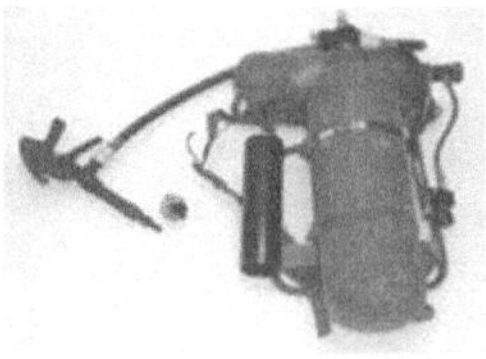

Figure 5-22. M33A1

MISSION ENHANCERS

5-25. Mission enhancers aid the commander in various tactical situations. Mission enhancers include various types of devices used for illumination, vehicle denial, traffic control missions, and force protection enhancement.

PORTABLE BULLHORN

5-26. The portable bullhorn (see *Figure 5-23*) provides squad leaders with the capability to project their voices above the noise and commotion created by crowds and mobs. The megaphone has a built-in siren for alarm purposes that allows the operator to use the device as an ordinary megaphone and siren alarm during emergencies. The bullhorn is portable and easy to carry. The acoustic range of the megaphone for a clear voice output is ¾ to 1 mile. The power output of the megaphone is rated at 15 watts with a maximum rating of 20 watts. The dimensions of the megaphone are 14.5 inches with a 9-inch diameter bell. It weighs 3.5 pounds without batteries. Battery requirements for the megaphone are eight dry-cell size C batteries.

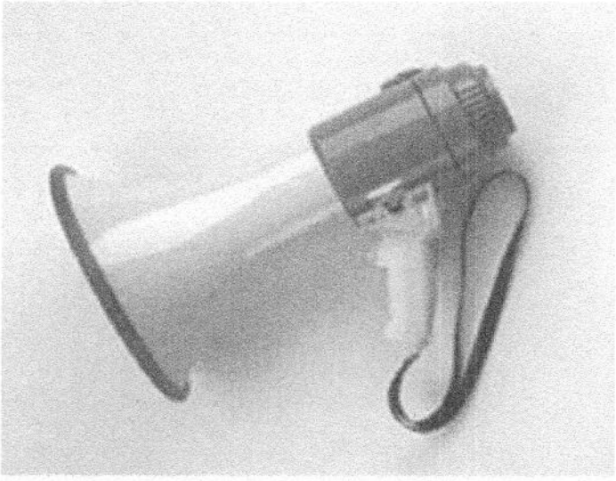

Figure 5-23. Portable Bullhorn

GROUND-MOUNTED BULLHORNS

5-27. Ground-mounted bullhorns (see *Figure 5-24,* page 5-14) are a critical communication enhancement device for conducting crowd control tactics. The bullhorn can facilitate communication with the crowd in conjunction with PSYOP support and assistance with the communication of commands to troops engaged in the crowd control process. The power output of the ground-mounted bullhorn is rated at 22 watts. The dimensions of the bullhorn are 13.75 by 9 by 14.75 inches, and it weighs 15 pounds.

INDIVIDUAL VOICE AMPLIFICATION SYSTEM (M7)

5-28. The M7 (see *Figure 5-25,* page 5-14) is a critical communication enhancement device for conducting crowd control tactics using RCAs while wearing the M40 protective mask. The M7 is fitted to the M40 protective mask to facilitate oral communication and increase the user's ability to communicate using radios and other devices.

Figure 5-24. Ground-Mounted Bullhorns

Figure 5-25. Individual Voice Amplification System

INDIVIDUAL HIGH-INTENSITY SEARCHLIGHT WITH HOLSTER

5-29. The individual high-intensity searchlight (see *Figure 5-26*) is used primarily for illumination in crowd control operations at night. An individual can use it for general illumination of the operational area to pinpoint agitators and threats, reduce the ability of the rioters to see troop formations and actions, enhance tactical deception and actions, and enhance tactical deception techniques for units conducting crowd control operations.

HIGH-INTENSITY LIGHT

5-30. A high-intensity light (see *Figure 5-27*) is intended for use in low light or night conditions. A high-intensity light can project a beam that will enable the user to identify an individual person up to 1,900 yards away. It has an adjustable beam spread of 1° to 15° and will run continuously at maximum power for up to 45 minutes. The internal power supply requires 4 hours for a full charge.

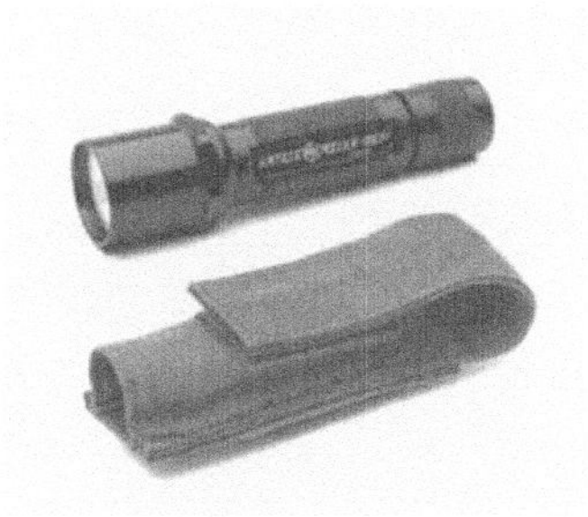

Figure 5-26. Individual High-Intensity Searchlight with Holster

Figure 5-27. High-Intensity Light

HIGH-INTENSITY LIGHT SYSTEM WITH 12-GAUGE SHOTGUN ATTACHMENT KIT

5-31. This 12-guage high-intensity light system (see *Figure 5-28,* page 5-16) is used to illuminate possible targets in crowd control situations where there is little or no light. It serves to blur or blind the aggressor's ability to see. It mounts easily to the end of the 12-gauge shotgun barrel.

40-MILLIMETER CARRYING POUCH

5-32. The 40-millimeter carrying pouch is made from durable canvas material, (see *Figure 5-29,* page 5-16) and can be easily affixed to load-bearing equipment (LBE) and/or a load-bearing vest (LBV) using Velcro straps. It is capable of carrying up to six 40-millimeter rounds.

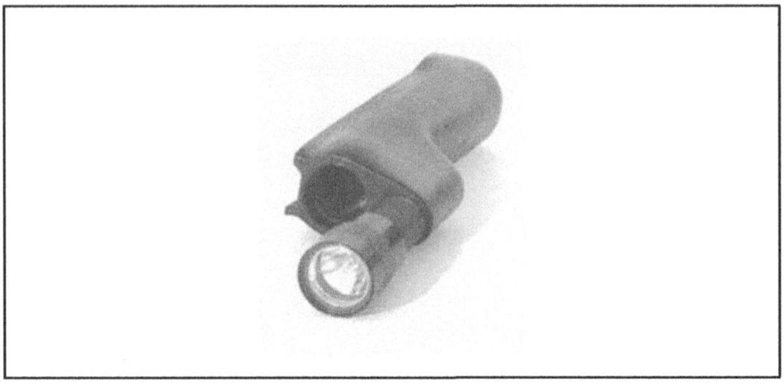

Figure 5-28. 12-Gauge High-Intensity Light System With 12-Gauge Shotgun Attachment Kit

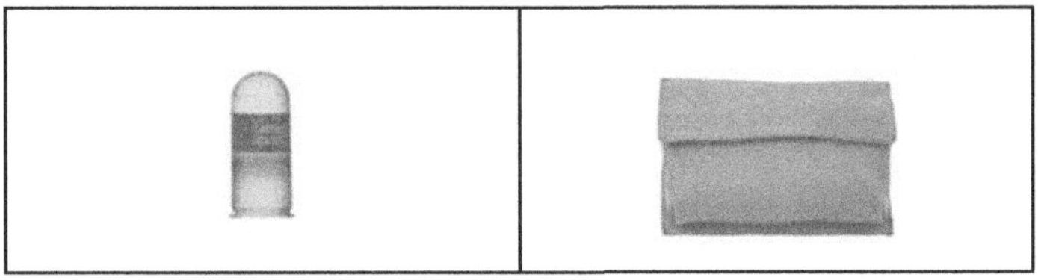

Figure 5-29. 40-Millimeter Carrying Pouch

12-GAUGE UTILITY POUCH

5-33. The 12-gauge utility pouch is made from durable canvas material (see *Figure 5-30*). It can be easily affixed to an LBE and/or LBV using metal clasps. The 12-gauge utility pouch is capable of carrying up to twenty-five 12-gauge shotgun rounds.

Figure 5-30. 12-Gauge Utility Pouch

12-GAUGE BUTTSTOCK CUFF

5-34. The 12-gauge buttstock cuff (see *Figure 5-31)* firmly attaches to the shotgun by sliding over the stock. Elastic straps aid in preventing the buttstock cuff from sliding or rolling. It holds up to seven 12-gauge cartridges, which give the soldier quick access to additional rounds.

Figure 5-31. 12-Gauge Buttstock Cuff

DIVERSIONARY AND/OR RUBBER BALL GRENADE POUCHES

5-35. The diversionary and/or rubber ball grenade pouch (see *Figure 5-32*) is an ammunition-carrying pouch made of durable canvas material. It is partitioned in a manner to provide the soldier with a means of holding up to six rubber ball grenades or six flash-bang grenades in separate compartments.

Figure 5-32. Diversionary and/or Rubber Ball Grenade Pouch

CALTROPS

5-36. The caltrop (see *Figure 5-33,* page 5-18) is a field-expedient area denial system. Securing approximately 50 caltrops to 550-type military cord allows a soldier to quickly toss and recover the system. Caltrops are used to deny vehicle and pedestrian movement in designated areas. The four prongs are approximately 2 inches tall and will puncture tires or boots. A soldier can easily employ the caltrop by scattering it in a designated area hidden by camouflage.

Figure 5-33. Caltrop

PORTABLE VEHICLE ARREST BARRIER

5-37. A portable vehicle arrest barrier (PVAB) (see *Figure 5-34*) is a lightweight, portable, manually-emplaced and recoverable-barrier entrapment device that is used (on command) to create a barrier that will stop vehicles being driven by known or suspected hostile forces with minimal damage to the vehicle and its occupants. Two individuals can easily emplace it in approximately 4 to 8 hours. It has the capability of stopping a light truck (up to 7,500 pounds) at speeds of up to 45 miles per hour with a stopping distance of less than 200 feet. It is not to be used as a stand-alone system. The PVAB may be used at dismount points, checkpoints, and roadblocks to prevent unauthorized access by wheeled vehicles into or out of areas.

Figure 5-34. PVAB

TRAINING DEVICES

5-38. Training devices that commanders can use to effectively prepare their soldiers are included in the NLCS. Training devices allow soldiers to learn and build confidence in the development of personal defensive skills. These are critical pieces of equipment because they allow soldiers to train at almost full speed, which provides reinforcement of realistic motor skills. Training in this manner has proven to be more effective than just striking out at the air.

IMPACT TRAINING SUIT

5-39. Impact training suits (see *Figure 5-35*) are made of closed-cell shock-absorbent foam that allows soldiers to hone their riot baton skills through realistic hand-to-hand training. Impact training suits absorb strikes inflicted by the current expandable and wooden riot batons. They provide protection for the head, face, hands, and legs without significant degradation of the wearer's mobility.

Figure 5-35. Impact Training Suits

TRAINING STRIKE BAGS

5-40. Training strike bags (see *Figure 5-36*) give the soldier a durable, yet absorbing surface to strike or kick for training. It also helps develops proficiency in open-hand control and riot baton techniques.

Figure 5-36. Training Strike Bag

PRACTICE GRENADE BODY (GG05)

5-41. The GG05 is blue in color and is easily identified as a practice grenade (see *Figure 5-37*). It is designed to replicate the live GG04 rubber ball grenade. It is made of durable rubber and can be hand-thrown or muzzle-launched from the 12-gauge shotgun. It can be reloaded.

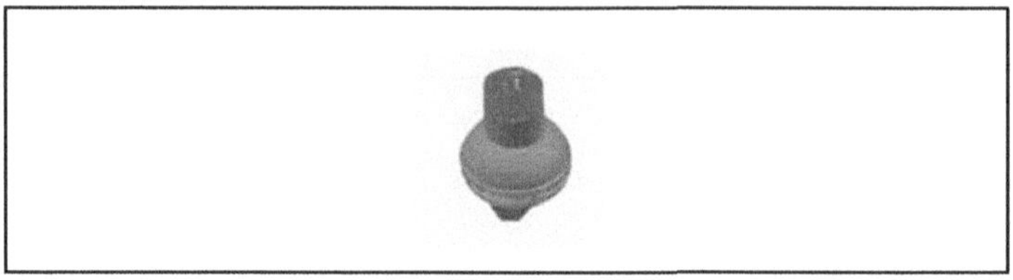

Figure 5-37. GG05

PRACTICE GRENADE FUSE

5-42. The practice grenade fuse (see *Figure 5-38*) is identified by its blue spoon and brown stripe. It easily screws into the body of the GG05. Once used, unscrew the practice grenade fuse from the body and reload it with a new fuse.

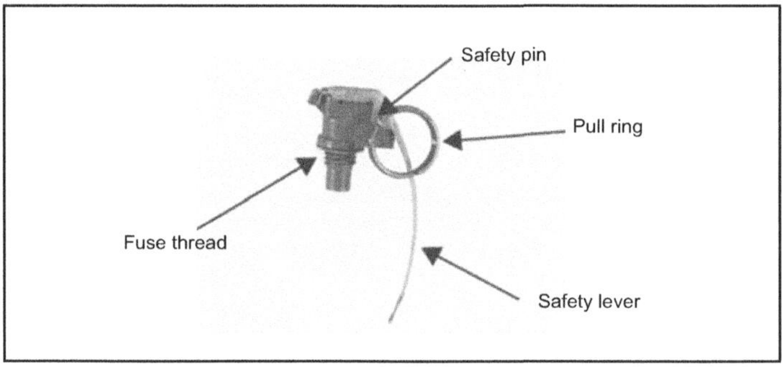

Figure 5-38. Practice Grenade Fuse

CORE CAPABILITIES

5-43. Core capabilities are those fundamental competencies that enable the commander to achieve the desired operational outcome. In the case of NLW, this includes providing a flexible means of response in order to protect friendly forces and/or influence the actions of potential adversaries and noncombatants. With the availability of the NLCS (see *Figure 5-39*) these goals can be achieved without resorting to lethal force and in a manner that will minimize collateral damage. The core capabilities associated with NL effects fall into the following two major categories:

NOTE: There are nine platoon equivalents within a light infantry battalion.

PLT Organization	Personnel and Weapons	Nonlethal Weapons & Equipment
PLT	7 personnel (5 M16/M4 and 2 M60/ M240)	3 riot face shields 3 sets of shin guards 3 sets of full-body restraints 3 individual voice amplification systems 3 individual RCA dispensers with carriers 1 portable bullhorn 1 ground-mounted bullhorn 1 high-intensity searchlight 1 set of caltrops 3 ballistic riot face shields 3 sets of ballistic shin guards 3 ballistic riot body shields 1 riot baton training suit 1 riot baton strike pad
SQD SQD SQD	3 personnel (3 M16/M4)	3 riot face shields 3 sets of shin guards 3 riot body shields 3 expandable riot batons with carriers 9 sets of cinch-strap restraints 3 individual voice amplification systems 3 individual RCA dispensers with carriers 3 individual high-intensity search lights with carriers (see *Figure 5-26*)
TM TM TM TM TM TM	24 personnel (12 M16/M4 and 6 M16/ M4 with M203 6 SAW)	24 riot face shields 24 sets of shin guards 100 sets of disposable flex-cuffs 24 individual voice amplification systems 24 individual RCA dispensers with carriers 6 expandable riot batons with carriers 11 grenade pouches 12 forty-millimeter grenade pouches 6 twelve-gauge ammunition pouches 6 twelve-gauge bitstock ammunition carriers

Figure 5-39. NLCS Distribution

- **Counterpersonnel.** NL counterpersonnel capabilities enable the application of military force with a reduced risk of fatalities or serious casualties among noncombatants or, in some instances, among enemy forces. There are several specific NL counterpersonnel capabilities to be explored. These include the means to influence the behavior and activities of a potentially hostile crowd, and the capability to bring a mob engaged in a riot under control. While there are many similarities in these two scenarios, each involves unique challenges, which may require radically different solutions. For more information, see *FM 3-22.40*.
- **Countermaterial.** NL countermateriel capabilities enhance operations by rendering equipment and facilities unusable without complete destruction. An NL countermateriel capabilities enable the

application of military force to defuse potentially volatile situations under circumstances in which more destructive conventional military means might prove counterproductive. For example, pre-emptive strikes against troublesome and aggressive nations may be politically unacceptable when only conventional weapons with the high risk of casualties are involved. With NL countermateriel capabilities, the enemy's ability to threaten its neighbors could be curtailed with less political risk by attacking only their weapons of war and their supporting infrastructure. For more information, see *FM 3-22.40*.

ADDITIONAL CAPABILITIES

5-44. NLW are only NL when fired within their parameters.

DANGER
If NLW are fired outside their parameters, they may cause serious injury or death.

5-45. If a situation turns from NL to lethal, an NL gunner only needs to change his point of aim and distance to inflict serious bodily harm and/or death to the threat. The closer the soldier is to the target, the greater the kinetic energy and the more severe the effects are to the threat.

EMPLOYMENT CONSIDERATIONS

5-46. NL munitions and equipment are not completely NL; they are NL by intent. The term NL does not guarantee zero mortality or nonpermanent damage. NLW can add flexibility to combat operations and enhance force protection by providing an environment in which friendly troops can engage threatening targets with a reduced risk of noncombatant casualties and collateral damage.

5-47. If the tactical situation dictates a NL response, a soldier who levels a weapon loaded with NL munitions must be trained in distance to the target and appropriate points of aim. Based on the soldier's understanding and training on the characteristics and capabilities of the round, they can easily discern and engage targets to the desired effect. If the tactical situation dictates a lethal response, soldiers can easily adjust the point of aim to a more vulnerable target location on the body and move in closer to the target. NL munitions can be used as a lethal response if the situation dictates and the intent of the soldier is to inflict death.

NONLETHAL WEAPONS CAPABILITIES IN FORMATIONS

5-48. The mix of NLW within the formation is based on METT-TC. For example, the first rank of the formation will have riot shields, riot batons, and their personal weapon (slung across their back [left to right] with the butt up and muzzle down or holstered). The second row will have a mix of shotguns and M203s. Shotguns and M203s may be used as area denial or point target weapons depending on the specific munitions used. The shotgun provides cover for the slower loading M203. The commander may move these weapons within the formation to meet the needs of the mission and to create distance between the formation and the crowd (see *Appendix D*).

Chapter 6

Civil Disturbance Formations

Control force formations, when properly employed and executed against a crowd, are some of the most practical methods of crowd control. The riot baton has been the weapon of choice for the control force. With the development of NLW, the commander now has more options at his disposal. Before NLW, the only other choices were chemical irritants, MWDs, and lethal force. Now commanders can create a standoff distance with NLW that reduces the risk of close fighting with the mob. During any crowd control operation, the commander must protect the control force with lethal overwatch. This overwatch may be in the form of marksmen and observer teams (sniper with observer) and a response force with lethal capabilities.

OVERVIEW

6-1. With advances in NLW, it is possible to create an effective standoff distance of 15 to 100 meters between the crowd and the control force formation. It is now possible to advance and disperse the crowd without coming into direct contact with it. Some NLW may also be used to assist in the identification and apprehension of instigators.

6-2. Control force formations are used to disperse, contain, and/or block a crowd. These formations are more effective in urban areas than in open areas, such as parks or fields. When control force operations are employed in urban areas, it is easier to split a crowd into smaller segments, isolate instigators, or funnel the crowd into the desired location by using buildings and other man-made structures. Commanders must be mindful of the fact that well-organized crowds may attempt to move to the flanks of a formation and/or get behind the formation to gain a tactical advantage.

6-3. Commanders must be aware of the limitations of control force formations. They are not the answer to all civil disturbance situations. Do not expose the formation to sniper fire and/or unnecessary violence simply for a show of force. Control force formations should be used when it is decided that a crowd poses a threat and must be dispersed or moved to a specified area. Before dividing a large crowd, the commander must consider that this might not solve the problem. It may worsen the problem by creating smaller elements that may engage the control force in small-mob tactics, such as sniping, looting, burning, and attempting to envelop the control force. Commanders must anticipate that breakaway crowds may engage in small-mob tactics. These tactics can be defeated by area control measures, such as building searches, saturation patrolling, and other tactics. After dispersing the crowd, the control force must not assume that there will be an immediate return to peaceful activities. The use of formations is only part of the total dispersal effort.

6-4. Commanders must assess and secure the areas of operations (AOs). This may be accomplished with the use of helicopter overflights and by visually securing rooftops, high elevations, and advantage points. Designated marksmen teams may be used to secure the rooftops and provide lethal overwatch for the control formation. The control force commander must be aware that designated marksmen teams are positioned on rooftops.

6-5. Crowd control formations and their support teams have more capabilities than just crowd dispersal. They also have the capability to apprehend and detain certain members of the crowd that the commander feels may instigate the crowd to further violence. Commanders must analyze the threat, determine the mission of the control force, and decide which control force formations will accomplish the mission according to METT-TC.

6-6. During CONUS operations, there are usually no enemy forces. However, commanders must analyze and prepare for any potential opposition. The formations discussed in this chapter are guidelines and may be adapted to fit any mission or situation. Whatever the modification, soldiers must stay in position and on line. Only through training and rehearsals will soldiers become proficient in basic formation movements and be able to adjust to changing situations. Soldiers must also be prepared to employ NLW. *Figure 6-1* is used as a key to identify soldiers within crowd control formations.

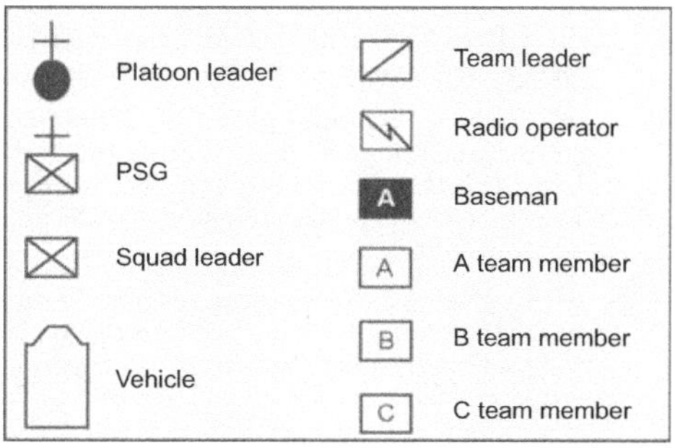

Figure 6-1. Symbols for a Formation Element

ACTIONS BEFORE MOVEMENT

6-7. Before moving the control force to the civil disturbance area and into a potentially hostile urban environment, the commander and his staff must review the preincident plans and expand them (as needed) based on current information. Because of the employment of forces against ambiguous threats,

IPB becomes critical. Information is the key to developing an appropriate response to civil disturbances.

6-8. Civil disturbance training must be part of the annual unit training plan. If a soldier is deployed to an area where civil disturbances are a recurring event, sustainment training must occur more frequently. Before moving the control force to the rally point, an intense rehearsal must be conducted by the leadership and the control force. A top-down review of the rules of engagement (ROE/rules for the use of force (RUF) must be accomplished. The ROE/RUF must be part of the unit training plan and each soldier must be trained to standard (see *Chapter 8*). The use of NLW (type specific) must be granted to the on-site commander from the approval authority.

6-9. Information is the key to developing plans for appropriate responses to a civil disturbance, and IPB is crucial to the success of the mission. During the planning phase (see *Chapter 2*), the information provided by the IPB will aid the staff in selecting various routes and rally points. The selection of ingress and egress routes must include a variety of ways in and out of the area. Direct and indirect routes and those with cover and concealment are necessary. The rally point must be cleared and secured by an advanced party capable of controlling the area.

RALLY POINT ACTIONS

6-10. Once at the rally point, the commander makes contact with the local police, civil authorities, or military officer. He then assesses the situation, decides the next COA, and issues orders, as appropriate. The platoon sergeant (PSG) forms the platoon into a column formation and places team leaders directly in front of their teams (see *Figure 6-2,* page 6-4). Team and squad leaders make last minute checks of their soldiers and await orders.

6-11. The commander places his lethal overwatch (designated marksmen [DM] teams) in areas that overlook the control force and the mob. They may be to the flanks and slightly to the front of the formation. It is their responsibility to protect the formation from lethal fire by constantly scanning the crowd.

6-12. When the commander receives word that DM teams are in place, he forms the control force into a crowd control formation and moves them quickly from the rally point to where the crowd is assembled. Depending on the situation, movement should be in a column formation or one of the three basic control force formations. Reserve forces are left at the rally point until ordered forward as additional lethal overwatch or to reinforce the formation.

CONTROL FORCE FORMATIONS

6-13. Control force formations, when properly employed and executed against a crowd of limited size, are one of the most practical methods of crowd control. Experience has indicated that the most frequently used control force

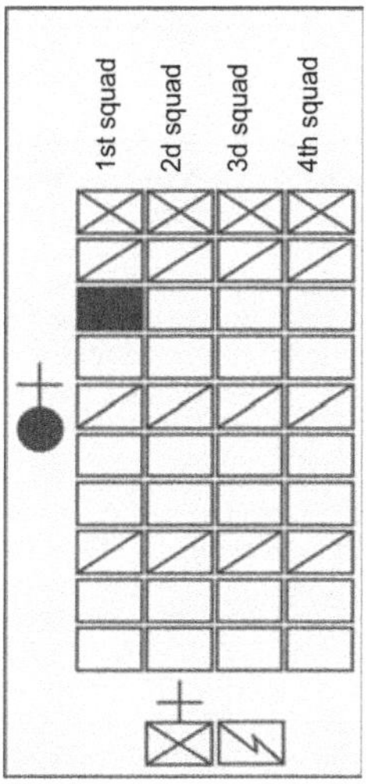

Figure 6-2. Column Formation

formations are the line, echelon, and wedge. The two lesser-used formations are the diamond and circle. Descriptions of all the formations are as follows:

- **Line formation.** The line is the basic formation and it is used more often because of its offensive and defensive applications. As an offensive formation, the line is used to push or drive crowds straight back, across an open area, or up a city street. As a defensive formation, the line is used to hold the crowd or to deny access to areas.
- **Echelon formation.** The echelon is an offensive formation used to turn or divert groups in open or built-up areas and to move crowds away from buildings, fences, and walls.
- **Wedge formation.** The wedge is an offensive formation that is used to penetrate and split crowds into smaller groups.
- **Diamond formation.** The diamond, when used as an offensive formation, is used to enter a crowd and is the formation of choice for extraction teams. As a defensive formation, the diamond is used when all-around security is required in open areas. The decision to use this formation is based on the conformation of the crowd.
- **Circular formation.** The circular formation is used for the same purpose as the diamond formation. The decision to use this formation is based on the conformation of the crowd.

6-14. There are many suitable variations of control force formations that may be employed, but appropriate commands and signals must be created to execute the formations. Because of their somewhat complicated nature and the coordination required for these formations, new variations must be trained, practiced, and rehearsed before they are used in a civil disturbance.

VEHICLES AND FORMATIONS

6-15. Vehicles may be employed with troops in control force formations, especially when a large riotous crowd is on hand. Although vehicles add strength to formations, certain precautions must be taken. Covering the windshield with sturdy, close mesh fencing and the standard safety glass will offer some level of protection to the occupants. Shields and/or mobile barriers may be built by mounting a wooden or metal frame strung with barbed wire across the front of a vehicle. Members of the formation should walk as near to the front corners of each vehicle as possible to keep rioters from attacking the sides and rear of the vehicles.

6-16. Up-armored HMMWVs, ASVs, or armored vehicles may be employed with the control force formation when determined by METT-TC. They present a strong psychological effect and offer protection for the occupants. Wheeled armored vehicles are best because they do not deface the pavement. Whenever vehicles are used in a crowd control formation, the commander must ensure that he is still able to see and control the formation. This may require him to occupy a position in a similar vehicle behind the formation or command subordinate leaders to move away from the front line and into line-of-sight (LOS) control. This will require additional hand-and-arm signals to allow the commander and subordinate leaders to communicate. Commanders can effectively use vehicles as primary communication tools by using horns and lights as signals to organize movement or actions. However vehicles are used, it takes a good measure of training and practice to be effective.

ELEMENTS OF A CONTROL FORCE FORMATION

6-17. Four elements make up the basic crowd control formation. They are as follows:

- **Base element.** The base element is made up of two ranks. The first rank is shield holders, while the second rank contains the NLW. This is the front line of the formation.
- **Support element.** The support element forms up in a column formation behind the base element and has a variety of uses. It may be used to replace base element members as needed, provide lateral or direct support, or perform extraction team operations. The support element helps the base element by performing the following three essential tasks:
  - **General support.** The general-support element is formed from an uncommitted squad in the platoon (generally the 2d squad). When a company is tasked as the control force element, one of the platoons becomes the general-support element. The element is in a column formation centered on and behind the main formation. From this formation, the general-support element can move to lateral or direct support, as needed.
  - **Lateral support.** The lateral-support element is used to protect the flanks of the formation. This is done by moving a set number of teams forward from the general-support element or by using the

end teams in the formation; however, this will make your formation smaller. Once in position, these teams become part of the formation, with their riot shields facing the flanks of the formation.

  - **Direct support.** When direct support is ordered, the general-support element moves forward as they form their own formation. While moving forward, the squad and team leaders from the control force formation step back and allow the riot shield holders from the support element to step behind riot shield holders of the original formation. This allows the support riot shield holders to form up between and behind the riot shield holders of the original formation. Now there are two lines of riot shield holders between the crowd and NLW firers. This formation is the strongest of the formations and requires more planning and practice to master.

- **Command element.** The command element contains several different members. A general configuration for the command element is the platoon leader, the PSG, a radio operator, a video recorder operator (if required), and an interpreter (if required). This element does not have a fixed location within the formation and moves about as needed.
- **Lethal overwatch element.** The lethal overwatch element is a team formed from reserve security forces.

6-18. The reserve support element is not part of the control force formation until it is brought forward from the rally point to join the formation. It remains until released by the formation commander. Lethal overwatch teams are formed from the reserve force. Once lethal overwatch teams are deployed and in position, they are under the control of the formation commander. Communication between the lethal overwatch team and the formation commander is a priority because he approves target selection and engagement.

WEAPON POSITIONS

6-19. At the commander's discretion, members of the control force formation may carry their weapons in any of the positions prescribed in *FM 3-21.5*. However, the primary method of carrying weapons in the control force formation is at the safe-port position (see *Figure 6-3*). It allows the soldier to control both ends of the weapon while moving in and out of the formation and advancing on the crowd.

6-20. High-port position (see *Figure 6-3*) is a position of complete readiness. It is used whenever troops are in contact with a crowd that is showing resistance or not withdrawing. This is a tiring position and is hard to maintain for extended periods. Commanders must rest their troops at every opportunity by using less tiring positions.

Safe-port arms | High-port arms

Figure 6-3. Weapons Positions

COMMANDS

6-21. Commands are given to the control force formation orally or with hand-and-arm signals. Oral commands for the on-guard position are given in one count. All other commands are given in two counts, a preparatory command followed by a command of execution. However, oral commands cannot be relied on completely, so commanders must plan to use hand-and-arm signals too. When vehicles are part of the formation, radios become an alternate method for relaying commands. To improve communication, commanders may use hand-and-arm signals (see *Figure 6-4,* page 6-8) with oral commands. The basic commands used in control force formations are outlined in *Table 6-1,* page 6-9. These commands can be added to or combined to form more complex commands, therefore forming more complex formations.

MOVEMENTS

6-22. Control force formations move in the same manner as regular formations. This allows the commander to more precisely maneuver the formation by commanding the number of steps it should move and in what direction, such as "Five Steps, Forward March." The normal rate of march for entering and leaving a control force formation is double-timing. The half-step march is used when the formation is in direct contact with the crowd. This slows the formation down and allows for better command and control. The squad and team leaders echo preparatory commands and provide cadence while in control force formations.

INTERVAL AND DISTANCE

6-23. Interval is the lateral space between elements. Distance is the space between elements in a column. The usual interval and distance between soldiers in control force formations is 30 inches. The interval and distance can be adjusted based on METT-TC. If the crowd has to be physically pushed back, a close interval is preferred. This allows the riot shields to be overlapped, creating a stronger wall. During peaceful demonstrations, the interval could increase to double arm width, allowing the formation to cover more area and still control the crowd. The interval and distance could change several times during the course of operations before the crowd is dispersed.

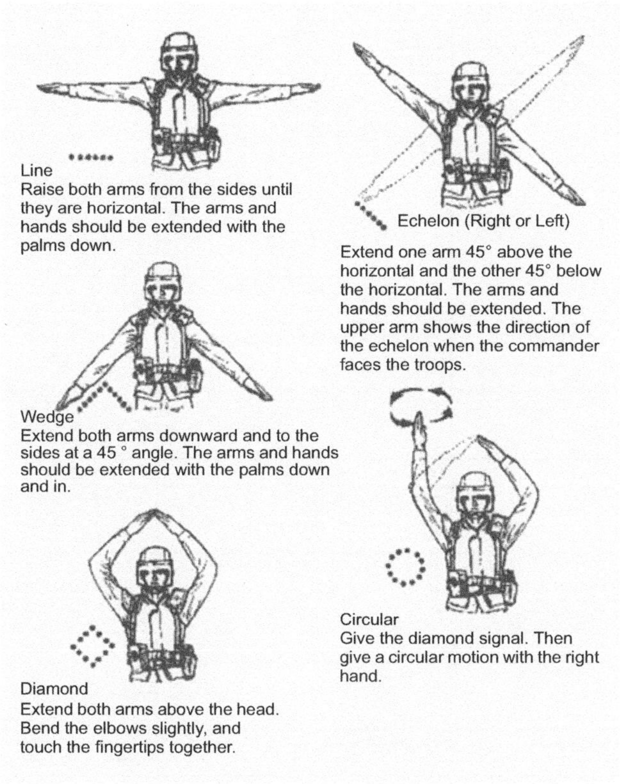

Figure 6-4. Hand-and-Arm Signals

Table 6-1. Basic Commands

Commands for	Command	Given by	Purpose
Formations	Platoon On Line	Platoon leader and/or commander	To form a line formation.
	Platoon Wedge	Platoon leader and/or commander	To form a wedge formation.
	Platoon Echelon	Platoon leader and/or commander	To form an echelon formation.
	Platoon Diamond	Platoon leader and/or commander	To form a diamond formation.
	Platoon Circle	Platoon leader and/or commander	To form a circular formation.
	Move	Platoon leader and/or commander	As a command of execution. **NOTE: The platoon leader and/or commander also identifies the location for the formation by pointing his arm to the desired location for the formation.**
Support	General Support	Platoon leader and/or commander	To place a specified unit in the rear of the base element.
	Lateral Support	Platoon leader and/or commander	To place support elements on the left or right flank of the base element.
	Direct Support	Platoon leader and/or commander	To move support elements forward to strengthen the base element.
	Open	Extraction team leader	To open a space in the formation.
	Up	Extraction team leader	To inform the extraction team to start moving back to the main formation.
Fire	Weapon System Gunners, Number of Rounds, Type of Round, and Prepare to Fire	Platoon leader and/or commander	To ready NL gunners to fire weapons; for example, M203 gunners, three rounds, area target, prepare to fire.
	Shield Down	NL gunners	To tell riot shield holders to get down on one knee and lower the riot shield to the ground.
	Shield Up	NL gunners	To tell riot shield holders to stand and raise the riot shield.
Weapons	Port Arms	Platoon leader and/or commander	To use when not in contact with the crowd.
	High Port	Platoon leader and/or commander	To tell NL gunners what position to hold their weapons in while the formation is in physical contact with the crowd. This allows for rapid targeting and firing of NL munitions.

EXTRACTION TEAM OPERATIONS

6-24. The extraction team is a squad (see *Figure 6-5*) that is generally formed from the support element, but could also be formed from reserve security forces. Once the team is formed, it is OPCON to the formation commander until the team's mission is complete and it returns to the rear of the formation. This team provides the riot control formation with the means to employ NL and lethal cover forward of the formation. It also provides the formation with the ability to remove and detain rioters downed by munitions.

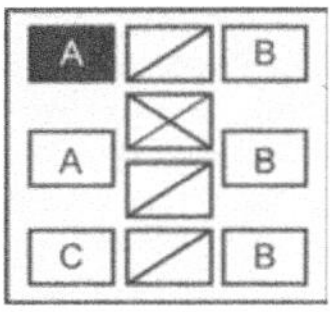

Figure 6-5. Extraction Team Formation

The extraction team—

- Extracts vehicles or personnel that are in immediate danger from the crowd.
- Detains and escorts downed rioters to the rear of the formation.
- Restrains and searches subjects.
- Covers confined or congested areas where a full riot control formation cannot be inserted.

6-25. The extraction team is usually a squad. While the recommended distance for deploying the extraction team from the control force formation is no farther than 10 meters, this may change based on METT-TC. The control formation protects the extraction team and provides immediate support, if necessary.

6-26. The extraction team may deploy from anywhere within the formation. When given orders to deploy, the squad leader of the extraction team gives the formation commands, identifies at least two personnel in the squad to conduct the search and apprehension of the subject, and leads the squad to the base line of the formation. He then extends his arm between the two riot shield holders and commands, "Open," while tapping the riot shield holders on the side of the shoulder. The two riot shield holders take one step backward and one step to the right or left. This clears a path for the extraction team. The extraction team leader then states the number of personnel leaving the formation, such as "Ten security personnel leaving." As the last extraction team member exits the formation he states, "Last Man." The two riot shield holders then return to their original position in the formation. The adjacent base element squad and/or team leaders pass the number of personnel in the team through the formation.

6-27. As the extraction team approaches an instigator or incapacitated demonstrator, the riot shield holders envelop the target and face in the direction of the nearest threat. NL gunners on the left and right cover their respective areas. The search and apprehension team immediately controls and restrains the target. To avoid confusion, one team member concentrates on controlling the subject while another team member applies a restraining system. When the search and apprehension team has the subject under control, the extraction squad leader sounds off with, "Up." Upon hearing the "Up" command, it is repeated by all extraction team members. Team members then grab a shoulder of the team member in front of them to backtrack into the formation. As the extraction team approaches the formation, the nearest base element leader will extend his arm and command, "Open," while tapping the riot shield holders on the side of the shoulder. The riot shield holders move one step backward and to the right or left to open the formation. As he enters the formation, the extraction team leader sounds off with the number of personnel entering the formation, such as "Ten security personnel and one demonstrator." The adjacent base element leader counts the number of personnel as they enter the formation. When the last man enters the formation, he states, "Last Man." The riot shield holders then return to their original places in the formation. The adjacent squad and team leaders in the base element pass information through formation that the extraction team has returned.

LETHAL OVERWATCH TEAMS

6-28. During a NL engagement, the use of a DM team provides confidence and safety to those facing the mob. If a lethal threat is presented, the DM team in an overwatch position (armed with a standard infantry rifle that is mounted with a high-powered scope) can scan the crowd, identify agitators and riot leaders for apprehension, or fire lethal fire if so ordered or warranted. It is also ideally suited for flank and countersniper operations.

6-29. The DM team sets the security overwatch and provides real-time intelligence and reporting that is vital to mission success. It provides coverage during the entire approach to the crowd, and its ability to select positions and provide cover fire (if and when warranted) is critical to the safety of the control force. Successful execution requires training, practice, and rehearsal with the control force formation.

6-30. The DM team is organized from a standard three- or four-man team. Each individual on the team is equipped with a standard-issue weapon. One individual has an infantry rifle with a scope, another has a set of binoculars, and two other individuals provide security for the team. Each team is equipped with a radio for communication with the HQ element of the control force formation.

6-31. The teams must have a visual advantage over the crowd to provide lethal protection to the formation. This is best accomplished by placing them on nearby rooftops, in the upper floors of buildings, and on hilltops. They must have the optical equipment to identify mob leaders, instigators, and/or individuals with weapons. Team integrity must be maintained. In organizations with three-man teams, one individual becomes the marksman, another the observer, and another the security. When teams have a 4th member, that individual becomes the video camera operator. This will help identify individuals for later apprehension and provides a record of events.

6-32. Two teams should be deployed to cover the flanks of the formation. Each team should be deployed to a position that is in front of the formation and in an overwatch position. As the formation moves forward, the teams leapfrog forward from the formation.

6-33. The reserve forces, located at the rally point, can provide another lethal force of some size that can be brought forward, as needed. Each soldier in the control force formation should have a sidearm (9-millimeter pistol) or an M16 or M4 rifle. The rifle is slung diagonally across his back with the rifle butt over his left shoulder and the muzzle below his right hip. Lethal munitions are in a separate ammunition pouch that is isolated from NL munitions. This is to prevent the mixing of NL and lethal ammunition.

6-34. Once the lethal overwatch teams are deployed, they are under the OPCON of the formation commander. The lethal overwatch element must have direct and secure communications with the formation commander. The number of teams deployed is based on METT-TC. Teams will leapfrog to keep ahead of the advancing control force formation with one team in the overwatch position at all times.

RESERVE FORCES

6-35. A large reserve of soldiers should be maintained during civil disturbance operations. Knowing that a large reserve force is available provides confidence and safety within the control force and helps prevent them from overreacting to provocative acts with disorderly and criminal elements in the crowd. When determining the number of reserve forces required, it should be resolved in favor of a large number. Lethal overwatch teams, forces to augment the control force, and apprehension teams come from the reserve forces.

6-36. Apprehension teams are a must when it is likely that a large number of people will be apprehended. These teams provide an organized response to what could be a chaotic situation. The teams can be organized at the team or squad level, depending on the number of apprehensions expected. These teams usually operate behind the base element. During peaceful demonstrations, these teams can be sent forward to remove demonstrators who refuse to disperse.

6-37. Each team consists of an apprehension, movement, and security element (added if the control force cannot provide security); a recorder; and a video or still photographer. The apprehension element is an officer or NCO and a civilian police officer (CONUS or OCONUS) or US marshal. It decides who will be apprehended, informs the subject of the reasons, supervises the handling of the subject, and ensures that the apprehension is properly conducted and documented. The apprehension team is responsible for the overall apprehension process, limiting the number of individuals appearing in court to document the apprehension. The movement element helps the apprehension element handle the subject. They move, restrain, and search the individuals being apprehended under the supervision of the apprehension element. The security element prevents the crowd from interfering with the mission. The recorder documents the apprehension on *DD Form 2708*. When possible, the video and/or still photographer recorder photographs each subject with the apprehension element. This provides the apprehension element with the necessary documentation for use in court. One individual may need to conduct many of the duties of an apprehension team based on METT-TC.

6-38. The video and/or still photographer record the situation before, during, and after the apprehensions. These images may also be used as pictorial documentation for use in court. It is possible that the presence of video and still cameras actively recording the situation can cause crowd members to be less prone to unruly or criminal acts.

6-39. The behavior of the selected individual to be apprehended is observed to decide if the subject is violent, cooperative, or passive. Unless there is good reason to believe otherwise, the subject must be considered dangerous. The mood of the crowd is observed to estimate what it might do. Is it only curious onlookers or does it support the individual targeted for apprehension? Will it interfere?

6-40. When apprehending an individual, the actions must be kept as simple as the situation allows. The element of surprise can give the apprehension team an advantage over the apprehended individual. It is best not to use surprise if the individual might panic and injure someone. Select an area for the apprehension that is to the advantage of the team and not the individual. Avoid apprehensions in crowded areas as these areas provide many avenues of escape and contain supporters of the apprehended individual.

6-41. All verbal commands given to the apprehended individual must convincing authority with short, clear, and distinct commands. Voice, attitude, and deportment must convey an expectation that the apprehended individual will comply.

6-42. The individual should be searched immediately; however, circumstances (violence or sympathizers for the individual being apprehended) may dictate temporarily postponing the search until the individual can be moved to a safer area. Flex-Cufs or hand irons may be applied.

SQUAD FORMATIONS

6-43. Civil disturbance training is a collective task. Using team and squad concepts, team leaders instruct their teams and squad leaders instruct their squads. Safety is key when using force on force training for the control force formation. Soldiers must understand that serious injuries may occur if they do not follow guidelines set by the commander.

6-44. The smallest formation is a squad formation, which is used in back of the main formation to cover side streets. Squad members must know their positions regardless of which control force formation they are in, squad to company.

6-45. Training must enforce soldiers' understanding of the need to stay informed for individual protection. Individual soldiers are much more vulnerable to attack when they break ranks and chase after crowd members. When an individual soldier breaks ranks he not only puts himself in danger, but the entire formation.

6-46. When moving a squad into a crowd control formation from a column formation, the squad leader takes one or more steps to the right of the squad and faces it. He gives the command and hand-and-arm signal for the formation he wants. Then he indicates where he wants the formation located by pointing in that general direction. The A team leader commands, "Follow Me." He places the baseman at the position indicated by the squad leader. The squad forms on the baseman position.

LINE FORMATION

6-47. The command for forming a squad in a line formation is "Squad Line Formation, Move." On the command of execution, the A team leader leads the baseman to the place indicated by the squad leader. The squad members align themselves in sequence with the baseman at normal intervals (see *Figure 6-6*). Based on the situation, the squad leader may designate a specific interval. He does this when he gives the preparatory command. If no interval is specified, the squad automatically forms using the normal interval.

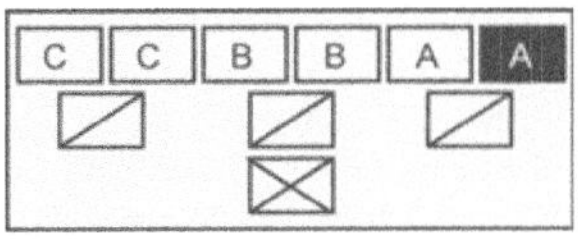

Figure 6-6. Squad Line Formation

ECHELON FORMATION

6-48. The command for having a squad form in an echelon formation is "Squad Echelon Right (or Left), Move" (see *Figure 6-7*). On the command of execution, the A team leader places the baseman at the location indicated by the squad leader. The squad members align themselves in sequence with the baseman, one pace to the side and one pace to the rear.

WEDGE FORMATION

6-49. The command for moving a squad formation into a wedge formation is "Squad Wedge, Move." On the command of execution, the A team leader moves the baseman to the place indicated by the squad leader. The A team lines up to the left of the baseman, one pace to the left and one pace to the rear of each preceding man. B team members align themselves with the baseman, one pace to the right and one pace to the rear of each preceding man. C team members align themselves in the same way to the left of the baseman (see *Figure 6-8*).

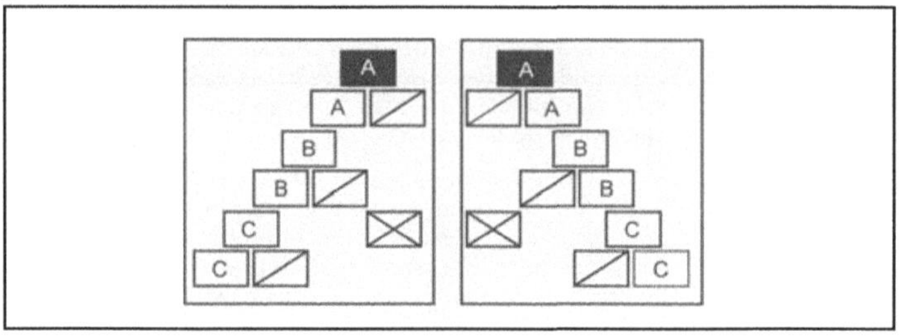

Figure 6-7. Squad Echelons Left and Right Formations

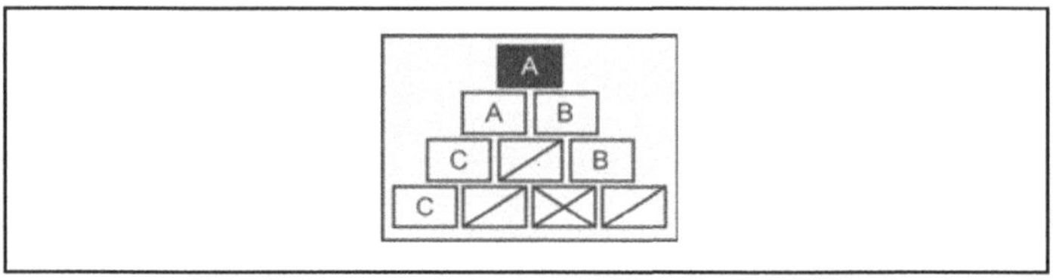

Figure 6-8. Squad Wedge Formation

ASSEMBLY

6-50. When assembling a squad from the line formation, the squad leader takes a position a sufficient distance to the rear of the squad and commands, "Squad Assemble." At the same time, he raises his right hand in the air and makes a circular motion. Squad members, who hear the command, automatically come to the port arms position. The A team leader and the baseman do an about-face movement.

6-51. The other squad members face toward the baseman. On the command of execution "Move," the squad leader points to the place where he wants the squad to assemble. The A team leader double-times to the designated spot, and the other members of the squad follow him. Team leaders fall into the file in their designated positions as they move to the AA. To assemble from a squad echelon formation, the steps are the same as a line formation.

6-52. To assemble from a squad wedge formation, the squad leader takes the same steps as for the line and echelon formations. B team members do a half left face, and the A team and C team members do a half right face. The C team pauses at the baseman position and allows the B team to clear the formation before moving to the AA.

DIAMOND AND CIRCLE FORMATIONS

6-53. The diamond and circle formations are used during extraction team operations. They are small formations used to penetrate the crowd or cover small areas. The decision on which formation to use is based on METT-TC and the ability of the squad to perform.

6-54. Commands for forming a squad in a diamond or circle formation is "Squad Diamond (or Circle), Move." On the command of execution, the A team leader moves the baseman to the place indicated by the squad leader. The A team lines up to the left of the baseman, covering the 8 to 12 o'clock positions. B team members align themselves to the right of the baseman, covering the 12 to 4 o'clock positions. C team members align themselves covering the 4 to 8 o'clock positions (see *Figure 6-9*).

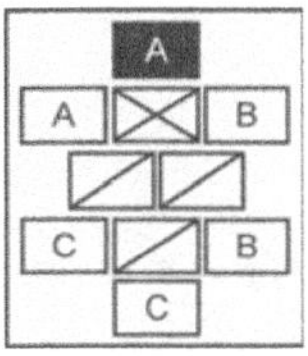

Figure 6-9. Squad Diamond or Circle Formation

6-55. To assemble from a squad diamond or circle formation, the squad leader takes the same steps as he would for a line formation. The A team leads the squad to the AA followed by the B team and then the C team. Teams fall into a column formation as they file back to the AA.

THREE-SQUAD PLATOON FORMATIONS

6-56. The platoon HQ consists of the following individuals: the platoon leader, the PSG, and a messenger or radio operator. When directed by the platoon leader, other individuals such as the selected marksman, interpreter, RCA disperser operators, and firefighters may augment the platoon HQ. In forming all control force formations from the column, the platoon leader moves to the right of the platoon and faces them when giving his commands. He gives the hand-and-arm signal and the preparatory command. On the command of execution "Move," he points to the approximate location where he wants the platoon to form. The platoon leader pauses between the preparatory command and the command of execution to allow each squad leader to issue a preparatory command to his squad.

Platoon Line Formation

6-57. The command for forming a platoon line is "Platoon on Line, Move." Immediately following the platoon leader's preparatory command, the squad leaders of the 1st and 3d squads give the command, "Follow Me." The leader of the 2d squad gives the command, "Stand Fast." On the platoon leader's command of execution, the 1st and 3d squads move forward to the point designated by the platoon leader. The 1st team member of the 1st squad is the baseman for the platoon formation. The 1st squad forms a squad line to the left of the baseman while the 3d squad forms a squad line to the right of the baseman and dresses with the 1st squad. After the 1st and 3d squads have cleared the platoon, the 2d squad leader gives the command, "Follow Me." The 2d squad splits and the A and B teams form a line to the right of the 3d squad.

The C team (with the squad leader) forms a line to the left of the 1st squad (see *Figure 6-10*).

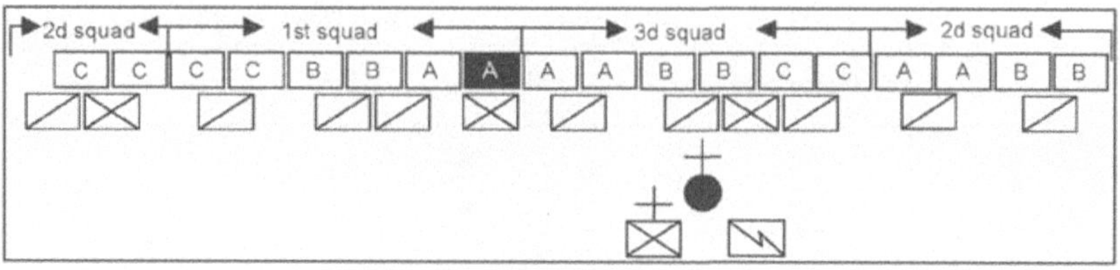

Figure 6-10. Platoon Line Formation

Assembling From a Platoon Line Formation

6-58. When assembling the platoon from the platoon line formation, the platoon leader and his HQ personnel take a position to the rear of the platoon. He then gives the command, "Platoon Assemble," while raising his right arm and making a circular motion above his head. Platoon members automatically assume the port arms position. The 1st squad leader and the baseman do an about face. The 3d squad leader moves to the head of his squad. All other members do an about-face movement toward the baseman. On the platoon leader's command of execution, the platoon leader points to where he wants the platoon to assemble. The 1st squad leader gives the command, "Follow Me," and moves his squad toward the AA. The squad moves to make a single file line to the baseman position followed by the C team of the 2d squad. As the 1st squad clears the baseman position, the 2d squad leader comes to a halt at the baseman position. The 3d squad leader gives the command, "Follow Me," and leads his squad to the AA followed by the A and B teams of the 2d squad. The A team leader of the 2d squad comes to a halt at the baseman position. Once the 3d squad has cleared the baseman position, the 2d squad leader gives the command, "Follow Me," and leads his squad to the AA. The 2d squad reassembles into their original column formation as they move to the AA. As the 1st squad comes to a halt (facing the crowd) in the AA, the 3d squad dresses with the 1st squad and leaves space for the 2d squad to return to its place in the formation. Once the platoon is formed into a column formation, HQ personnel take their position in the formation. The PSG then gives the command, "Countercolumn March," and moves the platoon to the rally point.

Platoon Line Formation With General Support

6-59. Whenever the command for establishing a crowd control formation contains the phrase "In Support" (without modification), it means that the support element is to remain in general support.

6-60. The command for forming a platoon line with the 2d squad in general support is "Platoon Line, 2d Squads in Support, Move." The 1st and 3d squads of the platoon line with general support execute the line, while the 2d squad remains in the column (see *Figure 6-11*). To assemble a platoon with a squad in general support, the procedure is the same as the platoon line, except that the 1st and 3d squads dress with the 2d squad.

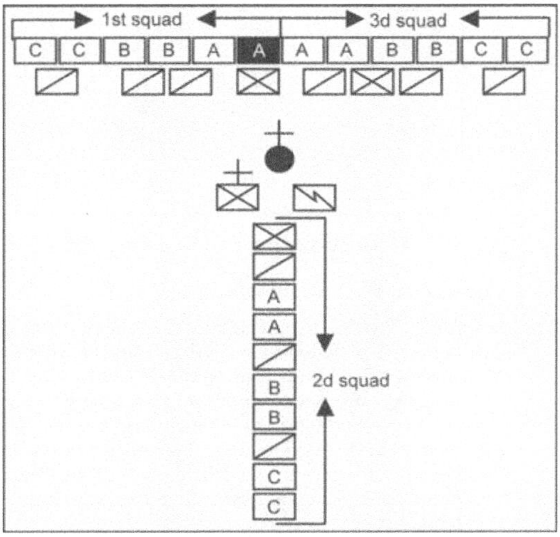

Figure 6-11. Platoon Line Formation With General Support

Platoon Line Formation With Lateral Support

6-61. The command for forming a platoon line with one squad in lateral support is "Platoon Line, 2d Squad in Lateral Support, Move." The 1st and 3d squads execute the line as before, and the 2d squad stands fast. After the line has been formed, the 2d squad leader gives the command, "Follow Me." The A and B teams move to the right flank while the squad leader and the C team move to the left flank of the formation. The 2d squad forms columns behind the last men on the line formation. If no direction is given, the 2d squad splits to cover both flanks. A squad may be used to cover one flank by giving the command, "2d Squad Right (or Left) Lateral Support" (see *Figure 6-12*).

6-62. The 2d squad may be moved from general to lateral support at any time by the platoon leader. The platoon leader commands, "2d Squad, Lateral Support, Move." To have the 2d squad join the line from either support position, the platoon leader commands, "2d Squad, Extend the Line, Move." The 2d squad leader commands, "Follow Me." On the command of execution, the squads set up individual lines, closing and dressing on the existing line. To assemble the platoon from a platoon line with lateral support, follow the same steps as for the platoon line.

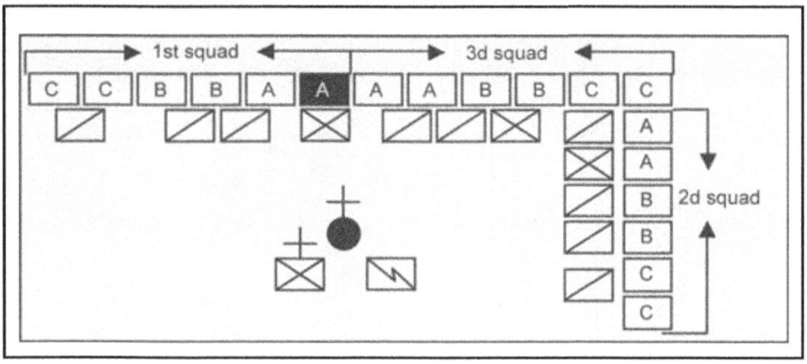

Figure 6-12. Platoon Line Formation With Lateral Support (Right)

Platoon Line Formation With Direct Support

6-63. The command for forming a platoon line with one squad in direct support is "Platoon Line, 2d Squad in Direct Support, Move." The 1st and 3d squads execute a line as before. The 2d squad executes a squad line directly to the rear and centered on the 1st line. The men in the supporting line cover the intervals between the men in the 1st line. To do this, the squad and team leaders of the 1st and 3d squads must take two steps to the rear to allow space for the 2d squad to move into place. After seeing the squad and team leaders move back from the line, the 2d squad leader commands, "Follow Me," and leads his squad into position (see *Figure 6-13*).

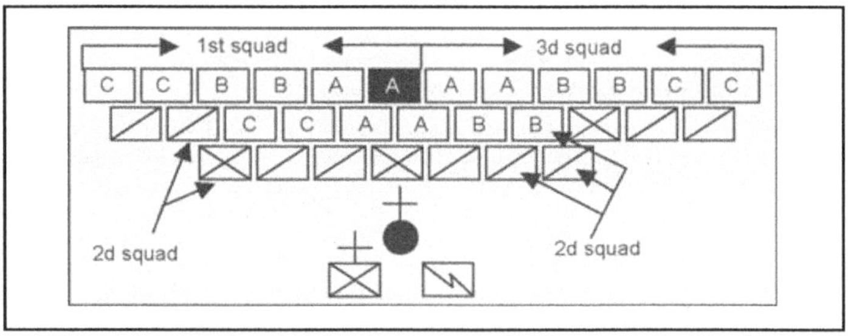

Figure 6-13. Platoon Line Formation With Direct Support

6-64. To assemble a platoon with direct support, the procedure is the same as the platoon line, except that the 2d squad moves to the AA first. To assemble the support squads from any position to general support, the platoon leader commands, "2d Squad in Support, Move." The 2d squad then returns to a column behind the line formed by the other two squads.

Platoon Echelon Right and Left Formations

6-65. The command for forming a platoon echelon right is "Platoon Echelon Right, Move." After the platoon leader's preparatory command, the squad leader of the 1st squad commands, "Follow Me." The squad leaders of the 2d and 3d squads command, "Stand Fast." On the command of execution, the 1st squad moves out and executes an echelon right at the location designated by the platoon leader. As the 1st squad clears the column, the 3d and 2d squads extend the echelon. The 2d squad performs support to this formation in the same ways it supports a line formation.

6-66. The command for forming a platoon echelon left is "Platoon Echelon Left, Move." The platoon echelon right is formed in the same manner as the echelon left (see *Figure 6-14*). *Figures 6-15; 6-16,* page 6-22; and *6-17,* page 6-23 show support formations in a left echelon.

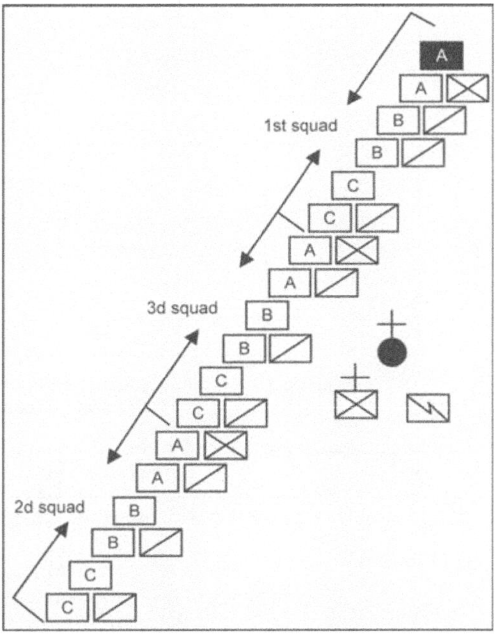

Figure 6-14. Platoon Echelon Left Formation

6-67. To assemble the platoon, the commander takes his position to the rear of the formation with the platoon HQ. On the preparatory command, "Platoon Assemble," the 1st squad leader does an about-face movement while all other

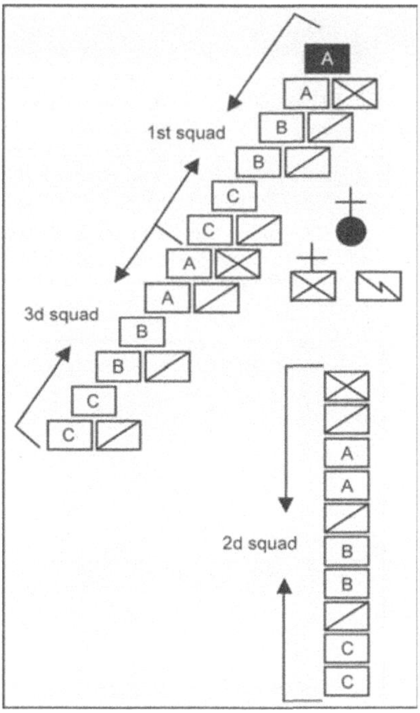

Figure 6-15. Platoon Echelon Left Formation With 2d Squad in General Support

members assume the port arms position and face the baseman. On the command, "Move," the 1st squad moves first, followed by the 3d and 2d squads. Each squad dresses with the 1st squad as it reaches the AA. The assembly of the echelon right formation is the same as the echelon left.

Platoon Wedge Formation

6-68. The command for forming the platoon wedge formation is "Platoon Wedge, Move." After the platoon leader's preparatory command, the squad leaders of the 1st and 3d squads command, "Follow Me." At the same time, the 2d squad leader commands, "Stand Fast." On the platoon leader's command of execution, the 1st and 4th squads move directly to the front. The 1st squad executes an echelon left on the baseman while the 3d squad executes an echelon right. When the last man of the 1st and 4th squads clear the 2d squad, the 2d squad leader commands, "Follow Me," and moves out to the left and right. The 2d squad splits and the A and B teams go to the right and the C

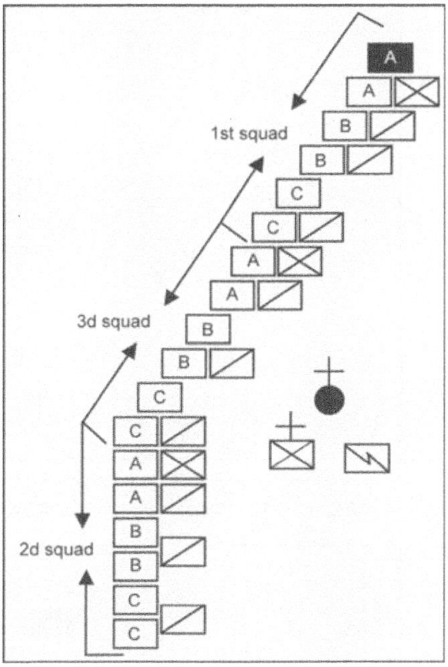

Figure 6-16. Platoon Echelon Left Formation With 2d Squad in Lateral Support

team goes to the left to extend the formation (see *Figure 6-18*). Assembly of the wedge formation is the same as the platoon line.

Platoon Wedge Formation With General Support

6-69. The command for forming a platoon wedge formation with support is "Platoon Wedge, 2d Squad in Support, Move." The 1st and 3d squads execute the wedge while the 2d squad remains in a column formation (see *Figure 6-19,* page 6-24).

Platoon Wedge Formation With Lateral Support

6-70. The command for forming a platoon wedge formation with lateral support is "Platoon Wedge, 2d Squad in Lateral Support, Move." The 1st and 3d squads execute the wedge while the 2d squad stands fast. After the 1st and 3d squads have cleared the column, the 2d squad leader commands, "Follow Me." If no direction is given, the 2d squad splits to cover both flanks. A squad

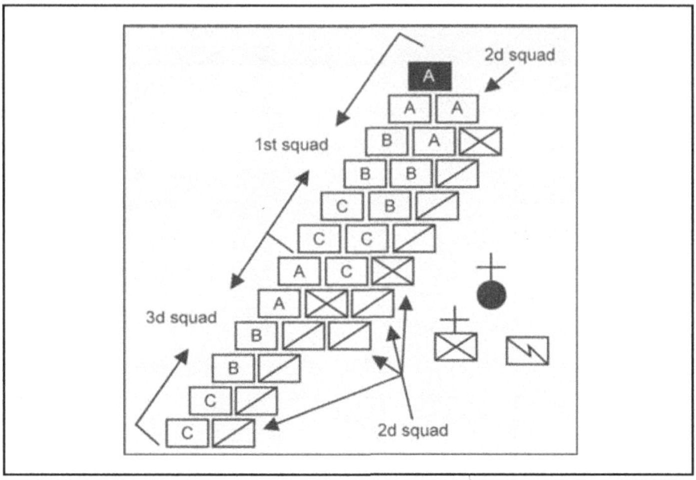

Figure 6-17. Platoon Echelon Left With 2d Squad in Direct Support

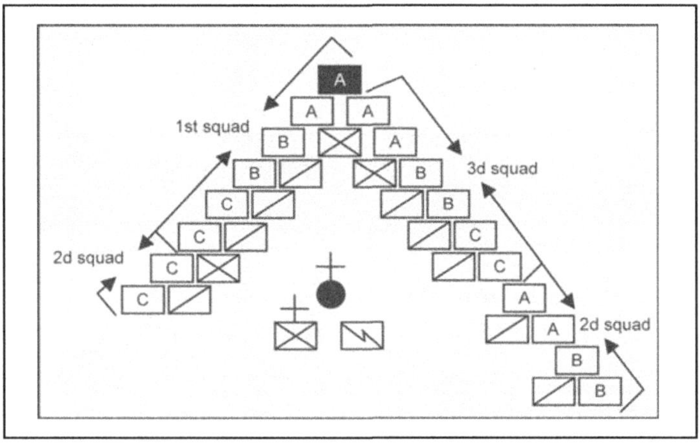

Figure 6-18. Platoon Wedge Formation

may be used to cover one flank by commanding, "2d Squad Right (or Left) Lateral Support" (see *Figures 6-20* and *6-21,* page 6-25).

6-71. To move the 2d squad from general support to lateral support, the platoon leader commands, "2d Squad, Lateral Support, Move." To have the 2d squad join the wedge from general or lateral support, the platoon leader

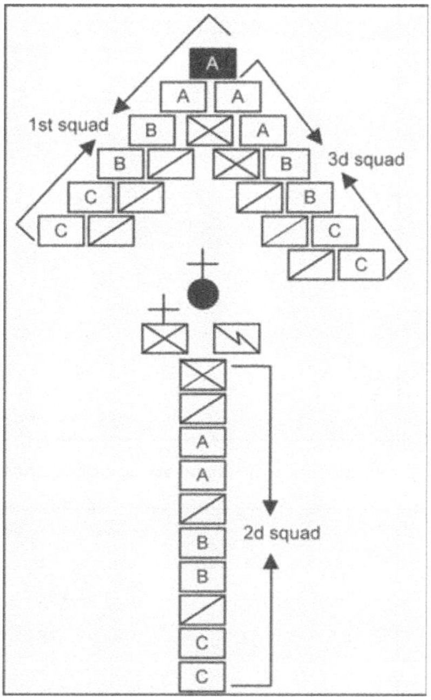

Figure 6-19. Platoon Wedge Formation With 2d Squad in General Support

commands, "2d Squad, Extend the Wedge, Move." To extend the wedge, the 2d squad splits and the teams move to the left and right, respectively.

Platoon Wedge Formation With Direct Support

6-72. The command for forming a platoon wedge formation with the 2d squad in direct support is "Platoon Wedge, 2d Squad in Direct Support, Move." The 1st and 3d squads execute a wedge. The 2d squad waits for the squad and/or team leaders to move to the rear, as in the line formation, before executing a wedge directly behind and centered on the main formation. The individuals in the supporting wedge cover the intervals between individuals in the leading wedge (see *Figure 6-22,* page 6-26).

Platoon Diamond and Circle Formations

6-73. These formations are used in open areas or where the crowd may be able to envelop the formation. The diamond and circle formations are primarily defensive in nature and difficult to maneuver. The decision about which formation to use is based on the ability of the platoon to perform them.

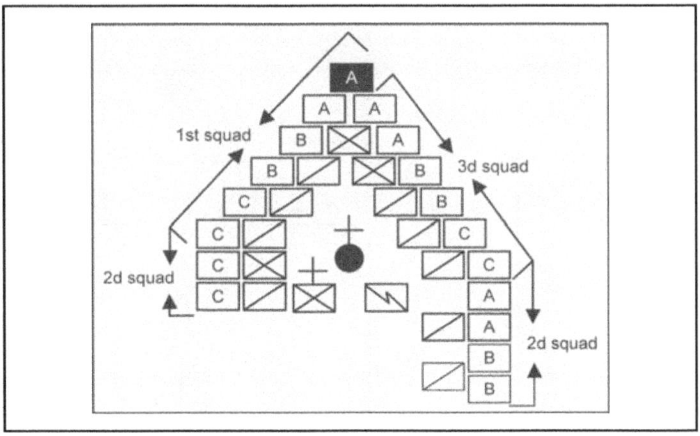

Figure 6-20. Platoon Wedge Formation With 2d Squad in Lateral Support

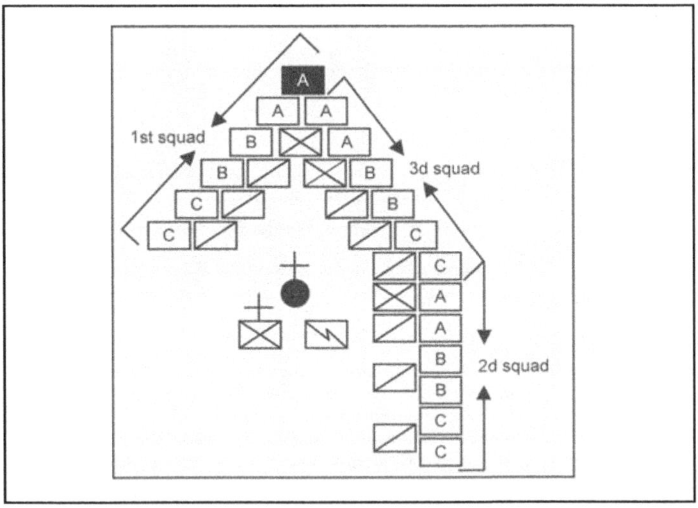

Figure 6-21. Platoon Wedge Formation With 2d Squad in Right Lateral Support

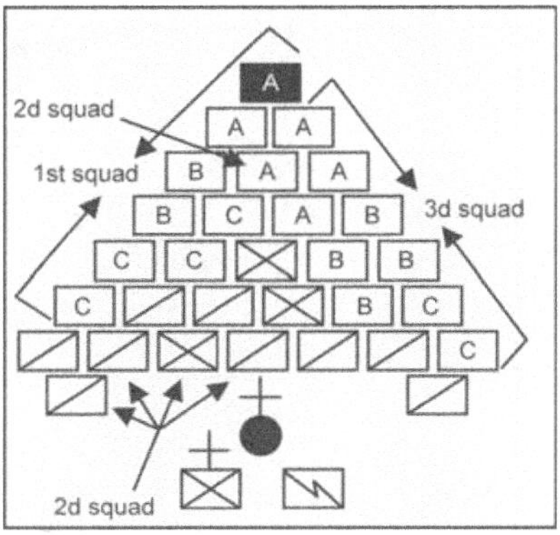

Figure 6-22. Platoon Wedge Formation With 2d Squad in Direct Support

The command for having a platoon form in a diamond or circle formation is "Platoon Diamond (or Circle), Move." On the command of execution, the 1st squad leader moves the baseman to the place indicated by the squad leader. The baseman becomes the 12 o'clock position while the 1st squad lines up to the left of the baseman covering the 8 to 12 o'clock positions. The 3d squad aligns itself to the right of the baseman, covering the 12 to 4 o'clock positions and the 2d squad aligns itself, covering the 4 to 8 o'clock positions (see *Figures 6-23* and *6-24*).

6-74. To assemble from a platoon diamond or circle formation, the HQ element takes the same steps as that of a line formation. The 1st squad leads the platoon back to the AA followed by the 2d and 3d squads. Squads fall into a column formation as they file back to the AA.

RELIEF ELEMENTS IN PLACE

6-75. During crowd control operations, the platoon leader may rotate his squads to give them a rest. He is not limited to using only those squads specified in the preceding examples as the lead elements of the formations. He can replace the base elements by commanding direct support to the 1st squad and then directing the 1st squad into general support. The support element (the 2d squad) then passes through the 1st squad. Once the 2d squad is formed directly behind the 1st squad, the 1st squad leader commands, "Right Face." The members of the 1st squad do a right-face movement and the 2d squad steps quickly though the 1st squad and forms the base formation. The 1st squad leader then commands, "Follow Me," and leads his squad to the general support position. This procedure is often necessary when protective masks are used and chemical irritants employed. The 1st squad then replaces the 3d squad in the same manner. Platoons participating in company operations are relieved in the same manner.

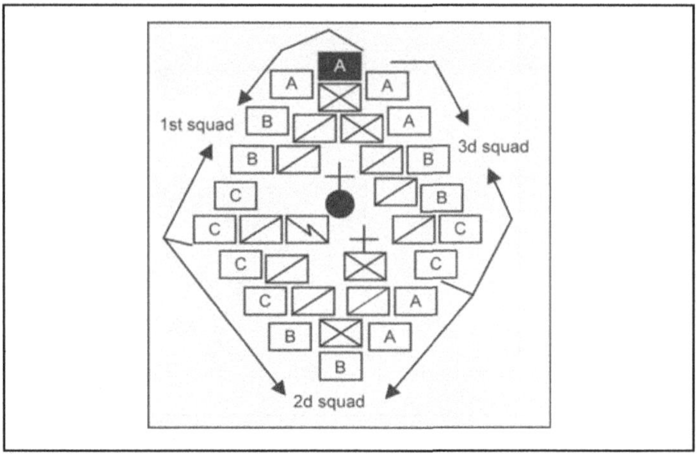

Figure 6-23. Platoon Diamond Formation

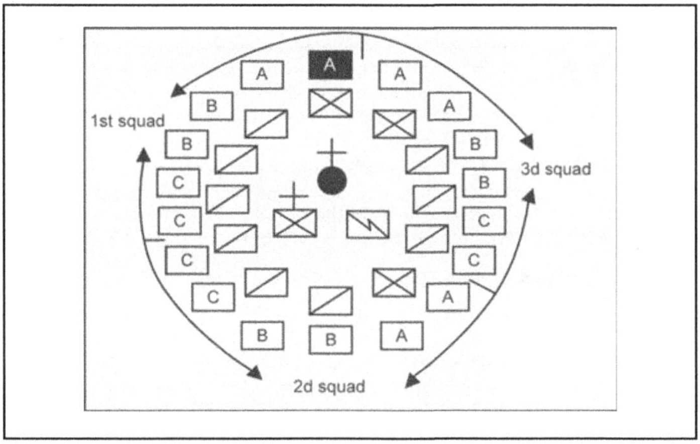

Figure 6-24. Platoon Circle Formation

FOUR-SQUAD PLATOON FORMATIONS

6-76. When forming control force formations from the column, the platoon leader moves to the right of the platoon and faces it when giving his commands. He gives the hand-and-arm signal and the preparatory command. On the command of execution, "Move," he points to the approximate location where he wants the platoon to form. The platoon leader pauses between the preparatory command and the command of execution to allow each squad leader to issue a preparatory command to his squad.

Platoon Line Formation

6-77. The command for forming a platoon line formation is "Platoon Line, Move." Immediately following the platoon leader's preparatory command, the squad leaders of the 1st and 4th squads command, "Follow Me." The squad leaders of the 2d and 3d squads command, "Stand Fast." On the platoon leader's command of execution, the 1st and 4th squads move forward to the point designated by the platoon leader. The 1st A team member of the 1st squad is the baseman for the platoon formation. The 1st squad forms a squad line to the left of the baseman and the 4th squad forms a squad line to the right of the baseman with all lines dressed to the right. After the 1st and 4th squads have cleared the platoon, the 2d and 3d squad leaders command, "Follow Me." The 2d squad forms a squad line to the left of the 1st squad. The 3d squad forms a squad line to the right of the 4th squad. All squads dress with the 4th squad (see *Figure 6-25*).

Platoon Line Formation Assembly

6-78. When assembling from the platoon line formation, the platoon leader and his HQ personnel take a position to the rear of the platoon. He then gives the command, "Platoon Assemble," while raising his right arm and making a circular motion above his head. Platoon members automatically assume the port arms position. The 1st squad leader and the baseman do an about-face movement. All other platoon members do an about-face movement toward the baseman. On the platoon leader's command of execution, he points to where he wants the platoon to assemble. He commands, "Follow Me," and moves his squad toward the AA. The 1st squad files to the baseman position, followed by the 2d squad. As the 2d squad clears the baseman position, the 4th squad leader then commands, "Follow Me," and leads his squad to the AA, followed by the 3d squad. As the 1st squad comes to a halt facing the crowd in the AA, the 2d squad assumes its position to the right of the 1st squad. As the 4th squad assembles, it dresses on the 1st squad, leaving space for the 3d squad to return to its place in the formation. Once the platoon is formed into a column formation, HQ personnel take their position in the formation. The PSG then commands, "Countercolumn, March," and moves the platoon to the rally point.

Platoon Line Formation With General Support

6-79. Whenever the command for establishing a crowd control formation contains the phrase "In Support Without Modification," it means that the support element is to remain in general support. The command for forming a platoon line with two squads in general support is "Platoon Line, 2d and 3d Squads in Support, Move." The 1st and 4th squads execute the line, while the 2d and 3d squads remain in the column (see *Figure 6-26,* page 6-30). To assemble a platoon with two squads in general support, the procedure is the same as that of the platoon line, except that the 1st and 4th squads dress on with the 2d and 3d squads.

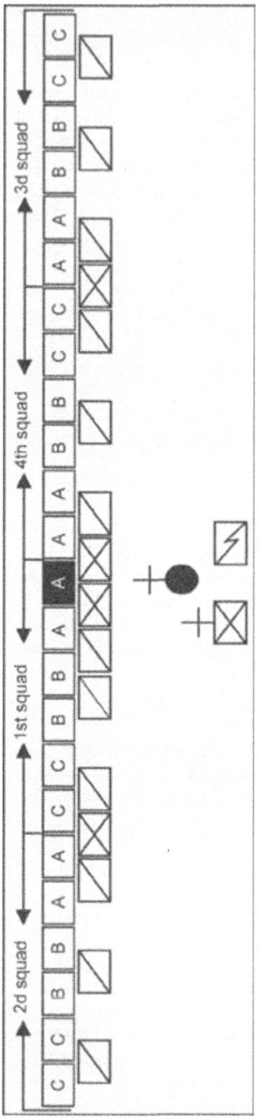

Figure 6-25. Platoon Line Formation With Four Squads

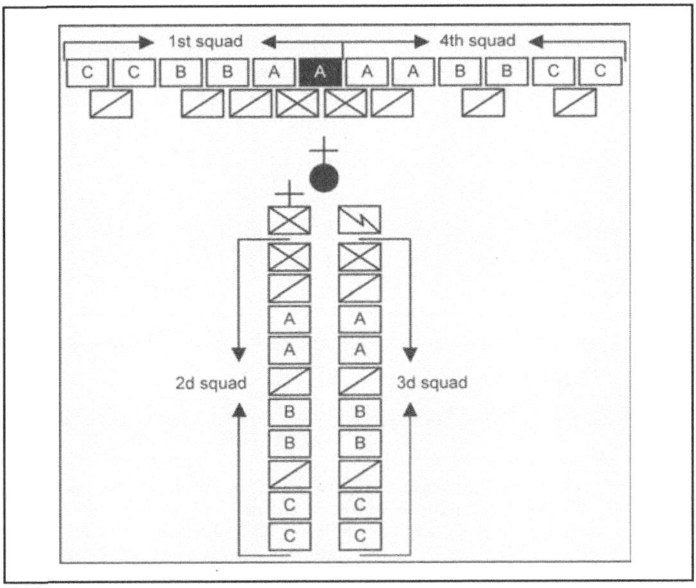

Figure 6-26. Platoon Line Formation With d and 3d Squads in General Support

Platoon Line Formation With Lateral Support

6-80. The command for forming a platoon line formation with two squads in lateral support is "Platoon Line, 2d and 3d Squads in Lateral Support, Move." The 4th and 1st squads execute the line as before, and the 2d and 3d squads stand fast. After the line has been formed, the squad leaders of the 2d and 3d squads command, "Follow Me." The 2d squad forms a column behind the last man of the 1st squad. The 3d squad forms a column behind the last man of the 4th squad (see *Figure 6-27*).

6-81. Squads may be moved from general support to lateral support at any time by the platoon leader. He commands, "2d and 3d Squads, Lateral Support, Move." To have the 2d and 3d squads join the line from either support position, the platoon leader commands, "2d and 3d Squads, Extend the Line, Move." The 2d and 3d squad leaders command, "Follow Me." On the command of execution, the squads set up individual lines, closing and dressing with the existing line. To assemble the platoon from a platoon line with lateral support, the procedure is the same as that of the platoon line, except lateral support squads face forward until they move to the AA.

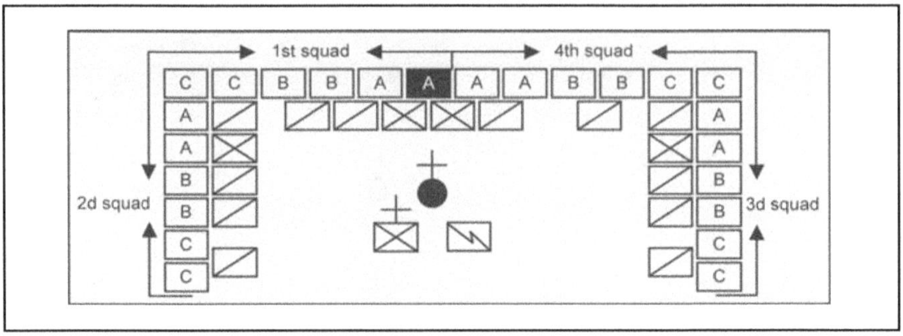

Figure 6-27. Platoon Line Formation With 2d and 3d Squads in Lateral Support

Platoon Line Formation With Direct Support

6-82. The command for forming a platoon line with two squads in direct support is "Platoon Line, 2d and 3d Squads in Direct Support, Move." The 1st and 4th squads execute a line as before. The 2d and 3d squads execute a 2d line directly to the rear and centered on the 1st line. The members in the direct support line cover the intervals between members in the baseline. To do this, the squad and team leaders of the 1st and 4th squads must take two steps to the rear to allow space for the 2d and 3d squads to move into place. After seeing the squad and team leaders move back from the line, 2d and 3d squad leaders command, "Follow Me," and lead their squads into position (see *Figure 6-28*).

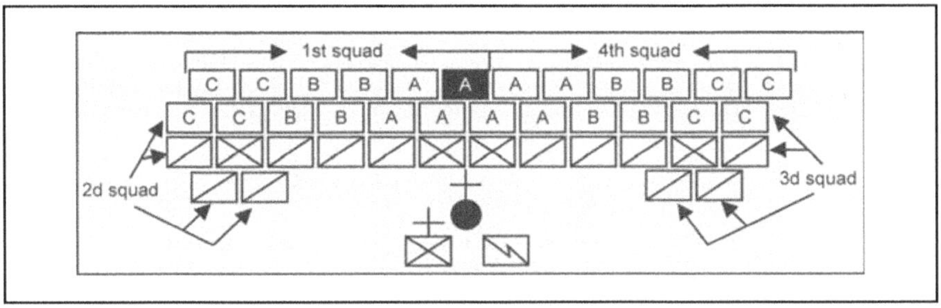

Figure 6-28. Platoon Line Formation With 2d and 3d Squads in Direct Support

6-83. To assemble the support squads from any position to general support, the platoon leader commands, "2d and 3d Squads in Support, Move." The 2d and 3d squads then return to a column behind the line formed by the 1st and 4th squads. To assemble the entire platoon, the procedure is the same as that of a platoon line, except that the 2d squad moves to the AA first, followed by the 3d, 1st, and 4th squads.

Platoon Line Formation With One Support Squad

6-84. The command for forming a platoon line formation with one squad in support is "Platoon Line, 3d Squad in Support, Move." The 1st squad moves out and executes a squad line. The 2d and 4th squads form a line on the side of the left and rights sides of the 1st squad. The 3d squad remains in column formation. The support squad may be used in lateral support on one or both sides of the formation or in direct support of any segment of the formation. To assemble the platoon, the procedure is the same as that of a platoon line, except that all squads dress with the 3d squad.

Platoon Echelon Right and Left Formations

6-85. The command for forming a platoon echelon right formation is "Platoon Echelon Right, Move." After the platoon leader's preparatory command, the squad leader of the 1st squad commands, "Follow Me." The squad leaders of the 2d, 3d, and 4th squads command, "Stand Fast." On the command of execution, the 1st squad moves out and executes an echelon right at the location designated by the platoon leader. As each squad clears the column, the next successive squad moves out and extends the echelon that has been formed. To assemble the platoon, the procedure is the same as that of a platoon line, except that all squads file back in order.

6-86. The command for forming a platoon echelon left formation is "Platoon Echelon Left, Move." The platoon echelon left is formed in the same manner as the echelon right, but in reverse order (see *Figure 6-29*). To assemble the platoon, the procedure is the same as that of an echelon right. See *Figures 6-30,* page 6-34; *6-31,* page 6-35; and *6-32,* page 6-36, for examples of echelon formations with support.

Platoon Wedge Formation

6-87. The command for forming the platoon wedge formation is "Platoon Wedge, Move." After the platoon leader's preparatory command, the 1st and 4th squad leaders command, "Follow Me." At the same time, the 2d and 3d squad leaders command, "Stand Fast." On the platoon leader's command of execution, the 1st and 4th squads move directly to the front. When the last members of the 1st and 4th squads clear the 2d and 3d squads, the 2d and 3d squad leaders command, "Follow Me," and move out to the left and right. The 1st squad executes an echelon left on the baseman and the 4th squad executes an echelon right. The 2d squad forms an echelon left on the last man of the 1st squad. The 3d squad forms an echelon right on the last man of the 4th squad. This completes the wedge formation (see *Figure 6-33,* page 6-36). To assemble the platoon, the procedure is the same as that of a platoon line.

Platoon Wedge Formation With General Support

6-88. The command for forming a platoon wedge formation with two squads in support is "Platoon Wedge, 2d and 3d Squads in Support, Move." The 1st and 4th squads execute the wedge while the 2d and 3d squads remain in column formation (see *Figure 6-34,* page 6-37).

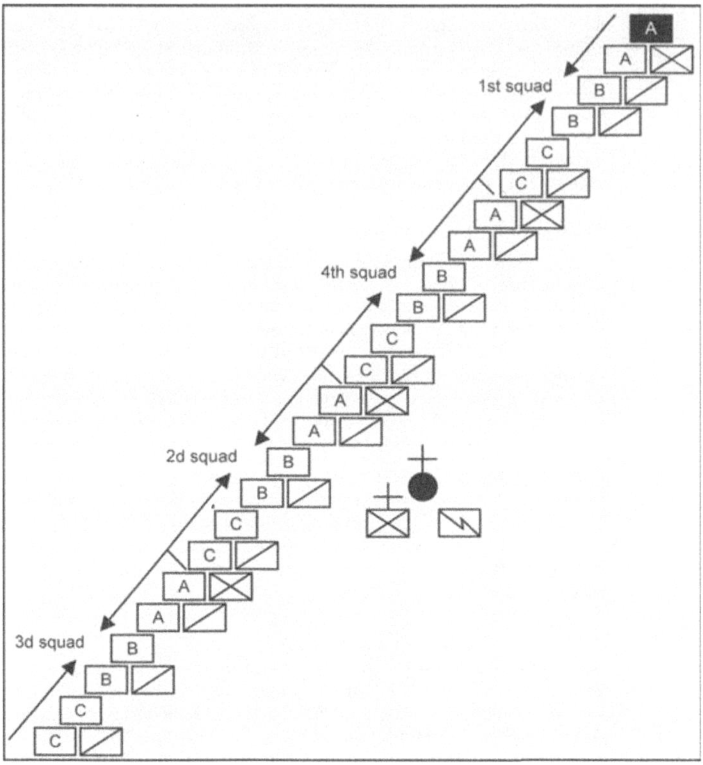

Figure 6-29. Platoon Echelon Left Formation With Four Squads

Platoon Wedge Formation With Lateral Support

6-89. The command for forming a platoon wedge formation with lateral support is "Platoon Wedge, 2d and 3d Squads in Lateral Support, Move." The 1st and 4th squads execute the wedge while the 2d and 3d squads stand fast. After the 1st and 4th squads have cleared the column, the 2d and 3d squad leaders command, "Follow Me," and move out to the left and right. The 2d squad forms a column formation behind the last man of the 1st squad. The 3d squad forms a column formation behind the last man of the 4th squad (see *Figure 6-35,* page 6-38).

6-90. To move the 2d and 3d squads from general to lateral support, the platoon leader commands, "2d and 3d Squads, Lateral Support, Move." To have the 2d and 3d squads join the wedge from general or lateral support, the

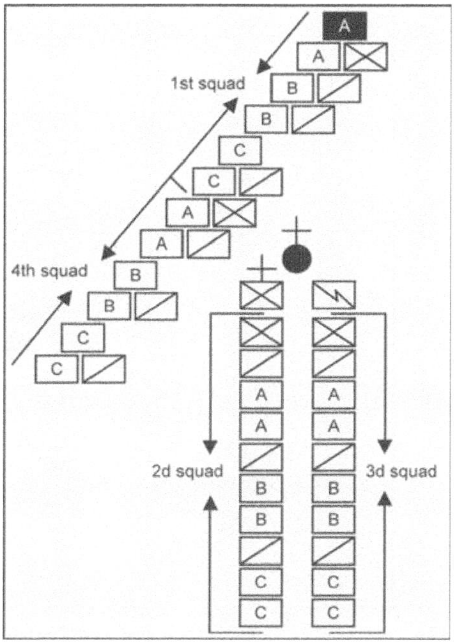

Figure 6-30. Platoon Echelon Left Formation With 2d and 3d Squads in General Support

platoon leader commands, "2d and 3d Squads, Extend the Wedge, Move." The 2d and 3d squad leaders command, "Follow Me," and move out to the left and right to extend the wedge on the 1st and 4th squads.

Platoon Wedge Formation With Direct Support

6-91. The command for forming a platoon wedge with two squads in direct support is "Platoon Wedge, 2d and 3d Squads in Direct Support, Move." The 1st and 4th squads execute a wedge. The 2d and 3d squads execute a wedge and close in on the leading wedge. The members in the supporting wedge cover the intervals between members in the leading wedge. To do this, the squad and team leaders of the 1st and 4th squads must take two steps to the rear to allow space for the 2d and 3d squads to move into place. After seeing the squad and team leaders move back from the line, the 2d and 3d squad leaders command, "Follow Me," and lead their squads into position (See *Figure 6-36,* page 6-38). To move the 2d and 3d squads back to general support, the platoon leader commands, "2d and 3d Squads Support, Move." Then the 2d and 3d squads move to their general support positions. To

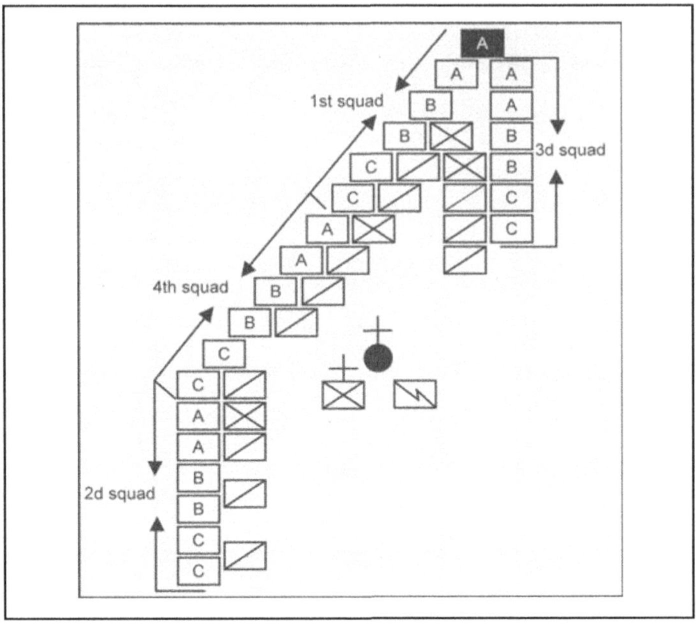

Figure 6-31. Platoon Echelon Left Formation With 2d Squad in Lateral Support

assemble a platoon from a platoon wedge with direct support, the procedure is the same as that of a platoon wedge, except that the 2d squad assembles first, followed by the 3d, 1st, and 4th squads.

Platoon Wedge Formation With One Support Squad

6-92. The command for forming a platoon wedge formation with one squad in support is "Platoon Wedge, 3d Squad in Support, Move." The 1st squad moves out and executes a squad wedge. The 2d and 4th squads form echelons left and right on the 1st squad. The 3d squad remains in the column formation.

Platoon Diamond and Circle Formations

6-93. The platoon diamond and circle formations are used in open areas or where the crowd may be able to envelop the formation. They are primarily defensive in nature and hard to maneuver. The decision concerning which formation to use is based on the ability of the platoon to perform them. The command for having a platoon form in a diamond or circle formation is "Platoon Diamond (or Circle), Move." On the command of execution, the 1st squad leader moves the baseman to the location indicated by the squad leader. The baseman becomes the 12 o'clock position. The 1st squad lines up to the

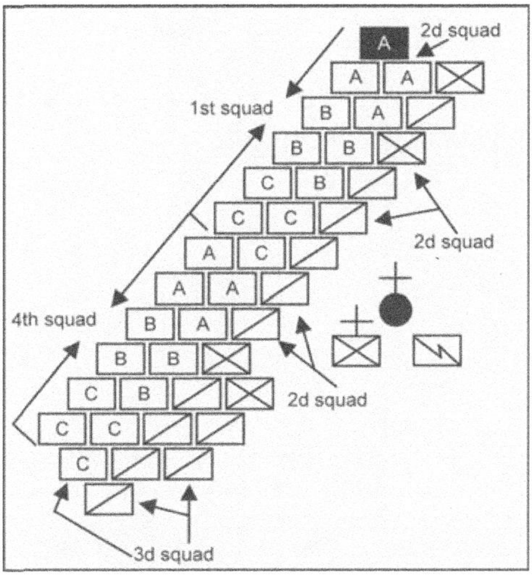

Figure 6-32. Platoon Echelon Left Formation With 2d and 3d Squads in Direct Support

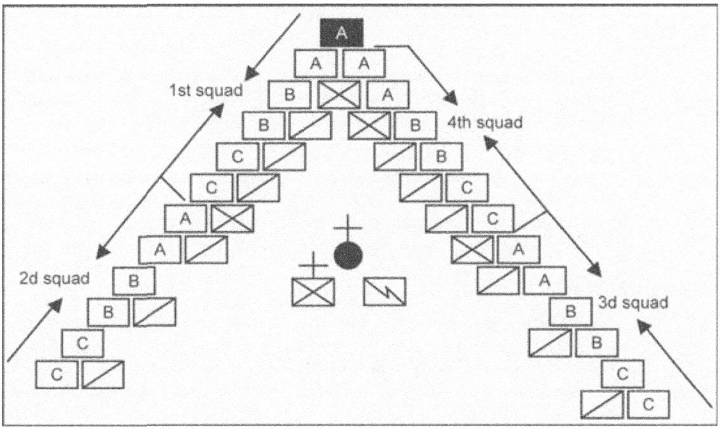

Figure 6-33. Platoon Wedge Formation With Four Squads

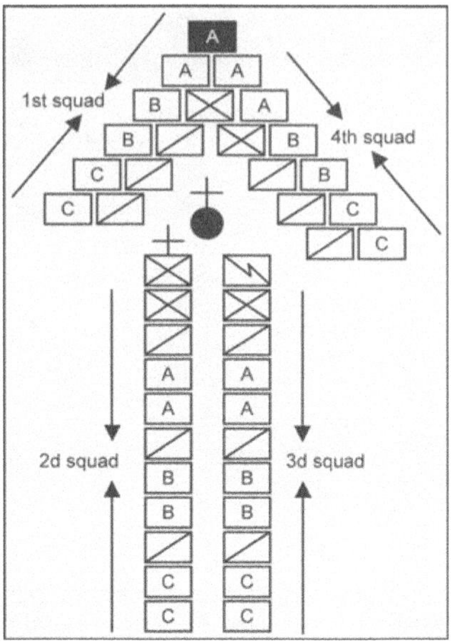

Figure 6-34. Platoon Wedge Formation With 2d and 3d Squads in General Support

left of the baseman covering the 9 to 12 o'clock positions. The 4th squad aligns themselves to the right of the baseman covering the 12 to 3 o'clock positions. The 3d squad aligns itself covering the 3 to 6 o'clock positions. The 2d squad aligns itself covering the 6 to 9 o'clock positions.

6-94. To assemble a platoon from a platoon diamond or circle formation, the HQ element uses the same procedures as that of a line formation. The 1st squad leads the platoon back to the AA followed by the 2d, 3d, and 4th squads. The squads fall into a column formation as they file back to the AA.

COMPANY FORMATIONS

6-95. When a company commander orders his company into control force formations from a column, he moves to the left near the head of the column so he can be seen by the platoon leaders. He then faces the company. As he gives his preparatory command, he points to where he wants the formation. If he does not indicate a location, the company forms immediately in front of the lead platoon.

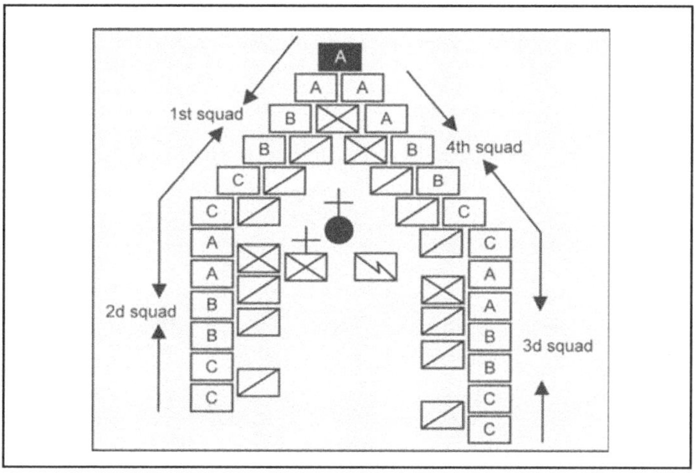

Figure 6-35. Platoon Wedge Formation With 2d and 3d Squads in Lateral Support

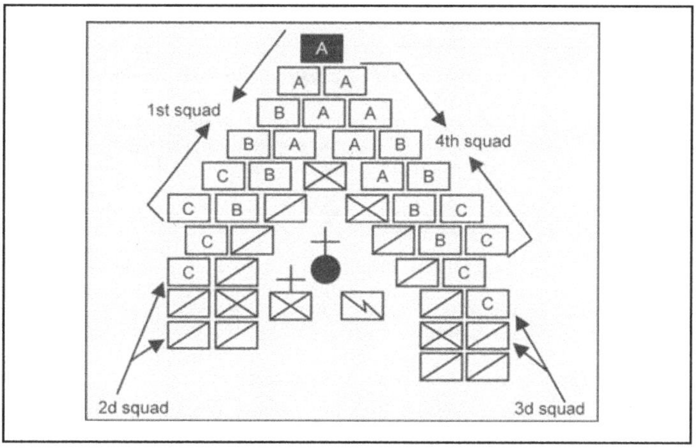

Figure 6-36. Platoon Wedge Formation With 2d and 3d Squads in Direct Support

Relieving Elements in Place

6-96. During crowd control operations, the platoon leader may rotate his squads to give them a rest. See *paragraph 6-76,* page 6-27.

Company Line in Depth Formation

6-97. The command for forming a company line in depth formation (see *Figure 6-37*) is "Company Line in Depth, Move." After the company commander's preparatory command, each platoon leader gives the command, "Platoon Line." The squad leaders follow with their commands to their squads to form the platoon line. On the command of execution, each platoon establishes a platoon line immediately to its front.

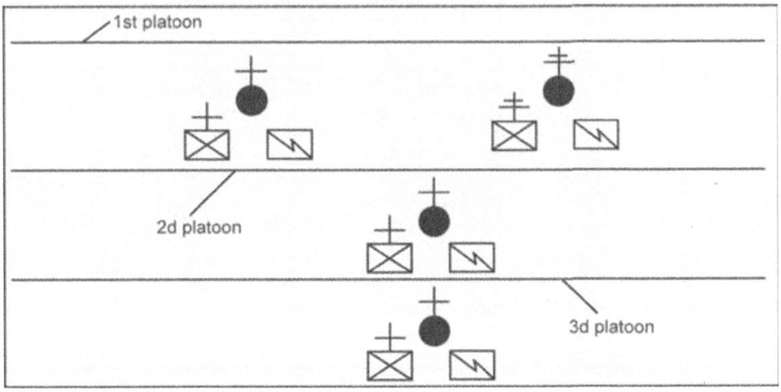

Figure 6-37. Company Line in Depth Formation

6-98. If the commander wants a more formidable formation, he commands, "2d Platoon, Direct Support, Move." The squad and team leaders from the 1st platoon take two steps backward. The 2d platoon moves forward and covers the intervals between the members of the lead platoon. The platoon leader of the 3d platoon then moves his platoon forward to occupy the position formerly held by the 2d platoon.

6-99. If the 3d platoon is called on to support the company line, it moves forward and takes up a position directly behind the 2d platoon in the same manner. Platoon leaders and PSGs help each other in controlling the company.

Company Line in Mass Formation

6-100. The command for forming a company line in mass formation is "Company Line in Mass, Move" (see *Figure 6-38,* page 6-40). The company line in mass is the same as a company line with both platoons in direct support. Instead of forming it in stages, the company commander indicates his desire for this formation in his initial command. The platoons each form a line, and the 2d and 3d platoons close on the 1st platoon without further command.

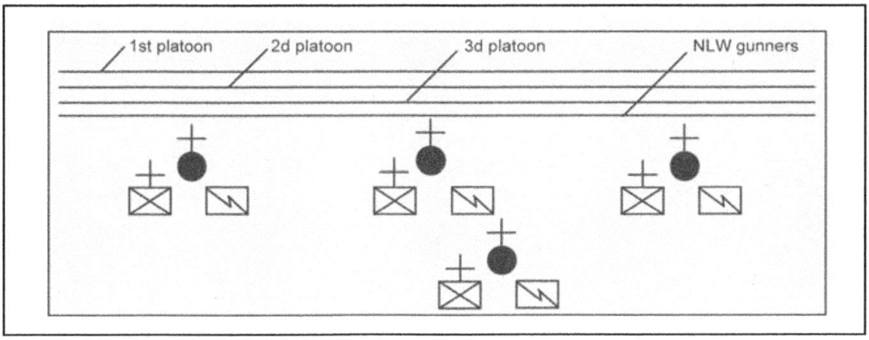

Figure 6-38. Company Line in Mass Formation

Company Line Formation With Support

6-101. In company formations, the 1st platoon usually forms the lead element and the 2d and 3d platoons provide support (see *Figures 6-39; 6-40;* and *6-41*). The support platoons can be employed in the same manner as the support squads in platoon formations. One platoon can be relieved from the lead element by another platoon while in formation by having a support platoon pass through the lead platoon.

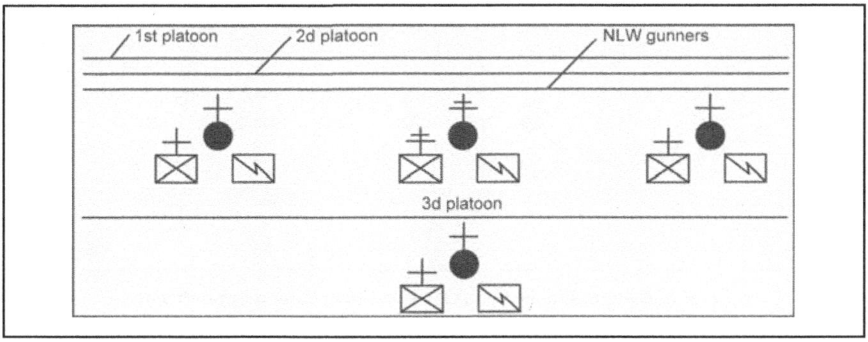

Figure 6-39. Company Line in Mass Formation With One Platoon in Depth

Company Echelons and Wedge Formations

6-102. Company echelons (see *Figures 6-42,* page 6-42*; 6-43,* page 6-43*; 6-44,* page 6-43*;* and *6-45,* page 6-44) and company wedge formations (see *Figures 6-46,* page 6-45; *6-47,* page 6-46; *6-48*, page 6-46; and *6-49*, page 6-47) are formed in the same manner and with the same variations as the company line formation. A company assembles from a crowd control formation in the same manner as a platoon or a squad does. However, due to the size of company

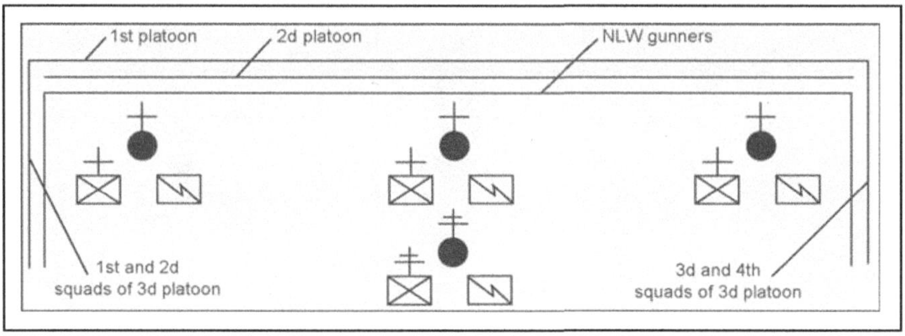

Figure 6-40. Company Line in Mass Formation With One Platoon in Lateral Support

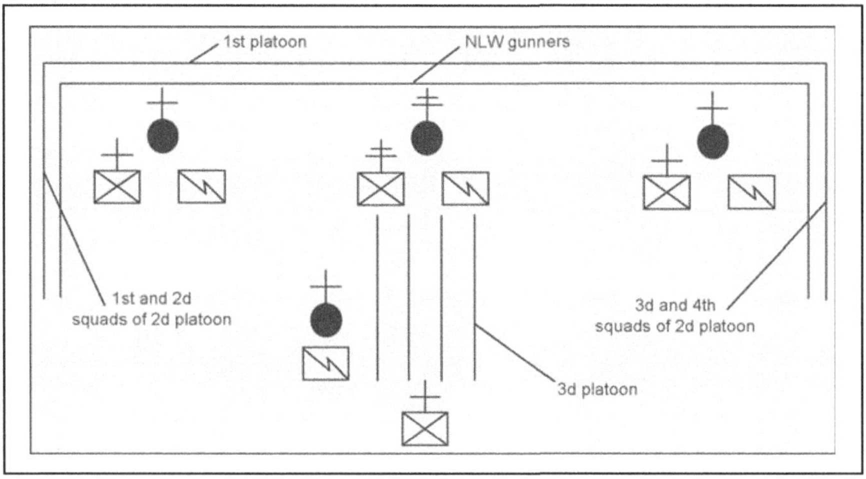

Figure 6-41. Company Line, One Platoon in Lateral Support and One Platoon in General Support

formations, the commander must consider the area that the company is operating in when he assembles the company. If he follows the guidelines of platoon assembly, he will have no difficulty. When assembling the company, the commander takes a position to the rear of the formation and commands, "Company Assemble." The platoon leaders take charge of their respective platoons. Usually, the 3d platoon assembles first, just behind the formation. After the 3d platoon is assembled, the 2d platoon assembles in front of the 3d platoon and the 1st platoon assembles in front of the 2d. All platoons assemble facing the commander. The 2d and 3d platoons then dress with the 1st platoon. As each platoon is assembled, they are ordered to port arms by the platoon leader to await further orders from the commander. If four platoons are used, the 4th platoon assembles first.

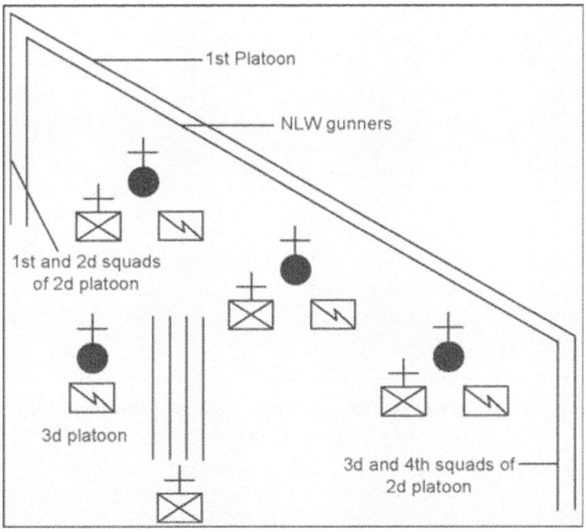

Figure 6-42. Company Echelon Right Formation With One Platoon in Lateral Support and One Platoon in General Support

Company and Platoon Formations With Vehicles

6-103. The commands for vehicles and foot troops are the same as those for foot troops alone. Because of the additional noise and distances involved, the company commander gives the hand-and-arm signal for the line. The command and the signal apply to the lead platoon only. The other platoons remain in general support until further directed (see *Figures 6-50,* page 6-48; *6-51,* page 6-49; *6-52,* page 50; *6-53,* page 6-50; *6-54,* page 6-51; *6-55,* page 6-52; and *6-56,* page 6-53).

6-104. The motor section moves out with vehicle number two moving to the right. Vehicle number two passes the lead vehicle and establishes the position for the center of the line. At the same time, the 3d and 4th vehicles swing out to the left and right. They form to the left and right and slightly to the rear of vehicle number two. The 1st and 4th squads of the lead platoon then move forward and form lines to the left and right, on the front end of vehicle number two. As soon as the troops are in position, the 3d and 4th vehicles close in on the squads and tighten the formation. The 2d and 3d squads of the lead platoon then form a line to the left and right of the 3d and 4th vehicles to complete the company line. The command vehicle (vehicle number one) takes

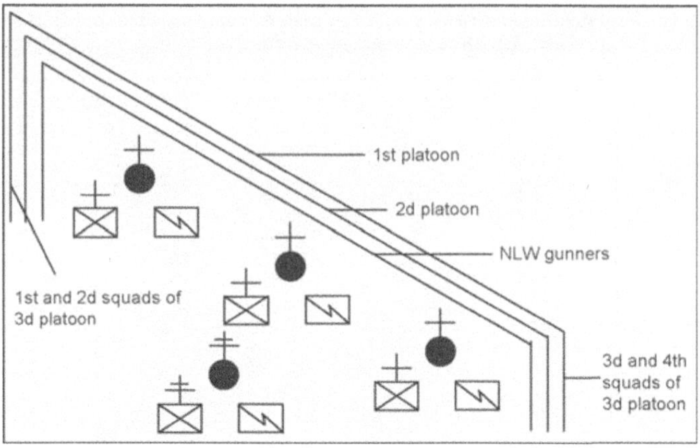

Figure 6-43. Company Echelon Right in Mass Formation With One Platoon in Lateral Support and One Platoon in General Support

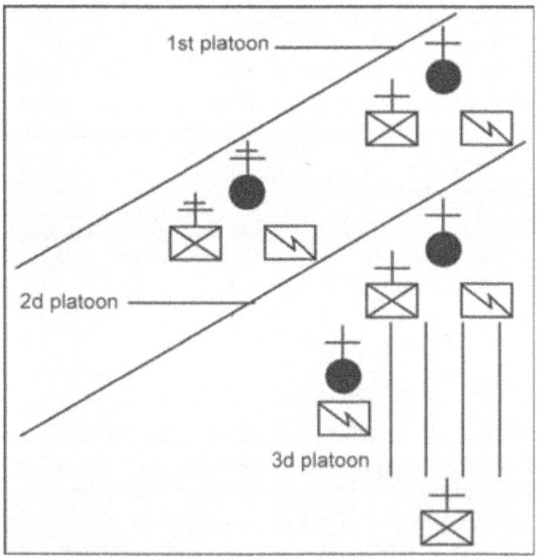

Figure 6-44. Company Echelon Right in Mass Formation With One Platoon in Lateral Support

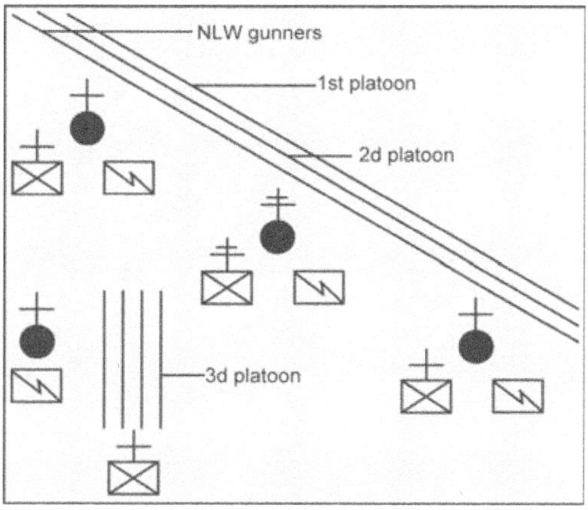

Figure 6-45. Company Echelon Right in Mass Formation With One Platoon in General Support

up a position behind the line where the commander can direct and control the unit. The 2d and 3d platoons move forward and are used in general, lateral, or direct support. In some instances, foot troops move into position first and vehicles join the formation, as needed.

6-105. To form a company echelon or a company wedge with vehicles, the company follows the same procedures used for company formations without vehicles. Vehicles are moved into the appropriate positions using procedures similar to those used for a company line.

6-106. To assemble troops and vehicles, the command vehicle moves to a position behind the formation. The commander faces the formation and gives the vehicle section the signal for assembly. Immediately, the vehicles return to their proper positions in column formation behind the command vehicle while the troops stand fast. The second in command then assembles the troops in the usual manner.

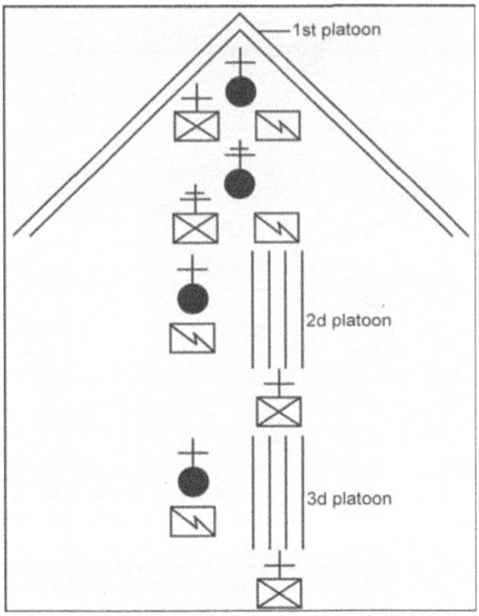

Figure 6-46. Company Wedge Formation With Two Platoons in General Support

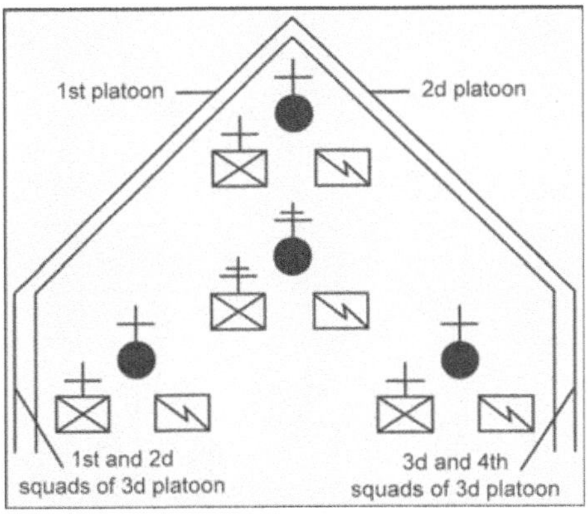

Figure 6-47. Company Wedge Formation With One Platoon in Lateral Support

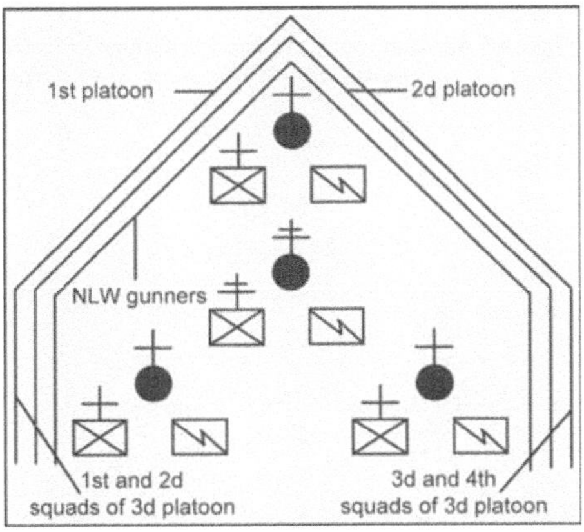

Figure 6-48. Company Wedge Formation With One Platoon in Lateral Support and One Platoon in General Support

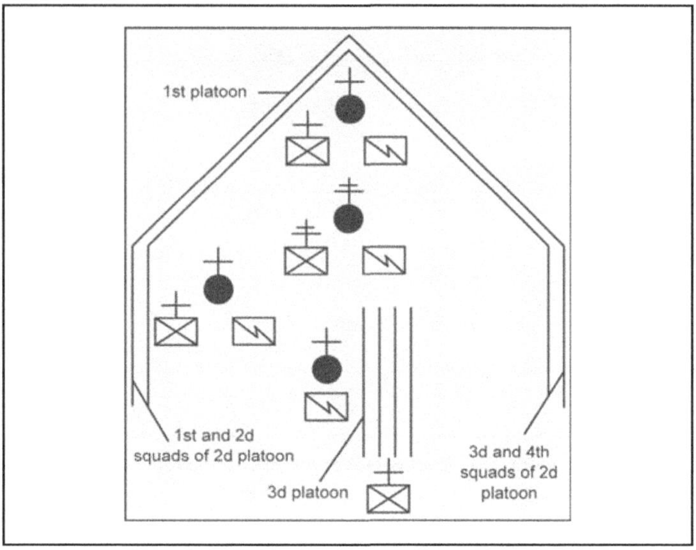

Figure 6-49. Company Wedge Formation With One Platoon in Lateral Support and One Platoon in General Support

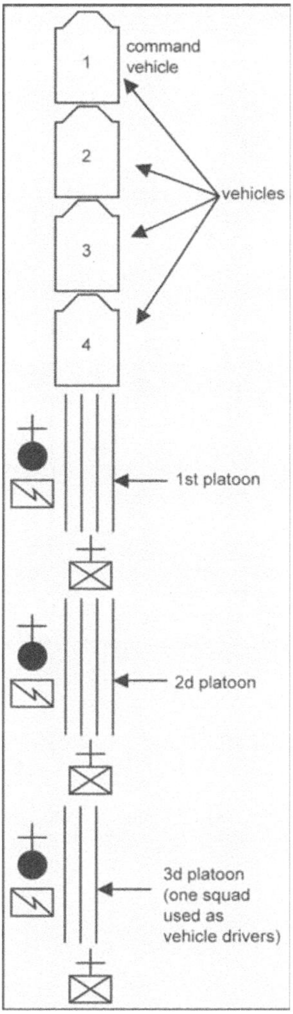

Figure 6-50. Company Column Formation With Vehicles

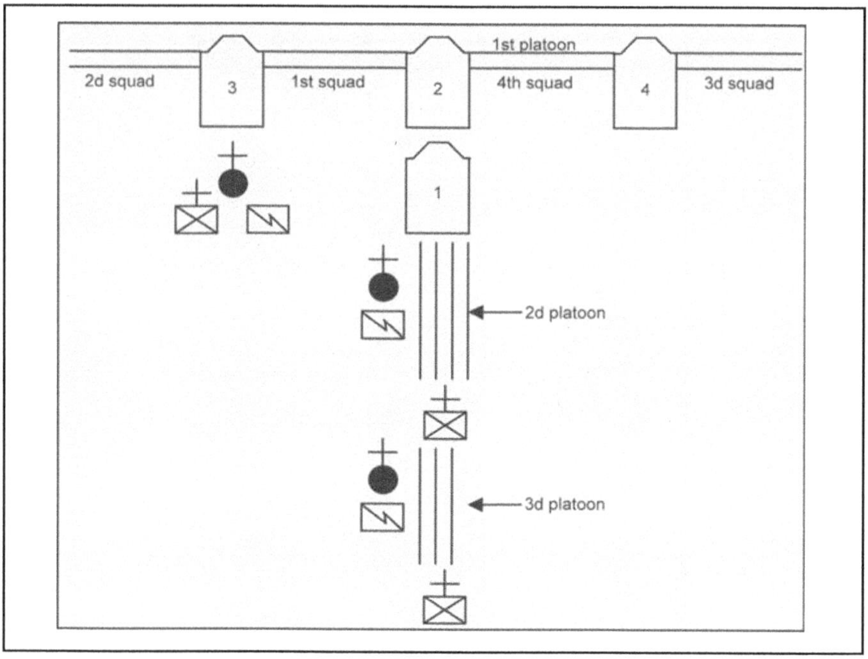

Figure 6-51. Company Line Formation With Vehicles and Two Platoons in General Support

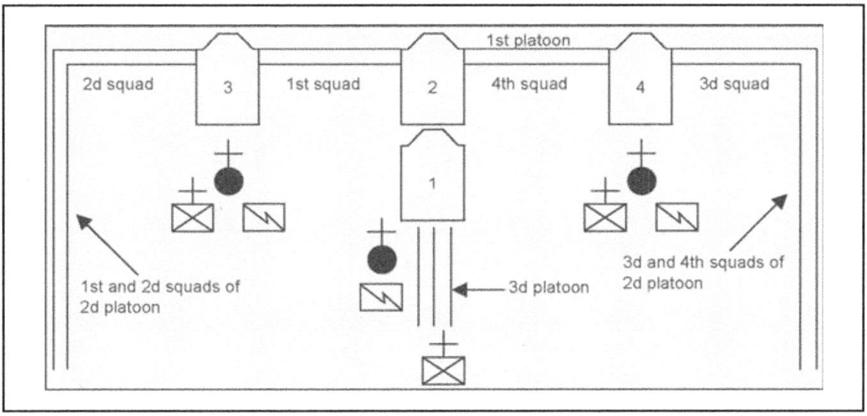

Figure 6-52. Company Line Formation With Vehicles, One Platoon in Lateral Support, and One Platoon in General Support

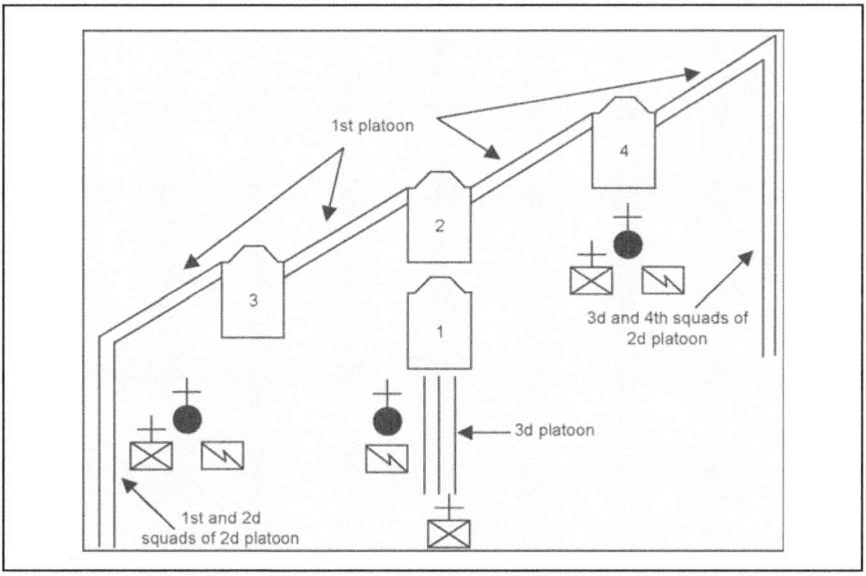

Figure 6-53. Company Echelon Right Formation With Vehicles, One Platoon in Lateral Support, and One Platoon in General Support

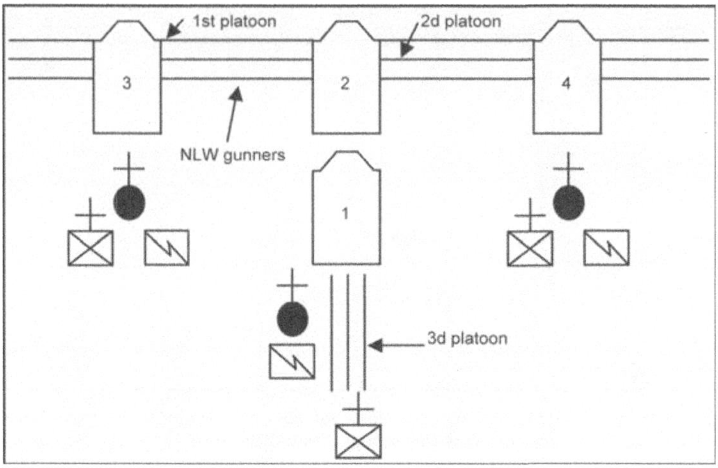

Figure 6-54. Company Line in Mass Formation With Vehicles and One Platoon in General Support

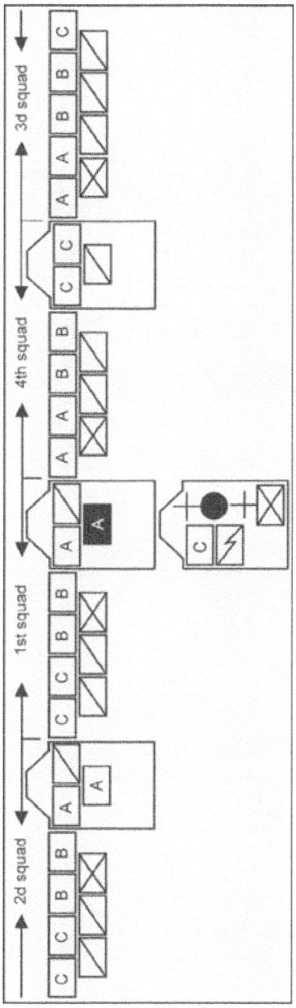

Figure 6-55. Platoon Line Formation With Vehicles

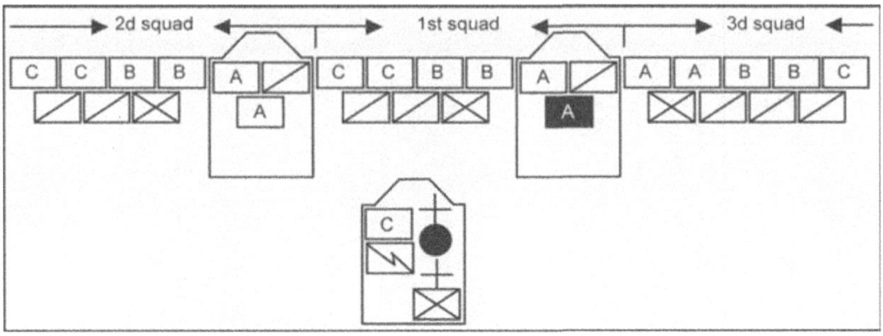

Figure 6-56. Platoon Line Formation With Direct Support and Vehicles

Chapter 7

Civil Disturbance Operations in Confinement Facilities

Civil disturbance doctrine is usually written for quelling disturbances that occur in large open areas where the size of the responding force or the type of formation employed is not hampered by a lack of space. However, in the military confinement environment, inmate disturbances often occur in small, closed areas, such as an inmate's housing unit, a gymnasium, a dining facility, or a chapel where maneuver space and equipment limit the size of the element that may be employed. The layout of each facility is different and presents its own unique problems when employing forces. This chapter covers the use of small military police teams, forced cell move teams (FCMTs), and the formations used to move an unruly inmate from one cell to another. Small generic riot control formations are discussed to provide the commander with a basic employment option that he can tailor to his facility to enable him to regain control of the affected area. The use of NL munitions and RCAs are incorporated into this discussion.

CROWD DYNAMICS WITHIN THE CONFINEMENT FACILITY

7-1. In confinement facility disturbances, inmates employ a number of tactics to resist control or to achieve their goals. Tactics may be unplanned or planned and nonviolent or violent. When a disturbance carries many purposes, it is likely that their tactics are well planned.

NONVIOLENT TACTICS

7-2. Nonviolent tactics range from name-calling to building barricades. Inmates may attempt to distract control force members by shouting at or ridiculing them or by using abusive language, obscene remarks, taunts, and jeers. The inmates' goals are to anger and demoralize the control force. They also want authorities to take actions that may later be exploited as acts of brutality.

VIOLENT TACTICS

7-3. Violent crowd tactics used by inmates are often extremely destructive and can include physical attacks on cadre, other inmates, and property. Use of violent tactics is limited only by the attitudes and ingenuity of the inmates, the training of their leaders, and the materials available to them. Inmates often commit violence with crude homemade weapons. If unplanned violence occurs, inmates will use mops, brooms, chairs, beds, or whatever else is on hand as weapons of violence. During planned violent disturbances, inmates can easily conceal makeshift weapons or tools to use against the control force. Inmates often erect barricades to impede troop movement and/or to prevent the control force from entering certain areas or buildings. Inmates can be expected to vent their emotions on individuals (cadre or inmates), the control force, their equipment, and/or the institution's property. Inmates direct dangerous objects like carts, barrels, liquids, and burning shoe polish at the control force.

GRIEVANCE PROTEST

7-4. A grievance protest may be organized as a riot. However, under normal circumstances, a riot of this type is not of an extremely violent nature. It may turn violent when leaders attempt to exploit any successes or weaknesses found in the control force.

UNORGANIZED RIOTS

7-5. Unorganized riots are characterized as being spontaneous in nature. However, they could be exploited and diverted by leaders at any subsequent stage of the riot, turning it into a different type. They may begin as an isolated assault against authority figures, or acts of civil disobedience. Under determined leadership, an unorganized riot could change to an organized riot.

MULTIPLE RIOTS

7-6. Multiple riots may happen when a disturbance occurs in one area of the confinement facility, causing others to riot in their areas. To control this situation with only one riot control force available, the force commander should first subdue the most violent riot. At the same time, he should employ some of his control forces to contain the other riots until the main control force is able to move to those locations to subdue them.

CADRE BEHAVIOR

7-7. The cadre is susceptible to crowd behavior. They become emotionally stimulated during tense confrontation with unruly and violent inmates. Commanders must be aware of the mood and attitude of the crowd and its effect on the control force cadre. To counteract the effects of crowd behavior on the cadre, the commanders must institute rigorous training and firm and effective leadership. This training must include a complete awareness and understanding of the use of force and the commander's intent. These are all necessary to offset the effect of crowd contagion upon the control force. Even with the best training and leadership, control force members must exercise individual and collective restraint.

PLANNING

7-8. Preplanning is the preparation and training conducted before a crisis occurs. The purpose of preplanning is to have plans and standing operating procedures (SOPs) in place so the cadre may react to an emergency and successfully contain and neutralize the situation. The preplanning process includes plan development, intelligence gathering, and training.

NOTE: Intelligence gathering is essential for the successful containment, neutralization, and/or prevention of a riot.

7-9. Once the control force team has been alerted to a disturbance within the confinement facility, leaders further develop preplans to fit the situation much the same way that DOD develops war plans.

7-10. During the planning process, leaders must be aware of the potential risks involved in quelling disturbances within a confinement facility. Careful planning minimizes collateral damage and risks to control force members and hostages.

7-11. The use of NLW and NL munitions must also be considered when developing plans. There must be strict accountability and control of RCAs, their employment, and other appropriate NL means (such as high-pressure water hoses). When using RCAs, plans must be flexible enough to accommodate changes in the situation and weather because it impacts the effectiveness of some agents. Planners must also consider the—

- Location of the disturbance.
- Estimated number of rioters.
- Accessibility of weapons, tools, or cleaning supplies.
- Inmate's military training.
- Hostage situation.

7-12. Based on an analysis of these factors, the commander makes an estimate of the situation. The estimate must be as thorough as time permits. Using the estimate, the commander considers COAs, selects RCAs, and determines the need for engineer support. In choosing a COA, consider the—

- Number of hostages.
- Number of inmates holding the hostages.
- Leader.
- Attitude and demeanor of the rioters.
- Location of the riot.
  - Inside areas of the confinement facility, such as the dining facility, work site, or housing wing.
  - Outside areas of the confinement facility, such as in exercise or work areas.
- Accessibility of weapons or material to make weapons.
- Accessibility of chemicals, such as cleaning supplies, cooking oils, and shoe polish.

USE OF CHEMICAL IRRITANTS

7-13. The use of chemical irritants can be a valuable NL tool for control force leadership to consider during the planning phase. These chemical irritants can drive a threat from an established, enclosed position or deny the rioters access to a certain area without long-lasting effects to those involved. The proper use of chemical irritants may prevent the control force from having to enter a dangerous area; however, improper use can cause injury, death, or property damage.

USE-OF-FORCE GUIDELINES

7-14. The installation commander authorizes the use of chemical irritants after the control force leader determines the type and dosage. The SOP outlines the procedures for securing authorization and provides guidance to help the control force leader make his decisions. The confinement facility commandant should inform the installation commander of the situation, since the installation commander is ultimately responsible. Team members must be aware of how chemical irritants affect personnel and plan accordingly. Inmates not involved must be removed from the area and the chemical cloud's path. The fire department and emergency medical services (EMS) should be on standby during operations involving chemicals.

PEPPER IRRITANT

7-15. The pepper irritant receives its name from varieties of the capsicum plant that is it made from. The active ingredient, Oleoresin Capsicum (OC), is extracted from the plant and micropulverized to make the irritant. It is neither a chemical nor a gas, but an all-natural organic substance. It is effective against individuals who are emotionally disturbed or under the influence of drugs or alcohol. It is effective against domestic and wild animals. OC is often used to divert the threat from making or continuing an assault. OC does not cause permanent damage. Its effects last 30 to 40 minutes after individuals are placed in fresh air.

7-16. OC can have various effects on people. Some effects are—

- Swelling of mucous membranes and the upper respiratory system; however, it does not shut down the system.
- Burning (intense) and discoloration (bright red) of exposed skin.
- Dilating the capillaries.
- Swelling of the eyelids.
- Burning and involuntary closing of the eyes.
- Coughing uncontrollably.
- Gagging.
- Gasping for air.
- Losing strength and coordination (temporarily).

OC has some disadvantages. They are that—

- The canister depressurizes over time if not checked regularly.
- Inmates may gain access to OC and use it against SRT members.
- The canister must be shaken on a regular basis.

DISSEMINATION OF CHEMICAL IRRITANTS

7-17. There are six methods for disseminating chemical irritants.

Pyrotechnic Method

7-18. In the pyrotechnic method of dissemination, the chemical irritant is placed in a canister with an inert material that is ignited when the device is activated. The agent is then carried into the air on the smoke particles of the inert material.

7-19. There are some disadvantages when using a pyrotechnic device indoors. These disadvantages are that the device—

- May cause a fire.
- Has a very slow saturation time (40 to 45 seconds to burn completely).
- Can be thrown back by the threat, even while burning.

Bursting Method

7-20. In the bursting method of dissemination, the irritant and an inert powder are placed in a saw-toothed canister. After a small detonation, the canister splits and expels the irritant in a cloud.

7-21. The bursting method has its advantages and disadvantages. Some advantages are that—

- There is no risk of fire.
- There is no risk of the canister being thrown back.
- The contents disseminate quickly.
- The canister is effective, easy to carry, and easy to control.

Some disadvantages are that—

- There is a slight possibility of fragmentation when detonated.
- The fuse head may separate from the canister.
- Only 50 to 90 percent of the irritant may be disseminated.
- The direction of the cloud formation depends on where the device is detonated.

Aerosol Method

7-22. In the aerosol method of dissemination, the irritant is suspended in an inert liquid that is located in the rear portion of the device. When the projectile penetrates the structure, the rear portion splits open and the irritant is disseminated in a mist. The aerosol method is most appropriate for tactical operations when chemical irritants are used before entry.

7-23. The aerosol method has its advantages and disadvantages. Some advantages are that it—

- Is easily carried.
- Is an excellent deployment system (uses 40-millimeter, 37-millimeter, or 12-gauge ammunition).
- Disperses the contents quickly.
- Is easily controlled and accurate (fin-stabilized).

NOTE: Practice firing is always necessary.

Some disadvantages are the possibility of—

- Injury to individuals if it is fired from close range.
- Damage to the structure if it strikes a weak portion.

Projectiles Method

7-24. Chemical-containing projectiles can be very effective when used correctly. A soldier must know which projectile to use in each unique situation.

7-25. **Thirty-Seven-Millimeter Projectile.** A 37-millimeter projectile can incapacitate anyone in an enclosed area of 4,500 cubic feet. It can penetrate the following:

- A sheet of ¾-inch-thick plywood at 25 yards.
- An automobile windshield at 33 yards.
- A hollow-core door at 58 yards.

7-26. **Forty-Millimeter Projectile.** A 40-millimeter projectile is more effective at greater ranges than a 37-millimeter projectile because of its rifled bore. It can incapacitate anyone in an enclosed area of 4,500 cubic feet. It can penetrate the following:

- A 1-inch-thick plywood at 10 yards.
- An automobile windshield at 55 yards.
- A hollow-core door at 109 yards.

7-27. **Twelve-Gauge Projectile.** A 12-gauge projectile can be loaded into a shotgun with no modifications to the weapon. It can be fired as a single round, or it can be magazine-fed. The 12-gauge projectile can incapacitate anyone in an enclosed area of 1,000 cubic feet. The angle at which the projectile strikes the barrier can affect its penetration capability. It can penetrate the following:

- A sheet of ¾-inch plywood at 33 yards.
- An automobile windshield at 33 yards.
- A hollow-core door at 100 yards.

Fogging Method

7-28. The irritant is suspended in a fogging solution and dispersed with a fogging machine. It is not recommended for tactical operations.

7-29. The fogging method has its advantages and disadvantages. Some advantages are that it—

- Produces enough chemical irritant to cover an enclosed area of 100,000 cubic feet in 26 seconds.
- Disseminates quickly.
- Provides high-volume capabilities.
- Is effective for riot control or other crowd dispersal situations.

Some disadvantages are that it is—

- Noisy.
- Cumbersome.

Multipurpose Grenade Method

7-30. A multipurpose grenade can be hand-thrown or launched from a shotgun with an adapter. It has an extended shelf life of 6 years and an adjustable fuse delay of 2 to 5 seconds. The irritant is located in the cylinder portion of the device and is forced out the bottom of the grenade upon detonation.

> **WARNING**
> **A multipurpose grenade deployed incorrectly could result in injury. Safety rules should be observed at all times.**

RECORD OF EVENTS

7-31. Reporting procedures for serious incidents should be included in the plans. A record of events must be initiated to provide a basis for the preparation and submission of a formal report to higher HQ. As a minimum, include the—

- Time the incident was reported and by whom.
- Time the incident was reported to the facility commander.
- Time the control force was assembled.
- Time the control force entered the facility.
- Weather conditions as they relate to the use of RCAs.
- Number of cadre and inmates injured or killed, including how they were injured or killed and the medical attention given to them.
- Time the operation was completed.
- Time the riot control force restored order.

TRAINING

7-32. Guard units and associated teams must train on a regular basis in the five basic riot control formations. There must be a continuous training program established, to include—

- Interpersonal communication.
- Self-defense.
- Use of force.
- Use of a riot baton.
- Fatal areas of the body to avoid when using the riot baton.
- First aid.
- Emergency plans.
- Intelligence and counterintelligence techniques.
- The use of RCAs and the various methods of dispersing them.
- The use of NLW.
- The use and effects of high-pressure water.
- Riot control formations.

EQUIPMENT

7-33. The recommended basic riot gear used by the confinement riot control force is as follows:

- Riot baton.
- Riot shield.
- Helmet with face shield.
- Groin protector.
- Flak vest.
- Leather gloves.
- Shin protection.

RIOT BATON

7-34. The riot baton is an invaluable weapon in crowd control situations. A riot baton in the hands of a well-trained control force member is the most appropriate weapon, except with extremely violent crowds. The riot baton is not meant to replace NL munitions, but is instead used in situations in which NL munitions are not needed. Losing a riot baton to the crowd does not present a serious threat.

7-35. Use of the riot baton is based on the commander's appraisal of the situation and his choice of force option. Control forces committed with the riot baton must have teams equipped with RCAs, NLW, and lethal coverage. If the confrontation increases in intensity, the commander has these force options available and may rapidly engage the control force.

7-36. A riot baton is a versatile offensive weapon that varies in length from 26 to 42 inches with the optimum length being 36 inches. It is made of hickory and does not shatter or break easily. Both ends of the riot baton are rounded to prevent unnecessary injury. It is approximately 1¼ inches in diameter. Each riot baton has a hole drilled at the grip end with a leather thong threaded through the hole. The thong helps secure the riot baton to the user's hand.

RIOT SHIELD

7-37. The riot shield (see *Figure 7-1*) is an excellent piece of defensive equipment. It is used by control force personnel in the front and on the flanks of the formation.

7-38. The riot shield has its advantages and disadvantages. Some advantages are that—

- OC pepper spray or high-pressure water can be used to back the control force.
- It provides protection from thrown objects.
- It absorbs some of the impact of rioters hurling themselves at the control force.

The disadvantages of the riot shield are that it—

- Severely limits the use of the riot baton.
- May crack if hit with a heavy object.

Figure 7-1. Riot Shields

RIOT CONTROL FORMATIONS

7-39. The five basic formations for civil disturbance operations are the line, wedge, echelon, diamond, and circle. In the correctional environment the basic formations for disturbances are the line, wedge, or echelon with apprehension teams in support. Generally, the diamond and circle formations are not used in the correctional environment. The minimum size force that the commander should consider is an augmented platoon, with a two-to-one advantage over the rioters.

7-40. Correctional facility cadre must be proficient in all riot control formations. These are usually trained at the squad level and above, but performed by a platoon or company-sized unit. Squad line, echelon left, echelon right, and wedge formations form the basis for platoon and company formations. Each squad must be adept in the basic formations before practicing in platoon-sized or larger formations. For more detailed information on formations, see *STPs 19-95B1-SM* and *19-95B24-SM-TG*. In the correctional environment, the support platoon plays a key role by supporting the control formation with apprehension and equipment teams. See *Chapter 6* for an in-depth discussion on civil disturbance formations.

APPREHENSION TEAMS

7-41. Apprehension teams give the control force the ability to extract inmates. Inmates identified as leaders or agitators and inmates that are injured or no longer want to resist can be extracted from the crowd. Apprehension teams can also be used to enter confined areas where the formation cannot operate.

7-42. The apprehension team will consist of eight members and one leader. Each team will have four shield holders, two nonlethal gunners and two apprehension/search members. The squad leader controls the movement of the team and assists the team in exiting the control formation. He maintains communications with the formation and coordinates the team's operations with the formation. The squad leader also ensures that the team uses the minimum amount of force necessary to extract inmates.

7-43. The apprehension team may deploy through the middle of the formation or from around either of the flanks.

7-44. The apprehension team should be deployed within a distance of approximately 10 meters so that the formation can still provide immediate support. If the apprehension team is deployed farther than 10 meters, the formation must advance to close the gap.

7-45. The team will exit the formation in two columns; shield holders in the first and second ranks, nonlethal gunners in the third rank, apprehension/search members in the fourth rank and the leader last. Team members grip the shoulder or equipment of the member in front of them. The shield holders will envelop the inmate that is to be extracted and form a protective barrier. Once the shield holders have enveloped the inmate, the nonlethal gunners will cover their sides with their weapons while the apprehension/search members control and restrain the inmate. When the team has the inmate under control, the squad leader will give the order for the team to withdraw. The team withdraws in reverse order (apprehension/search members with the inmate, followed by nonlethal gunners and shield holders). Team members then move backwards, face the crowd and grip the shoulder or equipment of the member in front of them.

TEAM EQUIPMENT

7-46. The apprehension team uses the following equipment:

- Flex-Cufs.
- Restraint cutters.
- Pepper spray.
- Helmets with face shields.
- Groin protectors.
- Flak vests.
- Elbow pads.
- Leather gloves.
- Shin protectors.
- Knee pads.
- A video camera.

TEAM COMPOSITION

7-47. The apprehension team is composed of five team members. This team consists of four members and a squad leader. Team member one is the team leader.

Squad Leader

7-48. The squad leader assists the team in exiting the control formation and ensures that the team uses the minimum force necessary to subdue inmates. He is also responsible for signaling the PSG if more help is needed.

Team Member One (Team Leader)

7-49. Team member one's responsibility is to make initial contact with the inmate and use pepper spray and apply Flex-Cufs, as needed. He also carries a set of restraint cutters. Once the apprehension team has returned to the relative safety of the control formation, team member one searches the inmate.

7-50. Team member one receives the command to move forward and secure the inmate from the squad leader. This command is usually given by a signal and verbal command. Team member one then notifies the main force commander where his team exits the control force. Once the team is in position to exit the main formation, team member one signals the member of the formation in front of him. This is usually a prearranged signal, such as a tap on the right shoulder. The control force opens and the team exits the formation and makes contact with the assigned inmate. Team members then subdue and control the inmate. If pepper spray is needed, team member five shouts, "Pepper." This gives the other team members time to react to the command and spray the inmate. Team member one then applies Flex-Cufs to the inmate's wrists and ankles. Depending on the mission, the inmate may be left for follow-on security forces or carried behind the control force to be secured by security forces.

7-51. Should the mission require that the inmate be moved, team member one shouts, "Prepare to Move," this allows each member to grab an arm and position the inmate toward the control force formation. When the team is ready to move, team member two shouts, "Ready," alerting team member one that the team is ready for the last command. Team member one then commands, "Move." The team enters the control formation with the inmate. The inmate is searched by team member one and turned over to the security team.

Team Member Two

7-52. Team member two stands directly behind team member five with his hands on team member five's waist. He protects the left side of team member five by moving to team member five's left and helps to secure the inmate after team member one makes initial contact. He repeats team member five's commands and controls the inmate's head after the inmate is on the ground.

Team Member Three

7-53. Team member three stands behind team member two with his hands on team member two's waist. He protects the right side of team member five by moving to team member five's right and helps to secure the inmate after initial contact.

Team Member Four

7-54. Team member four stands behind team member three with his hands on team member three's waist. His task is to help subdue the inmate in any way possible. He usually concentrates on the inmate's body below the knees. He acts as the eyes and ears for the team during control of the inmate.

Team Member Five

7-55. Team member five gives commands to control the inmate.

FORCED CELL MOVE TEAMS

7-56. The intent of a forced cell move is to move an unruly and/or uncooperative inmate from one cell to another. This is a difficult task and must not be taken lightly. A forced cell move must be thoroughly planned, rehearsed, and resourced to ensure the safety of the team members and the inmate.

7-57. A number of reasons may compel the commander to consider a forced cell move. The inmate may refuse to eat, take medication, accept medical attention, get a haircut, keep himself clean, work, or move or rotate cells when ordered to. These are some of the reasons that the commander may order a forced cell move.

TEAM COMPOSITION AND EQUIPMENT

7-58. The FCMT is composed of six corrections cadre and two support cadre from military police and medical sections specially trained in forced cell movement. Primarily, the team consists of the following personnel:

- One officer in charge (OIC) (E7 or above).
- Five corrections cadre.
- One military police investigator.
- One medic.

7-59. The FCMT uses the following standard equipment:

- Helmets with face shields.
- Groin protectors.
- Flak vests.
- Elbow pads.
- Leather gloves.
- Shin protectors.
- Knee pads.
- A pinning shield.
- Restraints (hand or leg irons or Flex-Cufs).
- A video camera.

ACTIONS PRIOR TO TEAM EMPLOYMENT

7-60. The OIC gives an operations order (OPORD) to team member five (the team leader) of the FCMT. As a minimum, the OPORD should answer the following questions about the incident leading up to the move:

- Did the inmate fail to comply with orders?
- Did the inmate assault cadre or other inmates?
- Is the objective of the move to force a shower or haircut?
- Is OC pepper spray authorized for use?
- Has the inmate been moved before?

- What is the condition of the area where the inmate is currently located?
- Has the inmate tested positive for human immunodeficiency virus (HIV)?
- Does the inmate have a weapon or the materials needed to make one?
- Is the area barricaded?
- Has he smeared himself with body fluids or other slippery substances?
- Can you approach the inmate without risk to him and/or others?

7-61. The team is assembled for inspection by the OIC, and the team members are required to brief the OIC on their individual duties and responsibilities. A line formation is the customary formation for this inspection. It should take place out of site of the inmates.

7-62. Each soldier takes one step forward and briefs the OIC as shown in the following example: *"I am (rank and name), team member one. My mission is to pin the inmate using the minimum amount of force necessary. I carry the pinning shield."* He then faces so the camera can record the number printed on the back of his flak jacket and holds up the pinning shield [see *Figure 7-2*]).

Figure 7-2. Recording Team Members' Duty Position

INDIVIDUAL TEAM MEMBER RESPONSIBILITIES

7-63. Each individual team member has his own responsibilities during a forced cell move. Each team member must do his designated job while working as a team.

Team Member One

7-64. Team member one is the eyes and ears of the team. His responsibility is to carry the pinning shield up to the cell door and describe the scene with as much detail as possible (especially including anything that may cause harm to the team). The pinning shield is used to protect the team from body fluids thrown at the them by the inmate.

7-65. Team member one describes the scene in the cell to his teammates by turning his head to the right and speaking in a loud voice so that all team members can hear him. A slap on the right shoulder by team member two alerts him that all team members have heard him. If he receives a slap on the left shoulder from team member two, he repeats the situation report (SITREP) in a loud, clear voice. The decision to enter the inmate's area with or without the pinning shield is team member one's decision, unless the inmate is known to have a weapon. Should he decide to enter without the pinning shield, team member one then enters with both hands held approximately chest high. When team member one is unable to determine if the inmate has a weapon, the pinning shield is used. Charging through the open cell door, team member one drives the inmate to the bunk or ground with the shield. Once pinned, team member one secures the inmate's head and ensures that it is not twisted or turned in relation to its natural form. He may have to use the various pressure points on the inmate's head to help subdue him.

Team Member Two

7-66. Team member two steps forward and briefs the OIC as follows: *"I am (rank and name), team member two. My mission is to secure the inmate's right arm using the minimum amount of force necessary. I carry one set of hand irons."* He then holds the hand irons up for video camera taping (see *Figure 7-3*) and says, *"I also carry one canister of OC pepper spray to be used at the direction of the team OIC."* He then holds up the canister for video camera taping and faces so the camera can record the number on the back of his flak jacket (see *Figure 7-2,* page 7-13.)

Figure 7-3. Recording Team Members' Equipment

7-67. Team member two lines up directly behind team member one. He keeps contact with team member one until he can secure the inmate's strong arm. Team member two taps team member one on his right or left shoulder after being tapped by team member three. Once inside the cell, team member two secures the prisoners strong hand with wrist restraints and announces, "*Hands secure,*" to team member five.

Team Member Three

7-68. Team member three steps forward and briefs the OIC as follows: *"I am (rank and name), team member three. My mission is to secure the inmate's left arm using the minimum amount of force necessary."* He then faces so the camera can record the number on the back of his flak jacket (see *Figure 7-4*).

Team member three lines up directly behind team member two. Team member three keeps contact with team member two until he can grasp a portion of the inmate's body. He then concentrates on securing the inmate's weak hand and assisting team member two in applying hand restraints. Team member three taps team member two on his right or left shoulder after being tapped by team member four.

Figure 7-4. Recording Number on Flak Jacket

Team Member Four

7-69. Team member four steps forward and briefs the OIC as follows: "*I am (rank and name), team member four. My mission is to secure the inmate's right leg using the minimum amount of force necessary. I carry one set of leg irons.*" He holds the leg irons up for the camera to record (see *Figure 7-2,* page 7-13). He then faces so the camera can record the number on the back of his flak jacket (see *Figure 7-2*). Team member four lines up directly behind team member three. He taps team member three on his right or left shoulder after being tapped by team member five. Team member four keeps contact with team member three until he can secure a portion of the inmate's body. Team member four concentrates on securing the right leg of the inmate. It is his responsibility to apply the leg irons. Once the legs are secure, he announces, "*Legs secure,*" to team member five.

Team Member Five (Team Leader)

7-70. Team member five steps forward and briefs the OIC as follows: "*I am (rank and name), team member five. My mission is to secure the inmate's left leg using the minimum amount of force necessary. I carry a restraint key (hold up for the camera to view). I am also the team leader for this mission and it is my responsibility to ensure that the team uses the minimum amount of force necessary to subdue the inmate.*" He then faces so the camera can record the number on the back of his flak jacket (see *Figure 7-4*).

7-71. Team member five taps team member four on the right shoulder when he has heard the SITREP from team member one. If he does not hear the report or needs to hear it again, he taps team member four on the left shoulder. Team member five keeps in contact with team member four until he can secure a portion of the inmate's body. He secures the left leg and assists team member four in applying the leg irons. Once he has heard that the hands and legs are secure, team member five signals for the team to conduct an equipment check. The team members check the equipment in place.

NOTE: At no time is the inmate to be released by anyone.

Medic

7-72. The medic briefs the OIC in the following manner: *"I am (rank and name), the medic. My mission is to treat all injuries that may occur during the move.*" He follows the instructions given to him by the OIC/NCOIC. His mission is to observe the inmate for signs of injury before the forced cell move and assess the inmate's condition during and after the move.

Military Police Investigator

7-73. The military police investigator (MPI) is responsible for videotaping the entire forced cell move. This includes the OIC/NCOIC briefing the OPORD, each team member briefing his responsibilities to the OIC, and any medical treatment rendered to the inmate.

ACTIONS DURING THE FORCED CELL MOVE

7-74. There are many actions performed during a forced cell move. Each member of the FCMT must complete his task successfully to accomplish the forced cell move.

Officer in Charge or Noncommissioned Officer in Charge

7-75. The OIC, an E7 or above, is responsible for ensuring that the team uses the minimum amount of force necessary to subdue the inmate. He is also responsible for ensuring that all necessary precautions have been taken to ensure the safety of the team and the inmate. The OIC gives the inmate the initial orders to comply and informs him of the consequences of noncompliance. He does this by placing himself at the front of the inmate's cell (see *Figure 7-5*). The orders are not given until it can be enforced by the FCMT. The following prepared statement is read to the inmate by the OIC: "*I am (rank and name), OIC of this FCMT. At this time I am ordering you to lie face down on your bunk (or other designated location) with your head to the rear of the cell. Cross your hands behind your back and cross your feet as the team enters the cell and uses the minimum amount of force necessary to place you in restraints and move you to (give location). Failure on your part to comply with these instructions or to resist the FCMT will result in the use of force (OC pepper spray if applicable). Do you understand these instructions? If so, lie down on the bunk.*"

7-76. After reading the instructions to the inmate, the OIC faces the camera and states whether the inmate complied. He then waits 60 seconds and motions for the team to position itself in front of the inmate's cell (see *Figure 7-6*). The OIC directs the opening of the cell when the team is positioned at the entrance to the cell and prepared to enter. He positions himself so he can observe team entry to and departure from the cell. If OC spray is authorized, the OIC tells the team when to spray the inmate with OC spray. This task is done in silence. Communication between team members should be done through hand-and-arm signals, unless noted otherwise.

Figure 7-5. OIC Reading Compliance Statement to Inmate

Figure 7-6. OIC and FCMT

Team Member One

7-77. Once the team is positioned in front of the inmate's cell, team member one starts his report. At a minimum, he must answer the following questions and report them to the team:

- Does the inmate have a weapon, if so, what kind?
- Does the inmate appear dry?
- Is the inmate clothed or wearing extra clothing?
- Has the inmate smeared anything on his body?
- Can you see the inmate's hands?
- Is the cell barricaded or visibly booby-trapped?

- Is the cell dry?
- Are cell furnishings intact?

7-78. Based on the answers to the above questions, team member one must decide whether to enter the cell with or without the shield (see *Figure 7-7*). Team member one's ultimate goal is to pin the inmate on the bunk or floor.

Figure 7-7. FCMT Preparing to Enter a Cell

Team Member Two

7-79. Team member two keeps contact with team member one until he can secure the inmate's strong arm. If he cannot immediately secure the inmate's arm, he secures a portion of the inmate's body and progresses to the inmate's strong arm. Once the inmate's hands are secure, he announces, "*Hands secure*," to team member five (see *Figure 7-8)*.

Team Member Three

7-80. Team member three keeps contact with team member two until he can grasp a portion of the inmate's body. Team member three concentrates on securing the inmate's weak hand and assists team member two in applying Flex-Cufs. Team member three maintains control of the inmate's hands and looks at team member five for further instructions (see *Figure 7-8*).

Team Member Four

7-81. Team member four keeps contact with team member three until he can secure a portion of the inmate's body. Team member four concentrates on securing the inmate's right leg. He applies leg irons and when the inmate's legs are secure, he announces, "*Legs secure*," (see *Figure 7-8*).

Team Member Five (Team Leader)

7-82. Team member five keeps contact with team member four until he secures a portion of the inmate's body. Team member five concentrates on securing the inmate's left leg and assists with the application of leg irons. When he is informed that the hands and legs are secure, the team searches the inmate for weapons and contraband. Once the search is complete, team member five commands, "Prepare to Lift."

Figure 7-8. FCMT Securing an Inmate

Movement of the Inmate

7-83. Team member one positions his hands to control the inmate's head during the move, turning the inmate's head toward the wall. Team members two and three position themselves on the inmate's right and left sides. They grasp the wrist or forearm of the team member opposite them to form a platform for the inmate to lie on. Team member four positions himself on the inmate's right side even with the inmate's knees. This places him in position to control the inmate's legs. Team member five's position is on the inmate's left side and enables him to control the inmate's lower legs and feet. When in position, team member five ensures that all team members are in place before commanding, "Ready, Lift."

7-84. On the command "Lift," all team members simultaneously stand with the inmate. They should sandwich the inmate between them to facilitate control. Team member five then commands, "Prepare to Turn, Ready—Turn." The inmate is always turned toward the bunk. If a bunk is not available, team member five should specify which direction to turn the inmate. The inmate is then removed from the cell (see *Figure 7-9*, page 7-20).

7-85. Once the inmate is on the tier, team member five commands, "Prepare to Lower, Team—Lower." The team simultaneously lowers the inmate to the floor with the team keeping positive control of the inmate. The medic asks the inmate, "*Do you have any injuries at this time?*" (see *Figure 7-10,* page 7-20). If the inmate has no injuries, the medic looks at the camera and states, "*Inmate (inmate name) appears to have sustained no injuries.*" If the inmate has sustained injuries, the medic treats those injuries and completes a sworn statement. If OC was employed, the medic decontaminates the inmate using water and paper towels. The medic instructs the inmate to tilt his head to the side and flush his eyes with water for approximately 5 minutes. The medic then dries the inmate's face.

Figure 7-9. Moving the Inmate

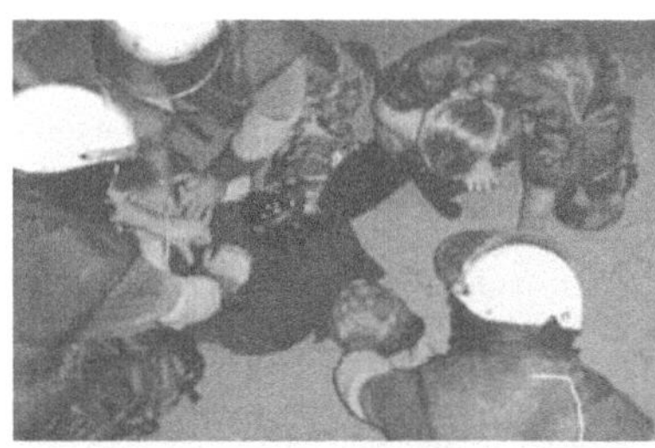

Figure 7-10. Treating Injuries

Placement of the Inmate in a Cell

7-86. Team member five orders an equipment check before returning the inmate to a cell. He then commands, "Prepare to Lift, Team—Lift." Team members position themselves exactly where they were when they removed the inmate from the cell. Next, team members simultaneously lift the inmate and return him to the predetermined cell. They move the inmate into the cell head first with his face toward the bunk. Once inside the cell, team member five commands the team to lower the inmate onto the bunk (see *Figure 7-11,* page 7-20). Team members then remove the inmate's restraints, keeping positive control of the inmate at all times. Team members use a nonverbal signal to notify team member five when the restraints are removed. He signals another equipment check. Team members then account for all equipment in place and prepare to exit the cell. The team exits the cell in the same manner as they entered, but in reverse order.

Figure 7-11. Placing Inmate in Cell

7-87. Team member four moves into position where he can control both of the inmate's legs while team member five moves toward the cell door. When team member four is ready to exit the cell, team member three moves to control the inmate's legs, team member two moves to where he can control both of the inmates arms, and team member one maintains control of the inmate's head. Team member four maintains contact with team member three at all times. When team member three is ready to exit the cell, team member two moves into a position where he can control the inmate's lower body and team member one moves into position to control the inmate's upper body. Using a prearranged, nonverbal signal, the team pulls team members one and two from the cell. The FCMT is prepared to reenter the cell if the inmate attempts to attack team members one and two as they exit the cell. The FCMT remains ready to reenter until the cell door is secured.

Chapter 8

Civil Disturbance Training

Training requires planning and preparation. Unit commanders structure their training program according to the mission essential task list (METL) (see *FM 7-1)* and expected contingencies. These training plans include prior training and the experience and proficiency of their unit. Leaders agree that the operational tempo and expected contingencies for the next 10 to 15 years will likely involve low-intensity military operations other than war (MOOTW). Therefore, civil disturbance training must be included in the commander's annual training program and must cover all aspects of that type of operation. To effectively function as a member of a control force, individual soldiers must be trained in civil disturbance operations. Training must be intensive, realistic, and sustained. Any lack of training (from basic soldier skills to leadership skills) can have an adverse effect to the point of endangering lives, especially the lives of soldiers. Leader involvement in the training process is essential for mission success in crowd control operations.

UNIT TRAINING STRATEGY DEVELOPMENT

8-1. Developing a unit training strategy begins with the mission analysis portion of the MDMP and identifying the GRM. Missions that require soldiers and units to enforce treaties or accords, protect the lives of civilians in uncertain or hostile environments (such as NEOs), or provide large-scale humanitarian assistance demand some sort of graduated-response criteria to maintain order and prevent uncertain environments from becoming hostile.

8-2. When developing training strategies, planners should focus training and evaluation on the echelon level above the target element (for example TF) or the unit identified as needing training.

8-3. Planners must agree on the intent of the GRM. The GRM is for most a training and rehearsal tool. It provides leaders with likely vignettes that can be incorporated into COA analysis, predeployment training, and rehearsals. The GRM becomes a handy reference during situations that require graduated responses.

TRAINING PLAN DEVELOPMENT

8-4. In developing a training plan for civil disturbances, many considerations should be addressed in terms of available resources. Considerations include the number of personnel to train, the equipment needed and its availability, the time available to train to standard, the location of where training will be executed, and the incorporation of SOPs (based on the theater of operations

[CONUS or OCONUS]). A personnel assessment is necessary to determine the training and experience level of soldiers. It is critical to identify and use soldiers with previous real-world experience and leaders that have graduated from the Interservice Nonlethal Individual Weapons Instructor Course (INIWIC) as primary instructors.

8-5. Commanders should structure the unit training program based on the METL, expected contingencies, prior training, and the proficiency of the unit. In developing a training program, use the seven-step approach. For in-depth instructions on training development, see *FMs 7-0* and *7-1*.

8-6. The objective is to train and evaluate units in the performance of civil disturbance procedures. Proficiency in performing civil disturbance procedures is reflected by a unit's ability to quickly and deliberately execute the disturbance control procedures required to prevent loss of life and property or to regain control of a civil disturbance in its assigned area of responsibility.

8-7. Training should focus on, but is not limited to—

- Instructors.
- Individual and collective drills.
- Leaders.
- Equipment.
- Weapons.
- Munitions.
- Live-fire range exercises.
- Qualification and familiarization.
- Soldier training exercises (STXs).
- Culminating exercises.

8-8. There are several successful methods to evaluate trained leader skills and trained soldier skills using scenarios, vignettes, lanes, or STXs. Flexibility is an important key in training plan development due to time restraints.

RANGE SETUP

8-9. The key to setting up a successful range for NL munitions is to ensure that they are in compliance with installation safety regulations. Kinetic energy is the force that pushes the projectiles down range. In most cases, kinetic energy will only carry these NL munitions projectiles 75 meters or less. This gives trainers flexibility as to the locations where a live-fire exercise can be conducted, especially in a field-expedient situation. *Figure 8-1* is an example for conducting a NL live fire range setup.

8-10. As illustrated in *Figure 8-1*, mark off 5-, 10-, 20-, 30-, 40-, and 50-meter distances from the targets with markers. Each of the NL munitions has minimum and maximum effective ranges within the distances of 5 to 50 meters. Impact power and accuracy are greater at close distances and depleted at farther distances. Targets can be anything from the standard E-type to more realistic three-dimensional manikins. Evenly divide your shooters into three groups as shown in *Figure 8-1*. Group one will initially be shield holders, group two will be shooters, and group three will be grenadiers.

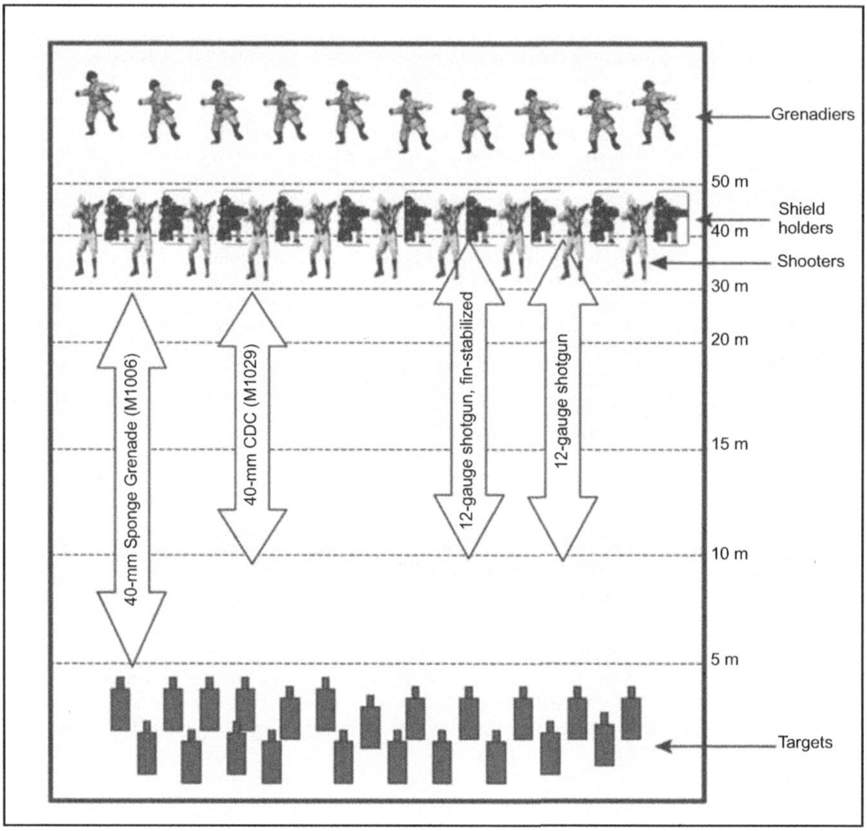

Figure 8-1. NL Munitions Range Setup

8-11. The formation should begin at the farthest effective range firing line (50 meters). After each volley of fire, the formation moves closer to the targets, but stays outside the closest effective range firing line (5 meters). For safety reasons, ensure that the formation stays online throughout each iteration.

8-12. When moving forward to each firing point, it is recommended that the formation always come to a halt before the fire command is given. It is also important to use the tap down drill method (see *Figure 8-1*) rather than having the shooters move forward and fire between the shield holders.

8-13. Range instructors can use verbal commands, such as, "Ready," or "Fire," for each volley. Shooters can be briefed to move forward, tap down the shield holder, and engage targets at the first explosion of distraction grenades thrown by grenadiers. In that instance, the command could simply be "Grenadiers Prepare to Throw, Throw."

8-14. The success of any civil disturbance operation is through effective communication, especially when incorporating the use of NL munitions. For effective training, use verbal commands in conjunction with effective, predetermined hand-and-arm signals during live-fire exercises.

8-15. Each iteration requires two phases of fire. During the first phase, half of the shooters will be firing NL, 40-millimeter rounds using an M203. The others will be firing NL, 12-gauge rounds using a 12-gauge shotgun. In the second phase, shooters switch weapons (safely). It is critical that line safeties are alert during this transition between phases.

8-16. Upon completion of each iteration, clear the weapons and the range. Establish a safe method and return all groups back to the farthest firing line. Once that is accomplished, group one rotates to become group three, group three rotates to become group two, and group two rotates to become group one for the next iteration. Using this method affords a safe and efficient way to get all shooters through the proficiency course. *Figure 8-2* is an example of how to fire each volley based on its characteristics, effectiveness, and capabilities.

Distance	Command	Subject	Actions
10 Meters	12-Gauge Shotgun, Two Point	Shotgun shooters	Load two M1013 rounds.
	M203, Two Point	M203 shooters	Load one M1006 round and have one readily available.
	Ready, Fire	Shotgun shooters	Tap down shields and engage designated targets.
		M203 shooters	Tap down shields, engage designated targets, reload, and reengage.
	Clear Range	Range safeties	Ensure that weapons are clear and safe.
	Safeties, Clear All Weapons	All shooters	Face to right, keep weapons pointed up and downrange.
	Shooters, Right Face, Move	All personnel	Step off shield holder, then shooters, and grenadiers on the command, "Move." Grenadiers reform at 50 meters.

Sample

Figure 8-2. Sample Live-Fire Exercise

Distance	Command	Subject	Actions
15 Meters	Shotgun, Two Point	Shotgun shooters	Load two M1013 rounds.
	M203, Two Point	M203 shooters	Load one M1006 round and have another available.
	Grenadiers, Prepare to Throw	Grenadiers	Make distraction grenade ready.
	Throw	Grenadiers	Throw grenade.
		Grenadiers	Throw distraction grenade.
	Ready, Fire	Shot gun shooters	Tap down shields and engage designated targets.
		M203 shooters	Tap down shields, fire at designated targets, reload with M1006 rounds, and engage designated target.
	Clear Range	Range safeties	Ensure that weapons are clear and safe.
	Three Steps Forward	All shooters and grenadiers	Echo preparatory command.
	Move	All shooters and grenadiers	Stay on line and step off three steps, counting each step aloud.
	Clear Range	Range safeties	Ensure that weapons are clear and safe.
	Seven Steps Forward	All shooters and grenadiers	Echo preparatory command.
	Move	All shooters and grenadiers	Stay on line and step off seven steps, counting each step aloud.

Sample

Figure 8-2. Sample Live-Fire Exercise (Continued)

Distance	Command	Subject	Actions
20 Meters	Shotgun, Three Point	Shotgun shooters	Load three M1013 rounds.
	M203, One Area, Two Point	M203 shooters	Load one M1026 round and have two M1006 rounds available.
	Grenadiers, Prepare to Throw	Grenadiers	Make distraction grenade ready.
	Throw	Grenadiers	Throw grenade.
		Grenadiers	Throw distraction grenade.
	Ready, Fire	Shotgun shooters	Tap down shields and engage designated targets.
		M203 shooters	Tap down shields, fire area at targets, reload with M1006 rounds, and engage designated target.
	Clear Range	Range safeties	Ensure that weapons are clear and safe.
	Shotgun, Two Point	Shotgun shooters	Load two M1013 rounds.
	M203, Two Point	M203 shooters	Load one M1006 round and have another available.
	Grenadiers, Prepare to Throw	Grenadiers	Make distraction grenade ready.
	Clear Range	Range safeties	Ensure that weapons are clear and safe.
	Three Steps Forward	All shooters and grenadiers	Echo preparatory command.
	Move	All shooters and grenadiers	Stay on line and step off three steps, counting each step aloud.

Sample

Figure 8-2. Sample Live-Fire Exercise (Continued)

Distance	Command	Subject	Actions
30 Meters	Shotgun, Two Area, Two Point	Shotgun shooters	Load two M1012 rounds and two M1013 rounds.
	M203, Two Area, Two Point	M203 shooters	Load one round M1026 and have one M1026 and two rounds M1006 available.
	Grenadiers, Prepare to Throw	Grenadiers	Make distraction grenade ready.
	Throw	Grenadiers	Throw grenade.
		Grenadiers	Throw distraction grenade.
	Ready, Fire	Shotgun shooters	Tap down shields, engage targets, and engage designated target with M1013 rounds.
		M203 shooters	Tap down shields, fire area at target, reload with M1006 rounds, and engage designated target.
	Clear Range	Range safeties	Ensure that weapons are clear and safe.
	Seven Steps Forward	All shooters and grenadiers	Echo preparatory command.
	Move	All shooters and grenadiers	Stay on line and step off seven steps, counting each step aloud.

Sample

Figure 8-2. Sample Live-Fire Exercise (Continued)

Distance	Command	Subject	Actions
40 Meters	Shotgun, Two Area, One Point	Shotgun shooters	Load two M1012 rounds and one round M1013.
	M203, One Area, One Point	M203 shooters	Load one M1026 round and have one M1006 round available.
	Grenadiers, Prepare to Throw	Grenadiers	Make distraction grenade ready.
	Throw	Grenadiers	Throw grenade.
		Grenadiers	Throw distraction grenade.
	Ready, Fire	Shotgun shooters	Tap down shields and engage targets with M1013 rounds.
		M203 shooters	Tap down shields, fire area at target, reload with M1006 rounds, and engage designated target.
	Clear Range	Range safeties	Ensure that weapons are clear and safe.
	Seven Steps Forward	All shooters and grenadiers	Echo preparatory command.
	Move	All shooters and grenadiers	Stay on line and step off seven steps, counting each step aloud.
50 Meters	Shotgun, One Area	Shotgun shooters	Load one M1012 round.
	M203, One Area, One Point	M203 shooters	Load one M1026 round and have one M1006 round available.
	Grenadiers, Prepare to Throw	Grenadiers	Make distraction grenade ready.
	Throw	Grenadiers	Throw grenade
		Grenadiers	Throw distraction grenade.
	Ready, Fire	Shotgun shooters	Tap down shields, engage targets, and engage designated target with M1013 rounds.
		M203 shooters	Tap down shields, fire area at the target, reload with M1006 rounds, and engage the designated target.
	Clear Range	Range safeties	Ensure that weapons are clear and safe.
	Seven Steps Forward	All shooters and grenadiers	Echo preparatory command.
	Move	All shooters and grenadiers	Stay on line and step off seven steps, counting each step aloud.

Sample

Figure 8-2. Sample Live-Fire Exercise (Continued)

NOTE: Shield holders are referred to as group one, shooters are referred to as group two, grenadiers are referred to as group three.

8-17. Currently, there are no qualification standards established in the Army for NL munitions. Area rounds, which contain multiple projectiles designed to affect multiple targets when engaged, only require approximately five rounds. This affords shooters an opportunity to realize the characteristics and capabilities of these rounds.

8-18. Single projectile rounds like the M1013 and the M1006 require a greater degree of proficiency. If situations call for the engagement of targets with these rounds, it is essential that trained shooters be able to hit what they are aiming at. For example, firing 10 rounds from various ranges can provide soldiers with the confidence to be effective shooters. A standard can be established that requires shooters to hit targets with at least 6 out of the 10 rounds used in the example to be considered proficient.

8-19. *Figures 8-3; 8-4,* page 8-11; *8-5,* page 8-21; and *8-6,* page 8-22, give examples of plans to follow if in a situation similar to the following scenario (see *Army Training and Evaluation Program (ARTEP) 19-313-10-MTP*):

A company commander is providing security patrols in support of a joint task force (JTF). Besides providing security patrols for a sector of responsibility, one platoon must always be designated as a quick reaction force (QRF) for civil disturbance operations. For the last 2 weeks the sector has been quiet and the commander has ordered all platoons to conduct civil disturbance training on a rotational basis.

A new order has been sent down to provide a squad-sized element to provide security for the future polling site for an upcoming election. This polling site has yet to open, but intelligence indicates potential problems as the election approaches.

On the day prior to the scheduled election, a crowd of approximately 25 to 30 demonstrators aggressively move toward the polling site, chanting slogans and yelling toward the security forces posted there. While some demonstrators begin throwing whatever they can pick up, the majority are simply yelling, locking arms, and blocking roadways. Hundreds of additional people, not part of the crowd, begin to gather to watch the activities as they unfold.

In an assessment of the demonstrators, it is determined that they intend to stay through the night into the morning of the election with the intention to block access to the polls. It is essential that the crowd is dispersed and that access to the polls is unrestricted in time for the election to begin the next morning.

Evaluation Scheme

The evaluation will consist of five lanes conducted over a 3-day period. Platoons will execute 24-hour operations during the evaluation. Any retraining will be done immediately following the execution of the lane if time permits.

The following tasks will be evaluated:

Lane 1	Cordon and search operations
Lane 2	Civil disturbance QRF
Lane 3	Security and defense of a critical site (polling site)
Lane 4	Night civil disturbance QRF
Lane 5	Night LP/OP (PIO)

Evaluation Schedule

Date and Time	1st Platoon	2d Platoon	3d Platoon	4th Platoon
	Lane 1	Lane 2	Lane 3	Lane 4
	Lane 2	Lane 3	Lane 4	Lane 5
	Lane 3	Lane 4	Lane 5	Lane 1
	Lane 4	Lane 5	Lane 1	Lane 2
	Lane 5	Lane 1	Lane 2	Lane 3

Sample

Figure 8-3. Evaluation Scheme

Unit Training Schedule and Firing Tables

Unit Training Schedule 1

Training Day 1

Time	Subject	Location	Uniform	Instructor	Notes
0800	Administrative time	Classroom	U1		
0830	GRM	Classroom	U1		
1000	ROE and use of force	Classroom	U1		
1200	Chow				
1300	MOOTW	Classroom	U1		
1500	Effective communication	Classroom	U1		

Administration:

Uniforms:

U1: BDU

U2: BDU and LBE with harness and canteens

U3: BDU and LBE with harness, canteens, flak jacket, and Kevlar helmet

U4: BDU and LBE with harness, canteens, flak jacket, Kevlar helmet, and assigned weapon

Sample

Transportation:

Weapons:

Ammunition:

Required Equipment for Training:

1. Projection screen
2. Video projector with CPU
3. Television
4. VCR

Miscellaneous:

1. All personnel are required to have a notebook and a pencil or pen.
2. Handouts must be prepared and passed out.

Figure 8-4. Sample Two-Week Plan

Training Day 2

Time	Subject	Location	Uniform	Instructor	Notes
0800	NLCS	Classroom	U1		
0900	PPCT and empty hand	Classroom	U1		
1000	PPCT and empty hand (Practical application)	PT field	U2		
1200	Chow				
1300	PPCT and empty hand (Practical application)	PT field	U2		
1700	COB				

Administration:

Uniforms:

U1: BDU

U2: BDU and LBE with harness and canteens

U3: BDU and LBE with harness, canteens, flak jacket, and Kevlar helmet

U4: BDU and LBE with harness, canteens, flak jacket, Kevlar helmet, and assigned weapon

Sample

Transportation:

Weapons:

Ammunition:

Required Equipment for Training:

1. Projection screen
2. Video projector with CPU
3. Training strike bags
4. Flex-Cufs

Miscellaneous:

1. All personnel are required to have a notebook and a pencil or pen.
2. Handouts must be prepared and passed out.
3. Medical personnel must be on location.

Figure 8-4. Sample Two-Week Plan (Continued)

Training Day 3

Time	Subject	Location	Uniform	Instructor	Notes
0800	PPCT and empty hand (Practical application)	PT field	U2		
1200	Chow				
1300	PPCT and empty hand (Practical application)	PT field	U2		
1700	COB				

Administration:

Uniforms:

U1: BDU

U2: BDU and LBE with harness and canteens

U3: BDU and LBE with harness, canteens, flak jacket, and Kevlar helmet

U4: BDU and LBE with harness, canteens, flak jacket, Kevlar helmet, and assigned weapon

Transportation:

Weapons:

Ammunition:

Required Equipment for Training:

1. Projection screen
2. Video projector with CPU
3. Training strike bags
4. Flex-Cufs

Miscellaneous:

1. All personnel are required to have a notebook and a pencil or pen.
2. Handouts must be prepared and passed out.
3. Medical personnel must be on location.

Sample

Training Day 4

Time	Subject	Location	Uniform	Instructor	Notes
0800	Riot baton training	Classroom	U1		
0900	Riot baton training (Practical application)	PT field	U2		
1200	Chow				

Figure 8-4. Sample Two-Week Plan (Continued)

1300	Riot baton training (Practical application)	PT field	U2		
1700	COB				

Administration:

Uniforms:

U1: BDU

U2: BDU and LBE with harness and canteens

U3: BDU and LBE with harness, canteens, flak jacket, and Kevlar helmet

U4: BDU and LBE with harness, canteens, flak jacket, Kevlar helmet, and assigned weapon

Transportation:

Weapons:

1. Expandable riot baton with carrier
2. Training riot batons

Ammunition:

Required Equipment for Training:

1. Projection screen
2. Video projector with CPU
3. Training strike bags
4. Flex-Cufs

Sample

Miscellaneous:

1. All personnel are required to have a notebook and a pencil or pen.
2. Handouts must be prepared and passed out.
3. Medical personnel must be on location.

Training Day 5

Time	Subject	Location	Uniform	Instructor	Notes
0800	Riot baton training (Practical application)	PT field	U2		
1000	Crowd dynamics	Classroom	U1		
1130	Chow				
1230	Crowd dynamics	Classroom	U1		
1330	CD formations	Classroom	U1		

Figure 8-4. Sample Two-Week Plan (Continued)

1430	RCAs and OC	Classroom	U1
1700	COB		

Administration:

Uniforms:

U1: BDU

U2: BDU and LBE with harness and canteens

U3: BDU and LBE with harness, canteens, flak jacket, and Kevlar helmet

U4: BDU and LBE with harness, canteens, flak jacket, Kevlar helmet, and assigned weapon

Transportation:

Weapons:

1. Expandable riot baton with carrier
2. Training riot baton

Ammunition:

Required Equipment for Training:

1. Projection screen
2. Video projector with CPU
3. Training strike bags
4. Flex-Cufs

Sample

Miscellaneous:

1. All personnel are required to have a notebook and a pencil or pen.
2. Handouts must be prepared and passed out.
3. Medical personnel must be on location.

Training Day 6

Time	Subject	Location	Uniform	Instructor	Notes
0800	RCAs and OC (Practical application)	PT field	U2		
1200	Chow				
1300	RCAs and OC (Practical application)	PT field	U2		
1700	COB				

Figure 8-4. Sample Two-Week Plan (Continued)

Administration:

Uniforms:

U1: BDU

U2: BDU and LBE with harness and canteens

U3: BDU and LBE with harness, canteens, flak jacket, and Kevlar helmet

U4: BDU and LBE with harness, canteens, flak jacket, Kevlar helmet, and assigned weapon

Transportation:

Weapons:

Ammunition:

1. MK inert OC
2. MK 4 live OC disperser
3. M36 live CR disperser

Required Equipment for Training:

1. Five large clean trash cans with liners
2. Two water hoses with running water

Sample

Miscellaneous:

Medical personnel must be on location.

Training Day 7

Time	Subject	Location	Uniform	Instructor	Notes
0800	NL munitions employment	Classroom	U1		
1030	Chow				
1115	Safety brief	Range	U4		
1130	NL munitions (live fire)	Range	U4		
TBD	Evening chow				
TBD	Safety brief	Range	U4		
TBD	NL munitions night fire exercise	Range	U4		
TBD	Clean up and COB				

Figure 8-4. Sample Two-Week Plan (Continued)

Administration:

Uniforms:

U1: BDU

U2: BDU and LBE with harness and canteens

U3: BDU and LBE with harness, canteens, flak jacket, and Kevlar helmet

U4: BDU and LBE with harness, canteens, flak jacket, Kevlar helmet, and assigned weapon

Transportation:

Yes

Weapons:

1. M16A2 with M203
2. 12-gauge shotgun

Ammunition:

1. Fin-stabilized M1012
2. M1013 multipellet (rubber)
3. M1006 40-millimeter sponge grenade
4. M1026 40-millimeter multiball (rubber)
5. M84 diversionary grenade

Sample

Required Equipment for Training:

1. Fifteen to twenty human form targets
2. Eye and hearing protection
3. Fifteen to twenty riot shields
4. Two MREs per student
5. Video recorder
6. Three to five high-intensity lights
7. All students should be issued face shields, shin guards, and ammunition pouches

Miscellaneous:

1. Medical personnel must be on location.
2. Range safety officers must be available.
3. Primary and secondary means of communication with range control must be available.
4. Operational risk assessments must be conducted.

Figure 8-4. Sample Two-Week Plan (Continued)

Training Day 8

Time	Subject	Location	Uniform	Instructor	Notes
0800	CD formations (Practical application)	TA 1	U4		
1200	Chow				
1300	CD formations (Practical application)	TA 1	U4		
1700	COB				

Administration:

Uniforms:

U1: BDU

U2: BDU and LBE with harness and canteens

U3: BDU and LBE with harness, canteens, flak jacket, and Kevlar helmet

U4: BDU and LBE with harness, canteens, flak jacket, Kevlar helmet, and assigned weapon

Transportation:

Yes

Weapons:

1. Training weapons
2. Training riot batons

Ammunition:

Required Equipment for Training:

1. Eye protection
2. Fifteen to twenty riot shields
3. Video recorder
4. Three to five high-intensity lights
5. Face shields, shin guards, and ammunition pouches

Miscellaneous:

1. Medical personnel must be on location.
2. Primary and secondary means of communication must be available.
3. Operational risk assessment must be conducted.
4. Twenty to thirty OPFOR personnel must be available.

Sample

Figure 8-4. Sample Two-Week Plan (Continued)

Training Day 9

Time	Subject	Location	Uniform	Instructor	Notes
0800	Introduction to M5, PVAB, and M315	Classroom 1	U1		
1200	Chow				
1300	M5, PVAB, and M315 (Practical application/LFE)	Range	U4		
1700	COB				

Administration:

Uniforms:

U1: BDU

U2: BDU and LBE with harness and canteens

U3: BDU and LBE with harness, canteens, flak jacket, and Kevlar helmet

U4: BDU and LBE with harness, canteens, flak jacket, Kevlar helmet, and assigned weapon

Transportation:

Yes

Weapons:

Two complete M315s and installation kits

Ammunition:

1. Ten live MCCMs
2. Ten live M315 rounds of each

Required Equipment for Training:

1. Eye protection
2. Ten inert MCCMs
3. Two complete PVAB systems

Miscellaneous:

1. Medical personnel must be on location.
2. Primary and secondary means of communication must be available.
3. Operational risk assessment must be conducted.

Sample

Figure 8-4. Sample Two-Week Plan (Continued)

Training Day 10					
Time	**Subject**	**Location**	**Uniform**	**Instructor**	**Notes**
0800	Tactical consideration	MOUT classroom	U1		
0930	Safety brief	MOUT classroom	U4		
1000	Tactical exercise	MOUT Facility	U4		

Administration:

Uniforms:

U1: BDU

U2: BDU and LBE with harness and canteens

U3: BDU and LBE with harness, canteens, flak jacket, and Kevlar helmet

U4: BDU and LBE with harness, canteens, flak jacket, Kevlar helmet, and assigned weapon

Transportation:

Yes

Weapons:

1. Assigned weapons
2. Inert RCA/OC individual dispersers
3. Training riot batons
4. Fifty practice grenades

Sample

Ammunition:

Required Equipment for Training:

1. Eye protection
2. Ten inert MCCMs
3. Face shields, shin guards, and ammunition pouches
4. Fifteen to twenty full-length riot shields

Miscellaneous:

1. Medical personnel must be on location.
2. Primary and secondary means of communication must be available.
3. Operational risk assessment must be conducted.
4. A video recorder must be available.

Figure 8-4. Sample Two-Week Plan (Continued)

Sample Unit Training Schedule				
Training Day 1				
Time	**Event**	**Location**	**Instructor**	**Notes**
0730-0830	Force continuum	Classroom		L
0830-0930	Management of aggressive behavior	Classroom		L
0930-1030	ROE	Classroom		L
1030-1130	Crowd control	Classroom		L
1130-1230	Chow			
1230-1700	PPCT and empty hand	PT field		L, PA
Training Day 2				
Time	**Event**	**Location**	**Instructor**	**Notes**
0730-1130	Riot baton training	PT field		L, PA
1130-1230	Chow			
1230-1530	Riot baton training	PT field		PA
1530-1700	CD formations	Classroom		L
Training Day 3				
Time	**Event**	**Location**	**Instructor**	**Notes**
0730-1130	CD formations	PT field		PA
1130-1230	Chow			
1230-1700	NL munitions, NLW	Range		L, PA
Training Day 4				
Time	**Event**	**Location**	**Instructor**	**Notes**
0730-1130	RCAs and OC	Classroom and PT field		L, PA
1130-1230	Chow			
1230-1700	RCAs and OC	PT field		PA

Sample

Figure 8-5. Sample One-Week Plan

Training Day 5				
Time	**Event**	**Location**	**Instructor**	**Notes**
0730-1700	Tactical exercise	MOUT facility		OPS
NOTE: Appointed student leaders ensure that troops are fed and rotated for rest.				

Figure 8-5. Sample One-Week Plan (Continued)

Sample Unit Training Schedule				
Training Day 1				
Time	**Event**	**Location**	**Instructor**	**Notes**
	Shield and riot baton techniques	MOUT site		Individual and collective tasks
	NLCS and munitions	Range Classroom		Leader training
	Formation training	MOUT site		Individual and collective tasks
	Chow			
	Formation training	MOUT site		Individual and collective tasks
Training Day 2				
Time	**Event**	**Location**	**Instructor**	**Notes**
0730-1930, 12-hour operations	Live-fire exercise	Range		M203/12-gauge
	QRF exercise	MOUT site		Individual and collective tasks
NOTE: Once weapons and munitions familiarization is complete, shooters join the QRF exercise to be used. For live-fire range setup, see *Figure 8-1*.				

Sample

Figure 8-6. Sample Two-Day Plan

Appendix A

Metric Conversion Chart

This appendix complies with current Army directives which state that the metric system will be incorporated into all new publications. *Table A-1* is a metric conversion chart.

Table A-1. Metric Conversion Chart

US Units	Multiplied By	Metric Units
Cubic feet	0.0283	Cubic meters
Feet	0.3048	Meters
Feet per second	0.3048	Meters per second
Inches	0.0254	Meters
Inches	25.4001	Millimeters
Miles per hour	1.6093	Kilometers per hour
Pounds	0.4536	Kilograms
Yards	0.9144	Meters
Metric Units	**Multiplied By**	**US Units**
Cubic meters	35.3146	Cubic feet
Kilograms	2.2046	Pounds
Kilometers per hour	0.6214	Miles per hour
Meters	3.2808	Feet
Meters	39.3701	Inches
Meters	1.0936	Yards
Meters per second	3.2808	Feet per second
Millimeters	0.0394	Inches

Appendix B

Operations and Legal Considerations in the Continental United States

Providing military assistance to state and local governments to assist them in quelling a civil disturbance or riot requires close coordination through a host of state and federal agencies. It requires a thorough briefing of soldiers at all levels on what they can and cannot do with respect to law enforcement. Civil authorities must be briefed on the restrictions placed on federal forces by the *Constitution of the United States* and federal statutes and laws. The national guard (NG), when not in a federal status, operates under the control of the state governor and the adjutant general (AG). The NG has historically been the first military responder during emergencies.

FEDERAL INTERVENTION AND AID

B-1. Under the *Constitution of the United States* and *United States Code*, the President is empowered to direct federal intervention in civil disturbances to—

- Respond to state requests for aid in restoring order.
- Enforce the laws of the United States.
- Protect the civil rights of citizens.
- Protect federal property and functions.

B-2. Under the *Constitution of the United States*, each state is responsible for protecting life and property within its boundaries. State and local governments use their LEA forces to maintain law and order and to quell civil disturbances. However, if civil disturbance support requirements exceed the resources of a state, federal troops may be requested to help restore and maintain law and order.

B-3. The *Constitution of the United States* and federal statutes authorize the President to direct the use of federal armed troops within the 50 states and territories and their political subdivisions. The President is also empowered to federalize the NG of any state to suppress rebellion and enforce federal laws.

B-4. Federal assistance is provided to a state when the state has used all of its resources, including its NG, to quell a disorder and finds the resources insufficient. Usually, active duty federal forces are used to augment the NG of the requesting state. However, the President may choose to federalize the NG of another state and use them alone or with other forces to restore order.

B-5. The President may employ federal troops to ensure the execution of US law when a state opposes or obstructs US law or impedes the course of justice under those laws. The President may employ armed federal troops to suppress insurrection, domestic violence, unlawful assemblies, and conspiracy if such acts deprive the people of their constitutional rights and privileges and the civil authorities of a state cannot or will not provide adequate protection.

B-6. The president is also authorized to use armed federal troops to protect federal property and functions when the need for protection exists and local civil authorities cannot or will not give adequate protection. The right of the United States to protect all federal property and functions regardless of their locations is an accepted principle of our government.

B-7. As a temporary measure, federal military equipment and facilities may be loaned to law enforcement agencies of state and local governmental bodies for use during civil disturbances. These resources may also be loaned to state NG and non-DOD federal agencies. The requesting agencies are expected to provide enough resources of their own to minimize the need for US military resources. In addition, the loan of the resources must not conflict with US military needs.

ROLES AND RESPONSIBILITIES OF VARIOUS AGENCIES

B-8. Several agencies within the federal and state governments play a key role in supporting the request of local authorities for support during a civil disturbance.

SECRETARY OF DEFENSE

B-9. The Secretary of Defense retains approval for federal support to civil authorities involving the use of DOD forces, personnel, and equipment. Approval may also involve DOD support that will result in a planned event with the potential for confrontation with specifically identified individuals and groups or will result in the use of lethal force.

SECRETARY OF THE ARMY

B-10. The Secretary of the Army, as an executive agent for the Secretary of Defense, is the approval authority for federal emergency support in response to natural and man-made disasters (except weapons of mass destruction [WMD]). Military forces respond to directions and guidance from the director of military support (DOMS) (the action agent for the Secretary of the Army).

DIRECTOR OF MILITARY SUPPORT

B-11. The DOMS plans and executes the DOD domestic support mission to civil authorities. He is a general officer appointed by the Secretary of the Army. The DOMS is the DOD primary contact for all federal departments and agencies during DOD involvement in most domestic support operations.

STATE COORDINATING OFFICER

B-12. The governor appoints the state coordinating officer (SCO) to coordinate disaster operations for the state. The SCO is the primary point of contact for the federal coordinating officer (FCO) in facilitating disaster assistance.

DIRECTOR OF THE FEDERAL EMERGENCY MANAGEMENT AGENCY

B-13. The director of the Federal Emergency Management Agency (FEMA), on behalf of the President, appoints the FCO. The FCO coordinates the timely delivery of federal disaster assistance to the affected state and local governments and disaster victims. He works closely with the SCO to determine state requirements and to coordinate national level issues with the catastrophic disaster response group. There must be close and continuous coordination with FEMA to ensure the proper use of military assets in support of civil authorities.

DEFENSE COORDINATING OFFICER

B-14. The defense coordinating officer (DCO) operates under the authority of the President and is the DOD representative designated to coordinate on-the-scene activities with the FCO. FEMA and other federal agencies that request support from DOD go through the DCO for validation and resourcing.

NATIONAL GUARD

B-15. The NG, in state status (see *Title 32, United States Code {USC}),* is the governor's primary military response organization for emergencies and disasters. The NG responds and operates under state laws unless they are federalized, when they operate under the same constraints and laws as federal units.

UNITED STATES ARMY RESERVE

B-16. The US Army Reserve is capable of providing military support. This assistance and support may include the use of equipment and other resources, including units and individuals.

STATE AND LOCAL GOVERNMENT RESPONSIBILITIES

B-17. State and local government officials, operating under the authority granted by state constitutions and the *Constitution of the United States,* are responsible for daily safety and security issues that impact the quality of life of their citizens. State and local officials have primary responsibility for emergency preparedness.

STATE RESPONSIBILITIES

B-18. Many state government agencies are responsible for coming to the aid of state citizens.

Governor

B-19. A state governor is empowered by each state constitution and the *Constitution of the United States* to execute the laws of the state and to command the state NG while it is serving in state status. Governors are also responsible for issuing executive orders or emergency proclamations, such as declaring states of emergency and ensuring that state agencies plan for potential disasters and civil disorder.

B-20. Once a man-made or natural disaster occurs, the governor assesses its extent and determines if local government requests for assistance should be honored. If appropriate, the governor declares a state of emergency, activates the state emergency response plan, and may order the NG into state active duty. The governor assigns the NG its mission through the office of emergency service (OES)/emergency medical dispatching (EMD).

Plans, Operations, and Military Support Officer

B-21. Each state NG has a plans, operations, and military support officer (POMSO) who prepares and coordinates contingency plans for military support to civil authorities (MSCA) during disaster response and recovery operations. The POMSO coordinates training plans and exercises between the state NG and federal, state, and local emergency management agencies. The POMSO serves as the NG point of contact with DOD officials during a declared federal emergency or disaster.

Office of Emergency Services

B-22. Each state has a specific agency and/or office that coordinates emergency preparedness and serves as the governor's coordinating agency during an emergency. The titles of these offices vary from state to state. Some examples are the Division of Emergency Government, Emergency Management Agency, Department of Public Safety, or Office of Emergency Preparedness. This manual refers to this office using the generic term OES/EMD.

B-23. Generally, the OES/EMD is organized as a stand-alone office under the governor or aligned under the AG or the state police, and in some instances, the AG is the OES/EMD director. It operates the state emergency operations center and coordinates with state agencies and federal officials for assistance.

Adjutant General

B-24. The state NG is the governor's primary military response force in an emergency. The AG, through the state area command (STARC) (specifically the POMSO), develops and coordinates emergency response plans for disasters and emergencies. The AG is in command of state NG forces ordered into state active duty.

State Government Agencies

B-25. State government departments and agencies prepare emergency preparedness plans for their areas of specialization in support of the governor's plan. They also participate in emergency preparedness exercises.

LOCAL RESPONSIBILITIES

B-26. Mayors, city managers, local police, fire protection officials, county executives, sheriffs, prosecuting attorneys, and public health officials are some of the people responsible for law enforcement, safety, health, and fire protection on a daily basis. They are responsible for developing appropriate emergency response plans and responding to emergencies within their jurisdictions. Most local jurisdictions have an OES/EMD to plan and coordinate actions in an emergency. In many cases, local jurisdictions have mutual aid agreements with other jurisdictions that allow for firefighting, medical, and police assistance across jurisdictional boundaries. Once local officials determine that an emergency is beyond the scope of their resources or ability to respond, the senior local official is responsible for requesting additional assistance from the next agency in the assistance request process.

LEGAL CONSIDERATIONS AND CONSTRAINTS

B-27. The *Constitution of the United States*, laws, regulations, policies, and other legal issues limit the use of federal military personnel in domestic support operations. Any Army involvement in civil disturbance operations involves many legal issues requiring comprehensive legal reviews.

B-28. Under the *Constitution of the United States*, Congress has the authority to raise and support an army, provide and maintain a navy, and make rules for governing and regulating the land and naval forces. The *Constitution of the United States* places the military under civilian control and designates the President as commander-in-chief. Statutes provide for civilian leadership in the office of the secretary of defense, service secretaries, and various other civil authorities. Under the *Constitution of the United States*, the civilian government must preserve public order and carry out governmental operations within its territorial limits. The *Constitution of the United States* allows the use of the federal military to execute or enforce the law when necessary to protect federal or civilian property and functions. However, significant restrictions exist on employing federal military forces within the United States.

B-29. The unique capabilities of the military enable it to support federal, state, or local civilian agencies. In most circumstances, the DOD is one of many federal agencies reacting to a domestic emergency or crisis, playing a subordinate, supporting role to a lead civilian agency.

LAW

B-30. In the United States, civilian agencies provide for the needs of citizens. Civilian, federal, state, and local government and law enforcement agencies execute US laws. Laws governing the use of the military in domestic operations are complex, subtle, and constantly changing. Commanders involved in civil disturbance operations should staff plans, policies, programs, exercises, funding, operations, constraints, and limitations with their local SJA to ensure conformity with legal requirements. They should scrutinize each request for aid (whether it is for equipment, individuals, or training) to ensure that it conforms to statutory requirements. The following references are a snapshot of the law on military support to civilian authorities:

- *Section 331, Chapter 15, Part 1, Subtitle A, Title 10 USC.*
- *Section 332, Chapter 15, Part 1, Subtitle A, Title 10 USC.*
- *Section 333, Chapter 15, Part 1, Subtitle A, Title 10 USC.*
- *Section 334, Chapter 15, Part 1, Subtitle A, Title 10 USC.*
- *Section 1385, Chapter 67, Part 1, Title 18 USC.*
- *DOD Directive 3025.12.*

- *DOD Directive 3025.15*
- *DOD Directive 5525.5*
- *DOD Civil Disturbance Plan 55-2*
- *NGR 500-1/ANGI 10-8101*
- *FM 3-19.15*

INFORMATION OPERATIONS

B-31. Information superiority helps forces anticipate problems and requirements. It allows commanders to control situations earlier and with less force, creating the conditions necessary to achieve the end state. Public affairs, PSYOP, and civil military operations are activities that will allow the commander to control situations earlier and with less force.

B-32. Information is available from a multitude of sources, the primary being open sources, law enforcement, and the military. A diversity of sources is the best approach because it prevents biased behavior.

Open sources include—

- Libraries.
- Newspapers and news periodicals.
- Radio and television.

Law enforcement sources include—

- Local law enforcement agencies.
- National law enforcement agencies.

Military sources include—

- DOD intelligence community (most restrictive source).
- Local MI field offices.

B-33. The DOD intelligence community operates under limitations imposed by regulations and executive orders. Attempts to skirt these restrictions may violate regulations or federal statutes.

RESTRICTIONS

B-34. Strict limits are placed on collecting information related to a civil disturbance in order to protect the civil rights of people and organizations not affiliated with the DOD. Civil disturbance plans and materials must not include lists of groups or people not affiliated with the DOD. Information on civilians and civilian organizations can be collected only with specific authorization from the Secretary or the Under Secretary of the Army. Conditions for collecting information include the existence of threats against Army personnel, functions, or property (see *ARs 380-10* and *380-13).* Civil disturbance information (available in public documents) or open-source information may be collected. However, specific rules regarding its storage must be followed. Commanders must coordinate with SJA, MI, and USACIDC before collecting any such information.

B-35. Civilian law enforcement agencies may limit access to law enforcement intelligence. In such circumstances, the military may request that agencies release the unclassified version.

B-36. If the Department of Justice (DOJ) determines that federal intervention into a civil disturbance is likely, then information relating to the disturbance is provided to the Army Assistant Chief of Staff for Intelligence. The information is analyzed and then passed through channels to the TF commander for planning purposes.

B-37. Military intelligence collection efforts, except liaison, may begin only when DA orders. The orders will come through the CSA's personal liaison officer and the TF commander. Covert operations to gather information on non-DOD individuals and groups must be approved on an operation-by-operation basis by the Undersecretary of the Army.

B-38. Once DA approves the collection effort, MI elements establish and maintain liaison with the appropriate local, state, and federal authorities. Then they collect information pertaining to the incident and general situation. Their estimate of the situation is passed to the TF commander and DA.

POSSE COMITATUS

B-39. Generally, federal military forces may not give law enforcement assistance to civil authorities without conflicting with the *Posse Comitatus Act*. However, constitutional and statutory exceptions to this prohibition do exist. The recent emphasis on drug interdiction has led to an increase in those exceptions.

B-40. The *Judiciary Act of 1789* allowed US marshals to call upon the military as a posse comitatus. This continued until after the Civil War, when the federal government used the Army to execute reconstruction era policies. The southern states regarded the use of the military for this purpose as abusive and repressive, and in 1878, President Rutherford B. Hayes signed the original bill ending the practice.

B-41. The *Posse Comitatus Act* prescribes criminal penalties for the use of the US Army or Air Force to execute the laws of or to perform civilian law enforcement functions within the United States. DOD policy extends this prohibition to the US Navy and Marine Corps. Prohibiting the military from executing the laws means that military personnel may not participate directly in—

- An arrest; a search and seizure; a stop and frisk; or an interdiction of vessels, aircraft, or vehicles.
- A surveillance or pursuit.
- A civilian legal case or any other civilian law enforcement activity as informants, undercover agents, or investigators.

B-42. The *Posse Comitatus Act* does not apply to the following:

- Members of the NG when in state service.
- Members of a Reserve Component when not on active duty or active duty for training.
- DOD civilians, unless under the direct command and control of an active duty officer.

- A soldier when off duty and acting only in a private capacity.
- Soldiers taking action for the primary purpose of furthering a military or foreign affairs function of the United States; for example, enforcing military justice, maintaining law and order on military installations, and protecting classified materials.

CONSTITUTIONAL EXCEPTIONS TO THE POSSE COMITATUS ACT

B-43. Under its inherent authority, the US government is responsible for preserving public order and carrying out governmental operations within its territorial limits, by force if necessary. Under the *Constitution of the United States*, exceptions allow the use of military force to execute or enforce the law. Some of the exceptions are to—

- Protect civilian property and functions.
- Ensure preservation of public order and carry out government operations, by force if necessary.
- Protect civil rights where local authorities do not or cannot protect them.
- Protect federal property and functions.
- Quell civil disturbances and labor strife that rises to the level of civil disorder.

B-44. The President may order the armed forces to support state civil authorities suffering from an insurrection or civil disturbance. He must act personally by first issuing a proclamation calling on insurgents to disperse and retire peacefully within a limited time. This occurred when federal forces were called in to counter the Los Angeles riots in 1992.

Protect Civilian Property and Functions

B-45. A sudden and unexpected civil disturbance, disaster, or calamity may seriously endanger life and property and disrupt normal governmental functions to such an extent that local authorities cannot control the situation. At such times, the federal government may use military force to prevent loss of life or wanton destruction of property and to restore government functions and public order. This exception has rarely been used.

Protect Federal Property and Functions

B-46. The federal government may use military force to protect federal property and federal government functions when local authorities cannot or decline to provide adequate protection.

ARMED FORCES

B-47. Basic guidance for the interaction of the military, the Reserve Component, and civilian LEAs is given in *10 USC Chapter 18*. Guidelines on reimbursement and restrictions, direct participation in law enforcement activities, the use of information collected during military operations, and the use of military equipment and facilities are some of the topics covered. Title 10 prohibits the military from directly participating in arrests, searches, seizures, or other similar activity unless authorized by law (such as arrests on military property). The *Fiscal Year 1989 Act* and subsequent *National Defense Authorization Acts* have authorized the DOD to provide more support to LEAs in the counterdrug effort.

NATIONAL GUARD

B-48. For a description of how the secretary of defense may provide funds to state governments (including the District of Columbia, Commonwealth of Puerto Rico, and US territories) for civil disturbance operations performed by the National Guard when not in federal service, see 32 *USC*. The *Posse Comitatus Act* does not apply to the NG when not in federal service. Each state specifies how NG may be employed.

SUPPORT TO CIVIL LAW ENFORCEMENT

B-49. It is DOD policy to cooperate with civilian law enforcement officials to the greatest practical extent. However, cooperation must consistently meet the needs of national security and military preparedness, the historic tradition of limiting direct military involvement in civilian LEAs, and the requirements of applicable law. US military forces are never placed under the command of civilian law enforcement officers or nonfederalized NG. DOD policy concerning the provision of military support to LEAs, including personnel and equipment, are contained in *DOD Directive 5525.5.*

B-50. The Army assists civilian law enforcement by providing personnel, equipment, training, and expert advice within the limits of applicable law. Army NG units in state status (see 32 *USC*) provide the primary source of military assistance. They may be able to provide assistance to civil authorities in instances when federal units are precluded due to the restrictions of the *Posse Comitatus Act.*

ROLES OF THE NATIONAL GUARD AND FEDERAL FORCES

B-51. The preservation of law and order in the civilian community is the responsibility of state and local governments and law enforcement authorities. When a civil disturbance in a civilian community turns to widespread rioting that includes arson, looting, and acts of violence, the civil authorities may decide that they do not have the resources to quell the riot. They may then turn to the NG and federal forces to support the civil authorities in restoring law and order. Civil disturbances in any form are prejudicial to public law and order.

NATIONAL GUARD FORCES

B-52. The NG (as a state organization) responds to the governor according to state law for civil disturbance operations. NG regulations direct planning and training for the civil disturbance mission. During most civil disturbance situations, the NG will be the first military responder and will usually remain in state active duty status (according to 32 *USC*) throughout the operation. In extreme circumstances, the NG can be brought on federal service for civil disturbance operations when ordered to under the appropriate federal statute by the President. During the 1992 Los Angeles civil disturbance, the President federalized the California NG. Once the NG has been deployed to assist the civil authorities with a civil disturbance, it is too far into the deployment cycle to start determining what tasks the NG can do as a state or as a federally controlled force. It is the commander's responsibility not to accept missions that the unit is not adequately trained to do. The following are examples of appropriate missions for the NG:

- Manning traffic control points.
- Providing building security.
- Providing area security and patrols.
- Providing security at custody facilities.
- Providing security and escort for emergency personnel and equipment.
- Protecting sensitive sites.
- Transporting law enforcement personnel.
- Providing show of force.
- Dispersing crowds.
- Employing riot control munitions.
- Providing very important person (VIP) protection and escort.
- Providing quick reaction and reserve force.

B-53. Examples of inappropriate taskings:

- Negotiating hostage situations.
- Barricading suspects.
- Participating in criminal investigations.
- Acting as federal Army forces.

B-54. The commitment of federal troops to deal with domestic civil disturbances must be viewed as a drastic measure of last resort. Their role should never be greater than what is absolutely necessary under the circumstances. Commanders should take every measure to avoid the perception of an invading force. A JTF designated to respond to a civil disturbance should project the image of a restrained and well-disciplined force whose sole purpose is to help restore law and order with minimal harm to the people and property and with due respect for all law-abiding citizens.

B-55. The role of federal Army forces is to assist civil authorities in restoring law and order when the magnitude of the disturbance exceeds the capabilities of local and state LEA, including the NG. Under the provisions of the *Constitution of the United States* and selected federal statutes, the President may order the employment of the federal armed forces to aid local and state civil authorities to protect the Constitutional rights of citizens. Federal military forces may also protect federal facilities and installations in any state, territory, or possession.

B-56. The *DA Civil Disturbance Plan 55-2* gives guidance for Army forces directed to quell civil disturbances. In addition, *AR 500-50* outlines responsibilities, prescribes policies, and provides guidance for the use of Army resources in response to civil disturbances. It outlines statutory restrictions and designates staff and command responsibilities for planning and executing civil disturbance operations.

B-57. The JTF commander exercises control of all federal military forces (including NG in federal status) committed to assist civil authorities. Federal military forces remain under the military chain of command during civil disturbance operations. Federal forces will not be placed under the command of either civil officials or NG commanders in a nonfederal status. Civil authorities retain control of their state and local law enforcement agencies. The JTF commander establishes liaison with the senior representative of the attorney general (SCRAG) and other appropriate federal, state, and local civil authorities.

B-58. When a JTF is being assembled it must deploy with a provost marshal (PM) who will be the POC for situation assessment, operational information, and security matters. He also serves as a liaison with civilian and law enforcement agencies.

B-59. Federal military forces must be tailored to the specific civil disturbance situation. Sufficient combat support and combat service support units will be required to sustain the force throughout the deployment. Coordination with civil officials may allow the force to draw on resources available from state and local agencies. Close and continuous coordination between federal military forces and LEAs will provide the commander with the detailed and timely information required to employ and protect the force effectively.

B-60. In supporting *DA Civil Disturbance Plan 55-2,* intelligence personnel may conduct close and continuous liaison with LEAs and military police to ensure that their units receive the information needed to allow the commander to adequately protect the force. The JTF commander should staff intelligence support missions with his senior intelligence officer and legal counsel before approving the mission.

B-61. Federal military forces must be employed in tasks or missions appropriate to their organization and training; they must not be employed in ways that violate the legal restrictions in effect. Military forces may be used to disperse unlawful assemblies and to patrol disturbed areas to prevent unlawful acts. They may be used to assist in the distribution of essential goods and the maintenance of essential services. Forces may also establish traffic control points, cordon off areas, release smoke and obscurants, and serve as security or quick reaction forces. Certain types of missions (such as gathering intelligence on civilians) are always inappropriate for military forces during civil disturbance operations.

B-62. Requests for specific military missions are typically passed through a single state or federal law enforcement coordinating officer, as approved by the SCRAG. Validated requests are transmitted to the JTF commander and his HQ for staffing and coordination. Approved missions are assigned through the military chain of command to the appropriate element or unit for execution. Units and soldiers will not accept taskings or missions directly from law enforcement or civilian officials, except in a direct support relationship as approved and ordered through the military chain of command.

B-63. It is important that a military liaison be provided to each LEA HQ generating requests for support. This liaison can assist LEA officials in determining the types and quantities of military support to be requested. JTF HQ can facilitate this mission assignment process by providing LEAs with a detailed listing of the types of missions military forces may conduct.

B-64. The AO for a deployed unit should coincide with the jurisdiction or subdivision boundaries of the LEA it supports. This arrangement facilitates liaison and coordination between law enforcement and military chains of command.

LEAD AGENCY CONCEPT AND ROLE OF MILITARY

B-65. The DOJ is the lead federal agency for civil disturbance operations. The Attorney General's on-scene representative is the SCRAG. The DOD has designated DA as their executive agent for military assistance for civil disturbance (MACDIS) (see *FM 27-100.*)

JOINT TASK FORCE

B-66. Military support to civil authorities in disasters and domestic emergencies is a DOD (not a service component) responsibility. The supported commander may decide that the severity and scope of a disaster require a joint response. In civil assistance, as in operational level warfighting, the commander-in-chief (CINC) uses the different and complementary capabilities of each service to accomplish the mission. Such use requires knowledge of both the capability and the availability of all service component assets, to include their agencies and installation.

B-67. The CINC may establish a JTF to provide comprehensive military support. Establishing a JTF may provide the best DOD response to civil emergencies. A JTF is established to execute a specific mission limited in scope and duration. The JTF objective in civil emergencies is to deploy forces to the area rapidly to assist immediately in saving lives and safeguarding property and to continue providing assistance required by the federal response plan (FRP) and the particular situation.

B-68. The JTF is configured for each specific mission. In some civil emergencies, the JTF may require a greater proportion of combat service support type units and capabilities than in typical warfighting deployments. The JTF must be able to provide emergency assistance across all lines of support.

JOINT TASK FORCE CONTROL ELEMENT

B-69. Responsibility for controlling the civil disturbance resides with the civil authorities. The JTF is organized to support and lend assistance to them in their effort to return civil order to the community.

B-70. The control element consists of the command group and the crisis management team (CMT). The command group is made up of city, county, state, and military command personnel. They set policy and issue directives, coordinate the activities of civil and military support agencies, supervise the CMT, and coordinate with outside agencies. The CMT is made up of representatives of civilian and military staff sections. The CMT advises the command group and coordinates operations and support for the action element of the TF. The control elements are located with the emergency operations center (EOC) to facilitate information processing, resource management, and operational control. The JTF command element should colocate with the civilian command element at the EOC. This will expedite the exchange of information. If an EOC has not been set up, the commander may establish one while making provisions for including civil authorities to ensure a unity of effort.

B-71. Not all CMT members are located at the EOC. Some key people may use liaisons to represent them at the EOC. Key personnel can then research and discuss ideas freely, away from the confusion associated with crisis management. The liaison can transmit guidance and answers to the EOC. Some agencies may not be needed in the EOC at first, but they must be prepared for inclusion. The decision to include an agency in the EOC is based on the likelihood that they may have to take an action or a support role on the importance of the agency to the mission.

Appendix C

Urban Terrain Analysis

Of all the environments in which the Army may need to conduct operations, the urban environment confronts commanders with a combination of difficulties rarely found in other environments. The distinct characteristics of the environment result from the combination of man-made features and supporting infrastructure superimposed on the existing natural terrain and the density of civilians in close proximity to combat forces. The human dimension it represents is potentially the most important and perplexing for commanders and their staffs to understand and evaluate. Commanders at all levels must make extraordinary efforts to assess and understand their particular urban environment to plan, prepare for, and execute effective civil disturbance operations.

SIGNIFICANT FACTORS

C-1. Although urban areas throughout the world possess similar characteristics, no two are alike. The sprawl of Los Angeles, for example, bears little physical resemblance to New Delhi. The most significant factors affecting the uniqueness and complexity of each area are its societal characteristics. While the terrain can be extremely complex, information about this aspect, its potential effects on operations, and how it changes over time may be determined with some degree of certainty. However, the human dimension is more difficult to understand and assess, particularly its effect on military operations. The urban environment is not neutral—it helps or hinders. The commander who can best understand and exploit the effects of the urban environment has the best chance of success.

C-2. Whether a metropolitan area or a village, each urban environment has an identifiable system of urban components that constantly change and interact. This system of components consists of the terrain, the society, and the infrastructure that provides the critical link between the two (see *Figure C-1,* page C-2).

C-3. These systems are not separate and distinct categories but instead are overlapping and interdependent. A thorough analysis of these systems contributes to the commander's situational understanding and allows him to develop COAs that apply appropriate resources to control a crowd and prevent a major civil disturbance. This understanding allows commanders to engage decisive points critical to maintaining peace or restoring normalcy to the urban environment.

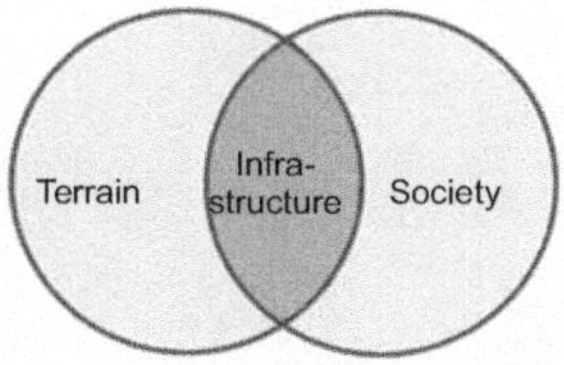

Figure C-1. Keys to Understanding the Urban Environment

C-4. The social aspects and the integrating infrastructure are the most difficult for commanders to understand and assess. These issues require greater dependence on LEAs and local governmental agencies for their information, knowledge, and expertise.

URBAN TERRAIN TYPES

C-5. The physical terrain is the most basic and easily recognizable aspect of the urban area; however, truly understanding it requires comprehending its multidimensional nature, urban and street patterns, urban form and function, and size. The infinite ways in which these factors (see *paragraph C-15)* can intertwine make it impossible to describe a typical urban area. However, these factors provide a framework for understanding the complex terrain within an urban environment. Although the terrain consists of both natural and man-made features, the latter dominates the physical characteristics of the urban environment and has significant effects on military systems and soldiers, tactics, and operations.

URBAN PATTERNS

C-6. The combinations of urban areas, lines of communications (LOCs), and natural terrain result in the identification of four urban patterns (see *Figure C-2*) that will influence civil disturbance operations. An integral part of each urban pattern is the hub or the central urban area around which outlying urban areas radiate. In SOSO, the hub is important, as it is often the economic, political, or cultural center of the surrounding area.

Satellite Pattern

C-7. This pattern consists of a central hub surrounded by smaller, dependent urban areas. LOCs tend to converge on the hub. The natural terrain throughout this pattern is relatively homogenous. Outlying or satellite urban areas often support the principal urban area at the hub.

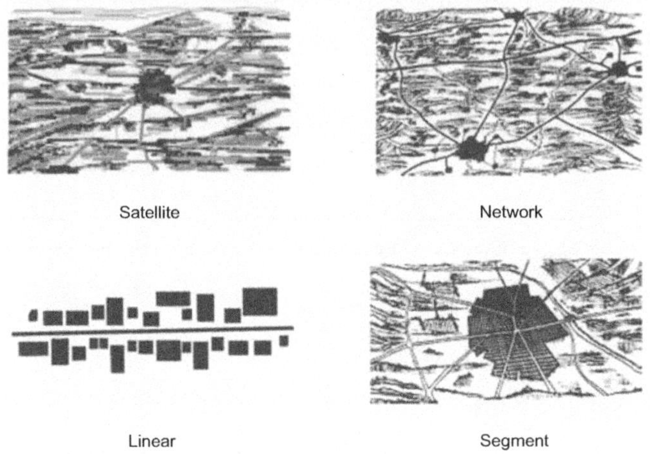

Figure C-2. Urban Patterns

Network Pattern

C-8. This pattern represents the interlocking of the primary hubs of subordinate satellite patterns. Its elements are more self-sufficient and less supportive of each other, although a dominant hub may exist. Major LOCs within a network are more extensive than in a satellite pattern and take a rectangular rather than a convergent form. The natural terrain within a network may vary more than in a single satellite array.

Linear Pattern

C-9. This pattern is a potential subelement of the previous two patterns. It may form one ray of the satellite pattern or be found along connecting links between the hubs of a network. Most frequently, however, this pattern results from the string of minor urban areas aligning a confined natural terrain corridor such as an elongated valley, a body of water, or a man-made communications route.

Segment Pattern

C-10. When dominant natural terrain (such as a river or man-made features [canals, major highways, or railways] divides an urban area, it creates a segmented pattern). This pattern makes it easier to assign areas of responsibility to units. However, if operations are in support of the city LEA, then units should be assigned to a police precinct to avoid crossing several precinct lines, as would happen when assigning units by dominant terrain features.

STREET PATTERNS

C-11. Patterns within the urban area result from the arrangement of the streets, roads, highways, and other thoroughfares. They are a result of the influences of natural terrain, the original designer's personal prejudices, and the changing needs of the inhabitants. The commander must understand the street patterns in the area in which his control force will operate. This will offer force protection, preventing the control formation from being outmaneuvered, flanked, or surrounded by a hostile crowd. Urban areas can display any of the three basic patterns or any combination of the three (see *Figure C-3).*

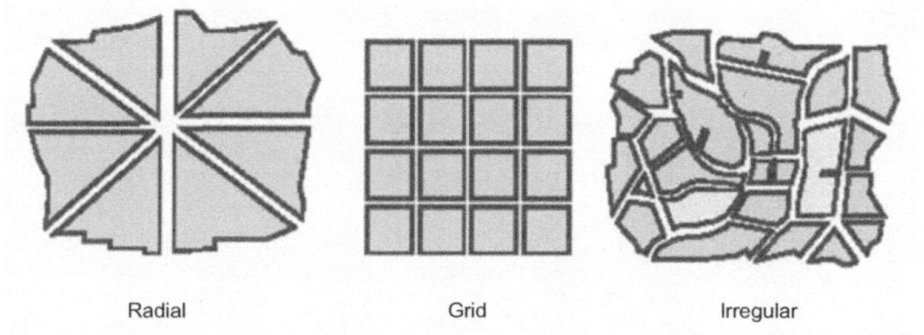

Figure C-3. Basic Internal Street Patterns

Radial

C-12. Societies of highly concentrated religious or secular power often construct urban areas with a radial design with all primary thoroughfares radiating out from the center of power. Terrain permitting, these streets may extend outward in a complete circle or may form a semicircle or arc when a focal point abuts a natural barrier such as a coastline or mountain. To increase mobility and traffic flow, societies often add concentric loops or rings to larger radial patterns.

Grid

C-13. The most adaptable and universal form for urban areas is the grid pattern (lines of streets at right angles to one another forming blocks similar to the pattern of a chessboard). A grid pattern can fill in and eventually take over an original radial pattern. Most urban areas, regardless of the original intent, plan, or vision, emerge from successive plans overlaid on one another.

Areas are well planned to fit with previous plans and others a haphazard response to explosive urban growth. The results may mix different patterns, blend patterns in symmetric combinations, or have no discernable geometric pattern (the irregular pattern).

Irregular

C-14. Irregular patterns can be specifically designed for aesthetic reasons (as in many suburban housing developments). They are sometimes used to conform to marked terrain relief (see *FMs 3-06* and *3-06.11*).

URBAN FORM AND FUNCTION

C-15. Throughout the world, urban areas have similar form and function. In form, urban areas contain like characteristics, readily divisible into distinct sections or areas. Functionally, they tend to be the center of population, finance, politics, transportation, industry, and culture.

- **Core.** The core is the heart of the urban area, the downtown, or central business district. Relatively small and compact, it contains a large percentage of the urban area's shops, offices, and public institutions.
- **Core Periphery.** As the name implies, the core periphery forms the outer edges of the core of the urban area. Generally, it has undergone less change than the core, resulting in buildings of uniform height (generally two to three stories in towns and five to ten stories in larger urban areas).
- **Industrial Area.** Industrial areas generally develop on the outskirts of the urban areas, where commercial transportation is easiest (along airfields and major sea, river, rail, and highway routes). The open/dispersed pattern of the buildings provides sufficient space for large cargoes, trucks, and materials-handling equipment.
- **High-Rise Area**. High-rise areas consist of multistoried apartments, commercial offices, and businesses separated by large open areas such as parking lots, parks, and individual one-story buildings. The automobile, mass transit systems, and improved road networks encourage these areas to grow and function further from the urban core.
- **Residential Area.** Residential areas can be found dispersed throughout the urban area; however, large suburban areas (or sprawl) normally form on the outskirts. The combined population of surrounding suburban areas often outnumbers that of the urban area. Proper and specific suburbs tend toward homogeneity based on ethnicity, religion, economics, or some other social aspect.
- **Commercial Ribbon Area**. Commercial ribbon areas are rows of stores, shops, and restaurants built along both sides of major streets that run through urban areas. These same types of areas often develop along the roads that connect one urban area to another (strip areas).

URBAN SOCIETY

C-16. Although complex, understanding the urban terrain is relatively straightforward in comparison to comprehending the multifaceted nature of urban society. As urban areas increase in size, they generally lose homogeneity; therefore, commanders must understand and account for the characteristics of a diverse population whose beliefs may vary. The behavior of civilian populations within an urban area is dynamic and poses a special challenge to commanders.

C-17. The center of gravity during an urban operation may be the civilian inhabitants themselves. The side that enjoys the support of the population has many advantages. To gain and retain this support, commanders must first understand the complex nature and character of the urban society. Second, they must understand and accept that every military action (or inaction) may influence the relationship between the urban population and Army forces, and by extension, mission success.

C-18. The identification of groups helps commanders focus on specific segments of the urban society to determine their beliefs and needs and how those beliefs and needs motivate them to future action (or inaction). Groups may be categorized based on race, religion, national origin, tribe, clan, class, party affiliation, education level, union memberships, occupation, age, or any other significant social grouping.

C-19. Commanders should consider political, economic, and historical factors in this analysis. These factors usually affect all groups to some extent and often provide the basis for many of their beliefs, needs (actual or perceived), and subsequent behavior. In most cases, training and discipline, grounded in cultural understanding and sensitivity, will help mitigate many of the potential adverse effects resulting from military and civilian interaction.

URBAN INFRASTRUCTURE

C-20. Urban infrastructures are those systems that support the inhabitants, their economy, and their government. They provide the link between the physical terrain and the urban society. Hundreds of systems may exist. Each system has a critical role in the smooth functioning of the urban area. All systems fit into the following five broad categories: communications and information, transportation and distribution, energy, commerce, and human services.

Appendix D

Practical Application

It is always best to derive lessons learned from practical operations, in the field. These are operations where a measure of success was acquired, along with important lessons learned. The following reflects an actual TF application of the doctrine highlighted in the contents of this manual.

OVERVIEW

D-1. The purpose of this manual is to provide TF units and elements with the latest guidance on providing crowd control and preventing civil disturbances.

D-2. Local civilians often gather in towns or along strategic locations throughout an AO to demonstrate grievances or other causes (for example, protesting the apprehension of a fellow citizen, policies of US forces, and policies of government services). Although civilians have the right to assemble, crowd gatherings are predominantly ethnically based, motivated, and influenced. Local civilians can apply for a permit to legally conduct a rally 72 hours in advance as long as their assembly is peaceful and unobtrusive to government services, freedom of movement, or the rights of others. On occasion, intelligence collection can indicate a possibility for violence. In these instances or during spontaneous crowd formations, the TF must provide forces in support of a safe and secure environment or to assist civilian police in preventing unlawful acts. If unlawful acts are committed, detain violators for adjudication under the legal system.

PROCEDURES

D-3. Take precautions to avoid becoming a catalyst to the civil disturbance. During lawful assemblies, the TF must make every effort to treat HN individuals with respect. This means that the TF must engage in peaceful gatherings using interpersonal communication skills. This ensures that demonstrators understand that TF personnel respect their right to lawfully assemble during peaceful demonstrations. When possible, TF engagement of a peaceful assembly should be unobtrusive.

D-4. Do not display civil disturbance equipment or weapons, overtly prepare defenses against a crowd, or aggressively violate crowd space, unless presented with information, intelligence, or negative indicators of pending violence.

NEGATIVE INDICATORS

D-5. Any response to negative indicators should be sequential, measured, and focused on deescalating rather than escalating the crowd.

NEGATIVE INDICATORS OF A POTENTIAL CIVIL DISTURBANCE

D-6. Predictive intelligence is the best indicator of a potential civil disturbance and may provide insight on internal planning, external influences, environmental conditions, or catalyst events that should influence TF planning and response.

D-7. Other indicators of a potential civil disturbance are people beginning to assemble during a scheduled or spontaneous rally. When people begin to assemble, leaders should pay particular attention to crowd dynamics. As the crowd demeanor or dynamics change from peaceful assembly to unruly behavior or unlawful acts, negative indicators should provide an advance warning and prompt a measured response. Negative indicators of a potential change in crowd dynamics and unlawful activities should prompt immediate countermeasures (see *Figure D-1*).

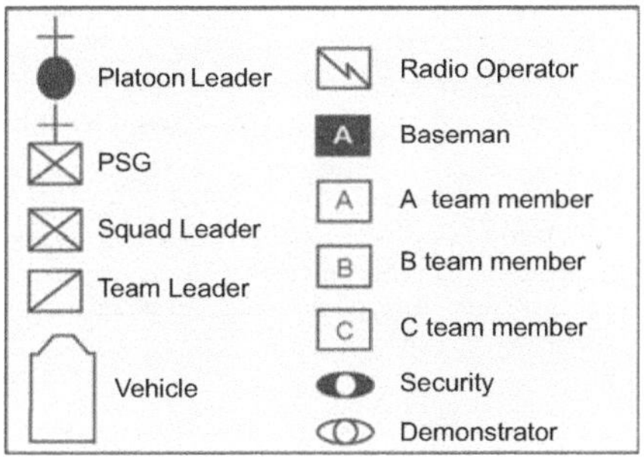

Figure D-1. Symbols for a Formation Element

Orientation

D-8. Crowd dynamics change from people milling around and talking among themselves in isolated groups to a more collective focus concentrated on a single objective (for example, TF personnel, specific agitators or magnets, a government figure, or other activity).

Massing

D-9. The crowd begins to mass or tighten into a large contiguous body from a loose formation to a static congregation located in a more concentrated area. Unchecked, massing can provide anonymity and a collective feeling of invulnerability. This change in group dynamics could provide ideal conditions for a civil disturbance (see *Figure D-2*).

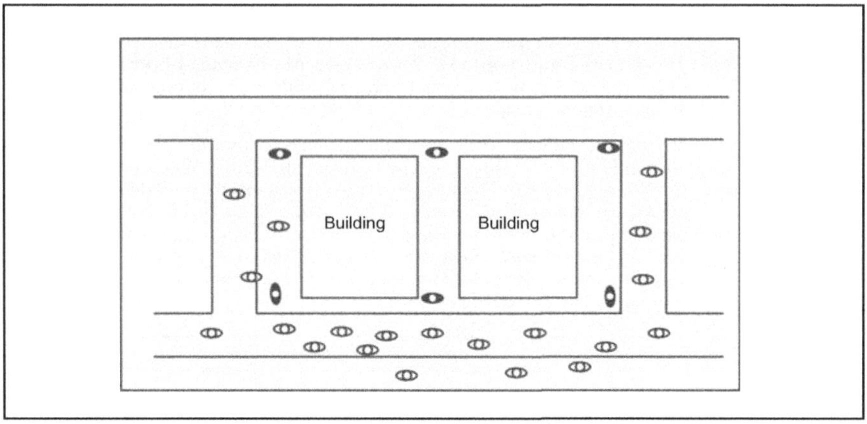

Figure D-2. Crowd Massing

Chanting

D-10. People may begin chanting slogans or yelling ethnically charged rhetoric or obscenities. Because chanting can indicate an increase in crowd intensity, interpreters should be positioned in strategic locations to interpret crowd rhetoric.

Signaling

D-11. Signaling occurs when members of the crowd position themselves to communicate and synchronize crowd actions tied to a plan with specific objectives. Signaling may indicate the intent to escalate the message through the use of violent or unlawful acts.

Blocking

D-12. The crowd may attempt to block public thoroughfares (for example, entrances and exits to public buildings, MSRs, community streets, or other traffic areas that may degrade freedom of movement or similar actions). This escalates the level of attention given to the crowd's message or prevents freedom of movement to opposing persons or groups.

Arming

D-13. Any signs that the crowd is arming itself with objects (for example, bottles, rocks, and sticks) is a clear indication of impending danger to the TF and other civilians. In fact, during a recent civil disturbance, some members of the crowd were seen handing out bottles to others in the crowd. Arming demonstrates intent toward violence is unlawful.

Outside Influence

D-14. The presence of people from outside the community may indicate more sophisticated planning and resources than would otherwise be expected from local community members. Outside people may feel insulated from responsibility for local actions and, therefore, provide a dangerous catalyst for a violent civil disturbance.

Agitators and Instigators

D-15. Agitators and instigators are common elements associated with civil disturbances. They often provide a catalyst to incite the crowd toward violence. Whether spewing rhetoric or committing unlawful acts, their continued presence will fuel potential strife.

Absence of Children

D-16. Recent civil disturbances have shown that even when children were initially present at a rally, they were absent before the civil disturbance erupted with violence. While the presence of children is not a guarantee of a peaceful assembly, the removal of children from a rally may provide early warning for a pending escalation of hostilities or a full-blown riot.

PREVENTION OF CROWD ESCALATION

D-17. If TF personnel believe, based on intelligence or negative indicators, that a crowd is planning to, or may spontaneously, riot they should immediately engage crowd participants using the preventive TTP measures.

D-18. The force executing these preventive TTP can include a number of civil disturbance teams and elements capable of executing the tasks shown in *Figures D-2,* page D-3; *D-3; D-4,* page D-6; and *D-5,* page D-8. These figures demonstrate a full range of crowd control teams and elements, to include camera teams with interpreters, checkpoints, QRF response, riot control formation, NLW teams, snatch and detain teams, blocking teams, and detainee evacuation, as required. *Figures D-2, D-3, D-4,* and *D-5,* illustrate the evolution of a crowd formation and the recommended positioning of teams and elements to engage people as they congregate. Included with each security team is a designated camera or video recorder.

D-19. Depending on the situation and crowd dynamics, it may not be necessary to establish all crowd control teams and elements, but they should be included in planning. Regardless of task organization, teams and elements must be prepared to execute the crowd control and civil disturbance TTP. The order to execute the following TTP should be tailored based on local conditions.

Early Contact

D-20. Engaging people as a crowd forms is essential to prevention. Whether based on intelligence or negative indicators, people should be engaged early, before the crowd masses as shown in *Figure D-2,* rather than waiting until after the crowd masses as shown in *Figure D-3.* Early contact will assist TFs in personalizing their interaction and ensuring that people understand their presence is known, documented, and can be linked to their actions.

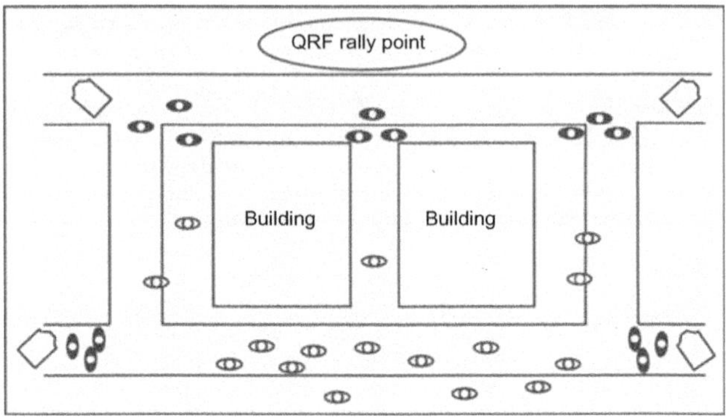

Figure D-3. Early Contact and Unobtrusive Crowd Control Measures

D-21. Cameras are one of the best tools for preventing civil disturbances. The use of cameras to document crowd formations, particularly lens-size groups (small enough to fit in a single photo), will personalize contact and eliminate feelings of anonymity among the crowd. People whose identity has been documented will think twice before committing unlawful acts. Even without documenting names, photo analysis can identify people associated with the civil disturbance, especially those who unlawfully participated as crowd instigators, agitators, or magnets or may commit unlawful acts.

D-22. As people begin to gather in and along streets, camera teams should be deployed as unobtrusively as possible to engage people before they congregate in mass. As depicted in *Figure D-3,* camera teams should consist of two soldiers with an interpreter and be employed in a widely dispersed pattern. Such small, dispersed camera teams are less likely to intrude on lawful rallies or create invasive friction that might become a catalyst to a civil disturbance.

D-23. As the crowd begins to demonstrate negative indicators for a possible civil disturbance (*Figure D-4,* page D-6), camera teams should become more invasive by actively engaging groups to take their pictures. Camera teams should focus cameras on instigators, agitators, or magnets (personnel that garner respect or compliance from the crowd). If a person is purposely avoiding the camera, even if his activities are lawful, the team should take special measures to obtain his picture. Each time a camera team takes a picture of a person or group of people, they must immediately engage them using the warning outlined in *paragraph D-26.*

D-24. Once the crowd has massed and displays negative indicators or begins to riot, camera teams should immediately move behind the riot formation line (see *Figure D-4*). It is essential, however, that camera teams continue to document the event, focusing on people who are committing illegal acts such as arming themselves, throwing objects, and assaulting TF personnel.

D-25. Immediately following a crowd control or civil disturbance operation, photos must be consolidated, processed, and submitted to the TF analysis and control element (ACE) for analysis. A near real-time analysis will provide immediate intelligence for ongoing operations to prevent subsequent civil disturbances or to detain violators from the previous operation.

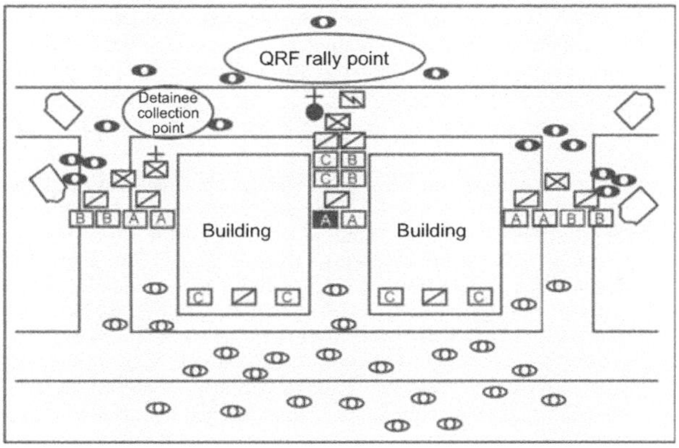

Figure D-4. More Obtrusive Engagement as the Crowd Demonstrates Negative Indicators

Crowd Warning

D-26. Once the team has taken a picture of a person or group of people, they should warn the crowd that as individuals they would be held accountable for their actions. Using an interpreter, the camera team leader will make solid eye contact with the person or group and state the following warning: *"We have your picture. You will be held accountable for your actions. If you commit an unlawful act, you will be detained, if not today, tomorrow. Blocking roads or access to public buildings is unlawful."*

Limit Crowd Size

D-27. Persons seeking to conduct a rally must apply for a permit through civilian authorities at least 72 hours prior to the meeting. If a rally has not been lawfully scheduled through civilian authorities, or if it has, but is demonstrating negative indicators and is turning into a civil disturbance, TF units and elements should attempt to limit the crowd or separate gatherings under 10 people (see *Figure D-3,* page D-5). Walking patrols can assist camera teams in breaking up large groups. Walking patrols should operate in conjunction with camera teams. This will ensure that camera teams can accurately document and provide the aforementioned warning to prevent further escalation of crowd size or behavior.

Clear Routes

D-28. Routes should be considered key terrain and must be controlled by the TF to retain the initiative. Note how a QRF divides in order to clear side streets between protected buildings in *Figure D-4.* Blocking routes or public thoroughfares is not only illegal, but severely restrictive to civil disturbance operations. Routes must remain open to allow movement of TFs and elements, whether it is for camera teams during early contact or for the QRF to respond during an actual civil disturbance. Because blocking routes is illegal, camera teams must document responsible persons for immediate or subsequent detention and legal processing.

Isolate the Area

D-29. As depicted in *Figure D-3,* page D-5, checkpoints should be established to isolate the area. While checkpoints can prevent people from entering the isolated area, they should never prevent people from exiting the area. When trapped, people may escalate the level of violence out of fear, anger, or desperation. When placing checkpoints, leaders should find a location that is both close enough to limit the number of persons being isolated, but far enough away that they will not draw a crowd from people within the isolated area. Checkpoints are essential to preventing others from joining the already gathered crowds. These checkpoints must provide storage room for processing vehicles in both directions; but when limited by space or other considerations, it should simply block traffic trying to enter the isolated area. Improve force protection measures in and around checkpoints as necessary. Strategically placed checkpoints can provide numerous countermeasures to prevent and control a civil disturbance or to identify and detain persons who may have committed unlawful acts during the riot. Use checkpoints to prevent potential participants from joining the crowd or civil disturbance. Checkpoints should also prevent any activity that may commingle ethnic groups. The checkpoint must process all people departing the area by documenting personal and vehicle information with the pictures of all occupants in front of or adjacent to the vehicle.

Establish Lethal Overwatch

D-30. The mentality of "taking the high ground" is just as important in civil disturbance operations as in any other type of operation. Although QRF assets on the ground have both NL and lethal capabilities, it is critical to ensure their safety by deploying marksman/observer teams as shown with the C-team designates on top of the buildings in *Figures D-4* and *D-5,* page D-8. Detain persons committing unlawful acts. The detention and legal processing of persons who commit unlawful acts, against TFs or local laws sends a clear message that all people will be held accountable for their actions. Immediately detain persons who are committing unlawful acts and are isolated or easily accessible, unless the detain operation would create a catalyst event or endanger the apprehension team. If it is not feasible to immediately detain them, a picture with relevant information should be forwarded for analysis and inclusion on the detain list for their detention at a later date and time. Remember, as stated in the warning, they will be detained, if not today, tomorrow. Detainees should be quickly relocated to the forward processing site for detainee evacuation (see *Figure D-5*). Detainee evacuation sites should be behind the control formations to prevent any interference from the local population. Sites should be positioned at two separate locations, such as at the end of a town or located along an MSR. All people who are detained based on

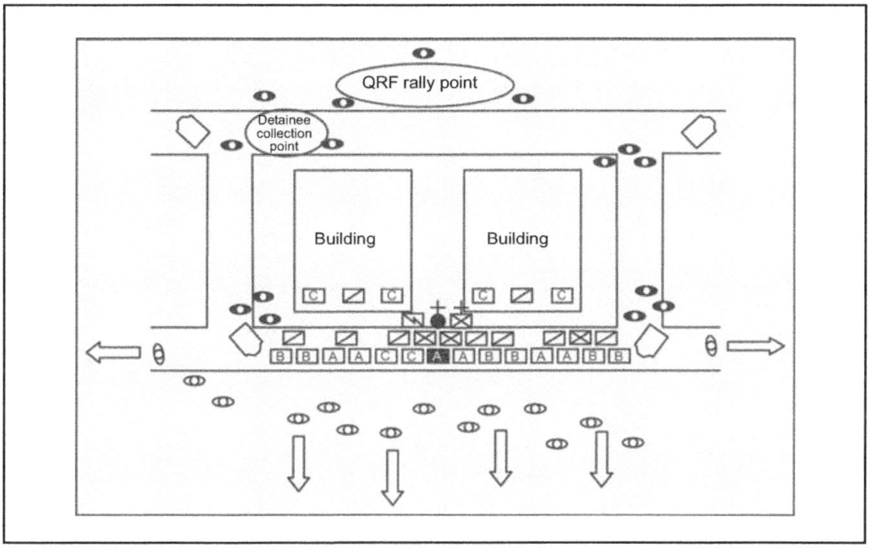

Figure D-5. Full Engagement as Crowd Unlawfully Demonstrates

solid evidence of a crime (for example, pictures of them armed with dangerous objects, blocking a road, assaulting TF members) will be turned over to local authorities for legal processing.

Detain List

D-31. When executing operations based on the detain list, do not show people the list or separate pictures from the list.

Quick Reaction Force Position

D-32. The QRF should be positioned with easy access to decisive points to interdict formations or gathering of crowds, support checkpoints, or provide sustained operations in support of early contact teams. Notice in *Figure D-1*, page D-2, how the QRF is positioned out of direct sight of the crowd until employment becomes necessary. QRF personnel are readily organized, equipped, and easily moved into position to move between buildings and through obstacles (see *Figure D-4,* page D-6).

D-33. At this point, if the crowd ignores verbal orders, NL munitions from appropriate standoff distances could be used to influence and motivate the crowd to comply. Use of NL munitions could prevent the necessity for QRF personnel to come into direct physical contact with the crowd. Early positioning of the QRF could become a catalyst event that may instigate a civil disturbance.

D-34. Assembling in mass (see *Figure D-5*), directly in front of the protected buildings, illustrates to the crowd that there is strength in numbers and the QRF is determined to disperse the crowd from the area. As a reminder, remove the flanking checkpoints that may block the dispersal of the crowd.

Glossary

A2C2	Army airspace command and control
AA	assembly area
AFL-CIO	American Federation of Labor-Congress of Industrial Organizations
AG	Adjutant General
AO	area of operations
ARTEP	Army training and evaluation program
AS	area security
ASV	armored security vehicle
ATTN	attention
BDU	battle dress uniform
CA	civil affairs
CD	combat development
CDC	crowd dispersal cartridge
CFR	Code of Federal Regulations
CINC	commander-in-chief
CMT	crisis management team
COA	course of action
COB	close of business
CONUS	continental United States
COTS	commercial off-the-shelf
CP	candle power
CPU	central processing unit
DA	Department of the Army
DCO	defense coordinating officer
DD	Department of Defense
DM	designated marksmen
DOD	Department of Defense
DOJ	Department of Justice
DOMS	director of military support
EMD	emergency medical dispatching
EMS	emergency medical services
EOC	emergency operations center
FCMT	forced cell move team
FCO	federal coordinating officer
FEMA	Federal Emergency Management Agency
FM	field manual

FRP	federal response plan
FSE	fire support element
FSO	fire support officer
G2	Assistant Chief of Staff, G3 (Intelligence)
G3	Assistant Chief of Staff, G3 (Operations and Plans)
G5	Assistant Chief of Staff, G5 (Civil Affairs)
G6	Assistant Chief of Staff, G6 (Information Management)
GOTS	government off-the-shelf
GRM	graduated response matrix
HIV	human immunodeficiency virus
HMMWV	high-mobility multipurpose wheeled vehicle
HN	host nation
HQ	headquarters
I/R	internment and resettlement
INIWIC	Interservice Nonlethal Individual Weapons Instructor Course
IO	information operations
IPB	intelligence preparation of the battlefield
JMC	joint military commission
JTF	joint task force
L&O	law and order
LBE	load-bearing equipment
LBV	load-bearing vest
LEA	law enforcement agency
LOC	lines of communication
LOS	line of sight
LP	listening post
MACDIS	military assistance for civil disturbance
MANSCEN	Maneuver Support Center
MCCM-GE	modified crowd control munition-ground emplacement
MCM	Manual for Courts-Martial
MDMP	military decision-making process
METL	mission essential task list
METT-TC	mission, enemy, terrain, troops, time available, and civilian considerations
MI	military intelligence
mm	millimeter
MO	Missouri
MOOTW	military operations other than war
MOUT	military operations on urbanized terrain
MPI	military police investigator
MRE	meal, ready to eat

MS	mobility support
MSCA	military support to civil authorities
MTP	mission training plan
MWD	military working dog
NCO	noncommissioned officer
NEO	noncombatant evacuation operations
NG	National Guard
NL	nonlethal
NLCS	nonlethal capabilities set
NLW	nonlethal weapons
OC	Oleoresin Capsicum
OCONUS	outside continental United States
OES	office of emergency services
OIC	officer in charge
OP	observation post
OPFOR	opposing forces
OPORD	operations order
PA	public address
PAO	Public Affairs Office
PIO	police intelligence operations
PM	provost marshal
PLT	platoon
PMO	Provost Marshal Office
POMSO	plans, operations, and military support officer
PPCT	pressure point control tactic
PPE	personal protective equipment
PSG	platoon sergeant
PSYOP	psychological operations
PT	physical training
PVAB	portable vehicle arrest barrier
QRF	quick reaction force
RCA	riot control agent
ROE	rules of engagement
RUF	rule for the use of force
S2	Intelligence Officer (US Army)
SCO	state coordinating officer
SCRAG	senior representative of the attorney general
SITREP	situation report
SJA	Staff Judge Advocate
SOFA	status of forces agreement

SOMA	status of mission agreement
SOP	standing operating procedure
SOSO	stability operations and support operations
SQD	squad
STARC	state area command
STP	soldier training publication
STX	soldier training exercise
TA	Training Area
TF	task force
TLP	troop leading procedure
TM	team
TPT	tactical psychological operations team
TRADOC	United States Army Training and Doctrine Command
TSOP	tactical standing operating procedure
TTP	tactics, techniques, and procedures
UCMJ	Uniform Code of Military Justice
UN	United Nations
UNGA	United Nations General Assembly
UNHCR	United Nations High Commissioner for Refugees
UNSC	United Nations Security Council
US	United States
USAMPS	United States Army Military Police School
VCR	video cassette recorder
VIP	very important person
WMD	weapons of mass destruction
WTO	World Trade Organization

Bibliography

AR 380-10. *Foreign Disclosure and Contacts With Foreign Representatives.* 6 June 2003.

AR 380-13. *Acquisition and Storage of Information Concerning Nonaffiliated Persons and Organizations.* 30 September 1974.

Code of Federal Regulations.

Constitution of the United States.

DA Form 2028. *Recommended Changes to Publications and Blank Forms.*

DA Form 4137. *Evidence/Property Document.*

DD Form 2708. *Receipt For Inmate or Detained Person.*

DOD Civil Disturbance Plan 55-2. *Operation Garden Plot.* 15 February 1991.

DOD Directive 3025.12. *Military Assistance for Civil Disturbances.* 4 February 1994.

DOD Directive 3025.15. *Military Assistance to Civil Authorities.* 18 February 1997.

DOD Directive 5525.5. *DOD Cooperation With Civilian Law Enforcement Officials.* 15 January 1986.

Executive Order 12333. *United States Intelligence Activities.* 4 December 1981.

Fiscal Year 1989 Act.

FM 3-06. *Urban Operations.* 1 June 2003.

FM 3-06.11. *Combined Arms Operations in Urban Terrain.* 28 February 2002.

FM 19.10. *Military Police Law and Order Operations.* 30 September 1987.

FM 3-19.4. *Military Police Leaders' Handbook.* 4 March 2002.

FM 3-19.40. *Military Policy Internment/Resettlement Operations.* 1 August 2001.

FM 3-21.5. *Drills and Ceremonies.* 7 July 2003.

FM 3-22.40. *Multiservice Tactics, Techniques, and Procedures (MTTP) For Tactical Employment of Nonlethal Weapons (NLW).* 15 January 2003.

FM 7-0. *Training the Force.* 22 October 2002.

FM 7-1. *Battle Focused Training.* 15 September 2003.

FM 27-100. *Legal Support to Operations.* 1 March 2000.

Judiciary Act of 1789.

MCM.

National Defense Authorization Acts.

NGR 500-1/ANGI 10-8101. *Military Support to Civil Authorities.* 1 February 1996.

Posse Comitatus Act. 1878.

STP 19-95B1-SM. *Soldier's Manual for MOS 95B Military Police Skill Level 1.* 14 January 2003.

STP 19-95B24-SM-TG. *Soldier's Manual and Trainer's Guide for MOS 95B, Military Police, Skill Levels 2/3/4.* 2 December 2002.

STP 19-95C1-SM. *MOS 95C, Corrections Specialist, Skill Level 1 Soldier's Manual.* 30 October 2003.

STP 19-95C24-SM-TG. *MOS 95C, Corrections Specialist, Skill Levels 2/3/4, Soldier's Manual and Trainer's Guide.* 30 October 2003.

Uniform Code of Military Justice.

USC, Title 32. *National Guard.*

USC, Title 10, Chapter 18. *Military Support for Civilian Law Enforcement Agencies.*

USC, Title 10, Subtitle A, Part I, Chapter 15, Section 331. *Federal Aid for State Governments.*

USC, Title 10, Subtitle A, Part I, Chapter 15, Section 332. *Use of Militia and Armed Forces to Enforce Federal Authority.*

USC, Title 10, Subtitle A, Part I, Chapter 15, Section 333. *Interference With State and Federal Law.*

USC, Title 10, Subtitle A, Part I, Chapter 15, Section 334. *Proclamation to Disperse.*

USC, Title 18, Part I, Chapter 67, Section 1385. *Use of Army and Air Force as a Posse Comitatus.*

Index

Made in the USA
Middletown, DE
04 November 2026

36555571R00265